SINGULAR SOLUTIONS AND PERTURBATIONS IN CONTROL SYSTEMS

A Proceedings volume from the IFAC Workshop, Pereslavl-Zalessky, Russia, 7 - 11 July 1997

Edited by

V. GURMAN
Program Systems Institute, RAS, Pereslavl-Zalessky, Russia

B. MILLER
Institute for Information Transmission Problems, RAS, Moscow, Russia

and

M. DMITRIEV
Program Systems Institute, RAS, Pereslavl-Zalessky, Russia

Published for the

INTERNATIONAL FEDERATION OF AUTOMATIC CONTROL

by

PERGAMON
An Imprint of Elsevier Science

UK	Elsevier Science Ltd, The Boulevard, Langford Lane, Kidlington, Oxford, OX5 1GB, UK
USA	Elsevier Science Inc., 660 White Plains Road, Tarrytown, New York 10591-5153, USA
JAPAN	Elsevier Science Japan, Tsunashima Building Annex, 3-20-12 Yushima, Bunkyo-ku, Tokyo 113, Japan

First edition 1997

Library of Congress Cataloging in Publication Data

A catalogue record for this book is available from the Library of Congress

British Library Cataloguing in Publication Data

A catalogue record for this book is available from the British Library

ISBN 9780080429328

Transferred to digital printing 2008

This volume was reproduced by means of the photo-offset process using the manuscripts supplied by the authors of the different papers. The manuscripts have been typed using different typewriters and typefaces. The lay-out, figures and tables of some papers did not agree completely with the standard requirements: consequently the reproduction does not display complete uniformity. To ensure rapid publication this discrepancy could not be changed: nor could the English be checked completely. Therefore, the readers are asked to excuse any deficiencies of this publication which may be due to the above mentioned reasons.

The Editors

IFAC WORKSHOP ON SINGULAR SOLUTIONS AND PERTURBATIONS IN CONTROL SYSTEMS

Sponsored by
International Federation of Automatic Control (IFAC)
IFAC Technical Committee on Optimization Methods

Co-sponsored by
International Association for the promotion of cooperation with scientists from the independent states of the former Soviet Union (INTAS)
Russian Basic Research Foundation (RBRF)
IFAC Technical Committees on
- Nonlinear Systems
- Control Design

Organized by
Russian National Committee of Automatic Control
Russian Academy of Sciences (Program Systems Institute and Institute for Information Transmission Problems)
State Committee of Higher Education of Russia (Russian Institute of Regional Problems and University of Pereslavl)

International Programme Committee (IPC)
Miller, B.M. (Russia) (Chairman)
Bars, R. (H)
Basar, T. (USA)
Bentsman, J. (USA)
Corless, M. (USA)
Dmitriev, M.G. (Russia)
Dragomir, T. (R)
Falcone, M. (I)
Gurman, V. (Russia)
(Chairman of Publishing Committee)
Isidori, A. (I)
Kirillova, F. (Belarus)
Kokotovic, P. (USA)
Leitman, G. (USA)
Rampazzo, F. (I)
Ryan, G. (UK)
Schneider, K. (D)
Troch, I. (A)
Vasilieva, A. (Russia)
Zolezzi, T. (I)

National Organizing Committee (NOC)
Dmitriev, M.G. (Chairman)
Fridman, L.
Kolmanovskii, V.
Melikyan, A.
Sesekin, A.

PREFACE

This volume contains the proceedings of the IFAC Workshop on SINGULAR SOLUTIONS AND PERTURBATIONS IN CONTROL SYSTEMS (SSPCS-97) held at Pereslavl-Zalessky, Russia, on 7 - 11 July, 1997. The Workshop was organized jointly by the Russian National Committee of Automatic Control, Program Systems Institute and the Institute for Information Transmission Problems of Russian Academy of Sciences, and University of Pereslavl, under the general sponsorship of IFAC.

SSPCS'97 was the 3rd event in a series of international workshops on Singular Solutions and Perturbations in Control Systems. The first two were held in 1993 and in 1995 in Pereslavl and were very successful. This event is the first of them under the sponsorship of IFAC. The objective of this workshop was to provide an international forum for the discussion of recent developments and advances in the fields of singular control problems, impulsive control, singular perturbations technique in control systems, computational problems and others.

The Workshop was devoted both to theoretical and applicative aspects of the so-called "nonclassical" problems in the area of control theory, such as problems with singular perturbations, problems with impulse and generalized controls, as well as other types of singularities. These problems arise in various areas of applications including: mechanics, information processing, medicine and economy, and at the same time they stimulate the development of new mathematical tools in classical theory of control and differential equations.

The contents of these Proceedings have been divided into seven sections:

The *Plenary session*, where two invited papers were presented. The first one by VASILIEVA, BUTUZOV and NEFEDOV was devoted to the asymptotic theory of contrast structures in nonlinear singulary perturbed problems, which were developed by authors in recent years. In the second paper by GURMAN the approach to optimization based on singularization of control systems was given.

In the section of *Impulsive and Singular Control Systems* the papers dedicated to the impulsive control theory were presented.

Various aspects of *Differential Equations with Singular Perturbations* are considered in the papers contained within that section.

In the section *Computational Problems* we have collected the papers devoted to the computational aspects of singular perturbated systems.

Problems of optimal control are presented in the papers in the section on *Dynamic Systems and Control.*

Various areas of applications, where the singular control problems arise, can be found in the papers from the section headed *Applications.*

The last section contains the papers devoted to *The Control of Systems with Incomplete Information.*

All papers included in the volume of Proceedings are given in the form presented by authors.

The financial support of the Workshop from Russian Basic Research Foundation and from International Association for the promotion of cooperation with scientists from the independent states of the former Soviet Union (INTAS) is gratefully acknowledged.

We hope that the Proceedings will be interesting for people working in the fields of control theory, differential equations, and applications.

Editors:
Chairman of Publishing Committee: **Vladimir I. Gurman**

Chairman of International Program Committee: **Boris M.Miller**

Chairman of National Organizing Committee: **Michael G.Dmitriev**

CONTENTS

PLENARY PAPERS

IMPULSIVE AND SINGULAR CONTROL SYSTEMS

DIFFERENTIAL EQUATIONS WITH SINGULAR PERTURBATIONS

COMPUTATIONAL PROBLEMS

DYNAMIC SYSTEMS AND CONTROL

APPLICATIONS

CONTROL OF SYSTEMS WITH INCOMPLETE INFORMATION

CONTRAST STRUCTURES AND EXCHANGE OF STABILITY IN SINGULARLY PERTURBED PROBLEMS*

A.B.Vasilieva, V.F.Butuzov, N.N.Nefedov

MSU, Faculty of Physics, Department of Mathematics,
119899 Moscow, Russia

Abstract: Recent results on nonlinear singularly perturbed problems which have solutions with internal layers are described. Two related cases – contrast structures and the case of exchange of stability – are considered.

Keywords: nonlinear system, singular perturbations, contrast structure.

1. INTRODUCTION.

This paper gives brief review of the results wich were obtained recently by the authors in nonlinear singularly perturbed systems. First part of the work describes some new results and problems of asymptotic theory of contrast structures for some classes of boundary value problems.

Second part of this paper containes recent results for different classes problems with exchange of stability where internal layers are a consequence of the intersection of roots of the degenerate problems.

2. CONTRAST STRUCTURES

This section of the paper presents results and problems of the asymptotic theory of contrast structures in nonlinear singularly perturbed problems which were developed by the authors during last years.

To define contrast structures we consider the following problem

$$\varepsilon^2 \Delta u = f(u, x, \varepsilon), \quad x \in D \subset R^2, \qquad u|_{\partial D} = g(x), \tag{1}$$

where $\varepsilon > 0$ is small parameter, Δ is the Laplacian. The spike–type contrast structure is the solution $u(x, \varepsilon)$ of the problem (1), which is close to some solution $\varphi(x)$ of the degenerate equation $f(u, x, 0) = 0$ everywhere inside the domain D with the exeption of a small neighborhood of some curve C where the solution $u(x, \varepsilon)$ significantly differs from $\varphi(x)$ (the solution $u(x, \varepsilon)$ has the spike).

The step-type contrast structure is the solution $u(x, \varepsilon)$ of the problem (1), which is close to two different solutions $\varphi_1(x)$ and $\varphi_2(x)$ of the degenerate equation on different sides of some curve C.

The main questions which were considered for the problem (1) and its extensions to another classes of DE include:

1. Existence and evolution of contrast structures.

2. Stability analysis of contrast structures.

3. Formation of contrast structures (investigation of the domain of attraction of the stationary solution).

The main extensions of the problem (1) were obtained for the parabolic problem

$$\varepsilon^2 (\Delta u - \frac{\partial u}{\partial t}) = f(u, x, t, ep) \tag{2}$$

and some classes of vector–case of (1) belonging to so-called Tikhonov's type systems (with "fast" and "slow" equations).

We develope a method of asymptotic expansion of contrast structures, which is based on a modification of boundary function method. Well–known boundary functions method is modified by the

*This work is supported by Grant $N^{\underline{o}}$ 96-01-00694 from the Russian Foundation for Fundamental Researches.

way, which gives a possibility to construct transition layer functions describing spike–or step–type contrast structures. The transition layer location, which is unknown in advance,is determined in the process of constructing of the asymptotic expansion.

In the most cases for PDE's investigation of existence, asymptotic estimates and stability of contrast structures is based on a new approach – asimptotic method of differential inequalities. The basic idea of this method is to construct upper and lower solutions by using the formal asymptotics. Related consideration of contrast structures can be found in (Vasilieva *et al.*, (1997a); Vasilieva *et al.*, (1997b)).

3. EXCHANGE OF STABILITY.

In this part we consider internal layers which arise in the case when the roots of the degenerate problem are intersecting. It appears that this phenomenon can lead to the specific behavior of the solutions. The investigation of this class of singularly perturbed problems was initiated by applications from chemical kinetics, where singularly perturbed differential equations describe fast bimolecular reactions. We demonstrate below our recent results for different classes of singularly perturbed problems with exchange of stability.

3.1. *Initial value problems.*

We demonstrate our results for some relatively simple models. Consider the problem

$$\begin{aligned} \varepsilon\frac{dx}{dt} &= f(x,t,\varepsilon)\,, \quad 0 < t \le T\,, \\ x(0) &= x^0\,, \end{aligned} \tag{3}$$

where $\varepsilon > 0$ is a small parameter and f is sufficiently smooth function, under the following conditions:

Condition 1. *In the domain under consideration the degenerate equation*

$$f(x,t,0) = 0$$

has two solutions $x = \varphi_1(t)$ *and* $x = \varphi_2(t)$.

Condition 2. *The curves* $x = \varphi_1(x)$ *and* $x = \varphi_2(x)$ *intersect at the point* $t = t_0 \in (0,T)$. *For definiteness we assume that* $\dfrac{d\varphi_1(t_0)}{dt} < \dfrac{d\varphi_2(t_0)}{dt}$.

Condition 3. *The root* $x = \varphi_1(t)$ *is stable for* $[0,t_0)$, *and the root* $x = \varphi_2(t)$ *is stable for* $t \in (t_0,T]$, *that is*

$$\begin{aligned} f_x(\varphi_1(t),t,0) < 0 \quad &for \quad t \in [0,t_0)\,, \\ f_x(\varphi_2(t),t,0) < 0 \quad &for \quad t \in (t_0,T]\,. \end{aligned}$$

The function

$$\hat{x}(t) = \begin{cases} \varphi_1(t)\,, & 0 \le t \le t_0\,, \\ \varphi_2(t)\,, & t_0 \le t \le T\,, \end{cases}$$

is called a composed stable solution.

Condition 4. *The initial value* $x(0) = x^0$ *lies in the domain of influence of the root* $x = \varphi_1(t)$.

Condition 5. *Composed stable solution satisfies the relations*

$$\varepsilon\frac{\partial\hat{x}}{\partial t} \le f(\hat{x}(t),t,\varepsilon)\,, \quad x^0 > \hat{x}(0) = \varphi_1(0)\,.$$

By using method of differential inequalities we have proved the theorem that $\hat{x}(t)$ is a limiting solution of the unique solution $x(t,\varepsilon)$ of the problem (3) and given the estimate of the difference between exact and composed stable solutions (Vasilieva *et al.*, (1996)).

3.2.

If $f(x,t,\varepsilon)$ has a special form

$$f(x,t,\varepsilon) = -x(x-\alpha(t))\,,$$

where $\alpha(t) < 0$ for $0 \le t < t_0$ and $\alpha(t) > 0$ for $t_0 < t \le T$. Then the solution of (3) exhibits more complicated behavior. It follows to stable root $x = 0$ until $t = t_0$ and continues to follow to unstable root $x = 0$ until point t^* which can be computed as the solution of the equation

$$\int_0^t \alpha(s)\,ds = 0\,.$$

Then the solution has a jump from unstable root $x = 0$ to stable root $x = \alpha(t)$. This result follows from the explicit presentation of the solution for this nonlinearity (Vasilieva *et al.*, (1996)). We call this case delayed exchange of stability.

3.3.

The extension the case 3.1 to Tikhonov's type systems is given in (Nefedov *et al.*, (1994)). More general types of nonlinearities in the case of delayed exchange of stability were obtained in (Nefedov *et al.*, (1996)). The generalization to BVP for scalar second order singularly perturbed ODE is given in (Butuzov *et al.*, (1997)). Resently more complicated Tikhonov's type boundary value problems for second order ODE's and scalar cases of problems (1) and (2) in the case of exchange of stability are also considered.

REFERENCES

Butuzov V.F. and Nefedov N.N. (1997), *Matematicheskie zametki* (in Russian, to appear).

Nefedov N.N. and Schneider K.R. (1996). Delayed exchange of stability in singularly perturbed systems. Weierstrass Institut für Angewandte Analysis und Stochastik, Berlin, Preprint $N^{\underline{o}}$ 270.

Nefedov N.N., Schneider K.R. and Schuppert A. (1994). Jumping behavior in singularly perturbed systems modelling bimolecular reactions. Weierstrass Institut für Angewandte Analysis und Stochastik, Berlin, Preprint $N^{\underline{o}}$ 137.

Vasilieva A.B., Butuzov V.F. and Nefedov N.N. (1997a). Asymptotic theory of contrast structures. *Avtomatika i Telemehanika*, $N^{\underline{o}}$ 7 (in Russian).

Vasilieva A.B., Butuzov V.F. and Nefedov N.N. (1997b). Contrast structures in singularly perturbed problems. *Fundamentalnaja i prikladnaja matematika* (in Russian, to appear).

SINGULARIZATION OF CONTROL SYSTEMS

Vladimir I.Gurman

Program Systems Institute, RAS
Pereslavl-Zalessky, Russia
E-mail: gurman@cprc.botik.ru

Abstract: Some effective methods developed in preceding studies of control systems representable by differential inclusions with unbounded right-hand sides (velocity sets) are generalized and extended on a more wide class of systems including ones with well bounded velocity sets when investigating their sliding modes as optimal solutions. It is attained by proper extension of the original inclusion to an unbounded one (what is called singularization) and further transformation to a control system of reduced order. As the result mutual compatible sufficient and necessary optimality conditions for sliding modes are deduced. The latter ones are local maximum conditions with respect to some multitude in state space as additional to usual maximum principle local maximum conditions with respect to some multitude in state space.

Keywords: impulses, optimal control, singular system, sliding mode

1. INTRODUCTION

In (Gurman, 1965; 1972; 1977; 1985), (Gurman et al., 1987, 1988) there was investigated the control system

$$\dot{x} = f(t, x, u), \; x \in \mathbb{R}^n, \; u \in U(t, x) \subset \mathbb{R}^p \qquad (1)$$

when its velocity set $V(t, x) = f(t, x, U(t, x))$ is unbounded. In this case its set of solutions contain those with impulses and discontinuous phase trajectories as some generalized ones, that are approximated by sequences of regular solutions. By this reason such a type of systems is called "singular". Under some general assumptions a singular system can be transformed to some equivalent (in integral sense) system of reduced order (*derived system*) what makes singular systems more simple and attractive in comparison with regular systems of the same order. Thus, there arise a quite natural idea of appropriate transformation of a regular system into a singular one at the expense of unbounded extension of the velocity set for the sake of succeeding investigation it by exact or approximate methods. What follows is development of this singularization idea. Let as notice that this approach in general should not be surprising for those who is familiar with investigations of problems in space flight control (Edelbaum, 1966; Krotov et al., 1969), laser technology (Krasnov, 1980) and other practical problems, where idealization of strong control actions as impulse ones proved to be effective. Here is an attempt to use the singularization for the theoretical study of much more wide class of problems (*degenerate problems* (Gurman, 1977)) that may include also those with well bounded controls.

2. DERIVED PROBLEMS FAMILY

The first step is standard transformation of (1) to the corresponding relaxed system

$$\dot{x} \in V_c(t, x) = \overline{\text{co}}\, V(t, x) \qquad (2)$$

that describes in terms of piecewise smooth solutions all usual and generalized solutions (sliding modes) of (1). Let us call *facet* any convex subset of V_c that does not contain relatively interior points of a convex subset having greater dimension.

It follows from our definition of a facet and the method of analytical description of convex sets by their support functions (Leichtweiss, 1980) that the whole class of facets of $V_c(t,x)$ can be defined as

$$F(t,x,p) = \text{Arg} \max_{v \in V_c(t,x)} p^T v, \quad p \in L_p(t,x)$$

where L_p is so called dual cone for V_c. In particular $F(t,x,p)$ may not depend on p within some subset of p, so $F(t,x,p) = F(t,x)$, and it we will be called *"isolated" facet*. If there exist a domain

$$B \subset \mathbb{R}^q, \quad q < n,$$

and a function $p(\beta)$, $\beta \in B$, such that $F(t,x,p(\beta))$ is Hausdorff continuous with respect to β in B then let us speak of a *continuous family of facets*.

A smooth $x(t)$ will be called *F-sliding mode* of (1) or (2) on $T = [t_I, t_F]$ if

$$\dot{x} \in \text{int}\, F(t, x(t)), \quad t \in T. \text{ [1]}$$

Let $F \subset V_c$ be a facet of dimension $k > 0$ and $P^{(k)}$ be its bearing plane with an interior basis $h = (h_1, h_2, \ldots, h_k)$. Consider the following auxiliary singular system

$$\dot{x} \in V_e(t,x) = V_c(t,x) \cup P^{(k)}(t,x). \tag{3}$$

Let the system

$$\partial x / \partial \tau_j = h_j(t,x), \; j = 1, \ldots, k$$

(basis system) have full integral

$$y = \eta(t,x) = (\eta^1(t,x) \ldots, \eta^{n-k}(t,x))$$

as a continuous and smooth function of t and x, so that

$$\eta_x h(t,x) = 0 \quad \left(\eta_x = \left(\frac{\partial \eta^j}{\partial x^i}\right)\right). \tag{4}$$

Then according to (Gurman, 1972, 1977, 1985) (3) can be equivalently (in integral sense) transformed to so called derived system.

$$\begin{aligned} &\dot{y} = \eta_x v + \eta_t, \quad v \in V_e(t,x), \\ &x \in Q(t,y) = \{x : y = \eta(t,x)\} \end{aligned} \tag{5}$$

with solutions as triples $(y(t), x(t), v(t))$ where $y(t)$ is piecewise smooth and $x(t)$, $v(t)$ are piecewise continuous. Also a piecewise continuous $x(t)$ will be consideredas a solution of (5) if there exist a piecewise smooth $y(t)$ and a piecewise continuous $v(t)$ such that the triple $(y(t), x(t), v(t))$ satisfies (5). Evidently (5) is an extension of (2) in the sense that any solution $x(t)$ of (2) is a solution of (5) (but not vice versa). Let us call (5) *F-derived system*. It has the following important property (A):

$$\dot{x}(t) \in P^{(k)}(t, x(t)) \tag{6}$$

for any smooth solution $x(t)$ of (5).

Indeed, any $v \in P^{(k)}$ can be represented as

$$v = v_0 + hw, \quad w \in \mathbb{R}^k$$

where v_0 is some arbitrary constant point of $P^{(k)}$.

Hence it follows from (5) that

$$\dot{y} = \eta_x v + \eta_t = \eta_x(v_0 + hw) + \eta_t = \eta_x v_0 + \eta_t$$

what is constant for any constant $t, x(t)$ because of (4). It means that (5) can be considered (in statics) as the equation of $P^{(k)}$. On the other hand $x(t)$ satisfies the equation

$$y(t) = \eta(t, x(t))$$

due to (5) so that

$$\dot{y} = \eta_t + \eta_x \dot{x}$$

that is $\dot{x}$ satisfies $P^{(k)}$ equation.

It follows from the property A that the set of pairs $(y(t), x(t))$ of all solutions of (5) with smooth $x(t)$ will be invariant when replacing V_e by V_c in (5).

Thus a family of F-derived systems can be generated, in accordance with all facets that satisfy the integrability conditions

$$h_m^T h_{lx} - h_l^T h_{mx} = 0, \quad l, m = 1, 2, \ldots, k$$

(that are equivalent to the well known Frobenius condition).

In these terms there are proposed global sufficient and new necessary optimality conditions for sliding regimes of (1) or (2), in the problem

$$t \in T = [t_I, t_F], \quad I = \Phi(x(t_F)) \to \inf, \tag{7}$$

$x(t_I)$ is fixed.

Consider two typical cases: the first one, when the problem can be represented in terms of "isolated" facet, and the second one, when some continuous parametric family of facets can be chosen to characterize the situation.

3. OPTIMALITY CONDITIONS: "ISOLATED" FACET

Let $F(t,x)$ be a k-dimension facet of $V_c(t,x)$ that generates the corresponding F-derived system. Define an extension of admissible set for

[1] Such a solution is often called "singular" regime for the system (2), and the term "sliding" is used to point out its generalized sense for the system (1). But this difference is not important here, so it will be used only one term for it to preserve the term "singular" for unbounded control systems.

the problem ((1), (7)) or ((2), (7)) taking the derived system (5) in the place of (2), and call the resulting substitute the *F-derived problem*. Any solution of this problem should satisfy the Pontryagin maximum principle, and in particular the condition of maximum of linear form $p^T v$ on V_c, where $p^T = q^T \eta_x$ and q is an ajoint $(n-k)$-vector for the derived problem. It follows from (3) that if for some q the maximum is reached at some point $v_0 \in F$ then all the facet F is the maximum set because it is apriori orthogonal to p.

In this case the F-derived problem solution should be a triple

$$(y_*(t), x_*(t), F(t, x_*(t))),$$

and $x_*(t)$ will be the desired ((2), (7)) — problem solution if

$$\dot{x}_* \in F(t, x_*(t)). \qquad (8)$$

But $\dot{x}_* \in P^{(k)}(t, x_*(t))$ apriori due to the property A, so the condition (8) is a condition for some point of a plane to be in a domain of this plane. Thus the following statement is true.

Theorem 1. Let $(y_*(t), x_*(t), F(t, x_*(t)))$ be a solution of the corresponding F-derived problem, $x_*(t)$ is piecewise smooth, and

$$\dot{x}_* \in F(t, x_*(t)).$$

Then $x_*(t)$ is a solution of the problem ((2), (7)), and optimal F-sliding mode if, in addition,

$$\dot{x}_* \in \operatorname{int} F(t, x_*(t)). \qquad (9)$$

Example 1. Degenerate quadratic problem with bounded control.

$$\dot{x}^1 = 2((x^2)^2 + (x^3)^2) + x^2 u^1 + x^3 u^2$$

$$\dot{x}^2 = x^3 + u^2, \quad \dot{x}^3 = x^2 + u^1, \quad |u^i| \le 1,$$

$$t \in [0, 1], \quad x^i(0) = a^i,$$

$$I = x^1(t_F) - x^2(t_F) x^3(t_F), \quad t = t_F.$$

Here the velocity set is convex: $V = V_c$, and it has 5 facets of dimension $k \ge 1$:

$$F_0 = V = \{\dot{x}\colon u^1 \in [-1; 1], u^2 \in [-1; 1]\} \ (k = 2),$$

$$F_{1,2} = \{\dot{x}\colon u^1 \pm 1,\ u^2 \in [-1, 1]\} \qquad (k = 1),$$

$$F_{3,4} = \{\dot{x}\colon u^1 \in [-1, 1],\ u^2 = \pm 1\} \qquad (k = 1).$$

Consider the corresponding derived problems and their solutions. For the facet F_0:
$\dot{y} = (x^2)^2 + (x^3)^2, \quad y = x^1 - x^1 x^3, \quad I = y(t_F)$;
$x_*^i(t) = 0,\ i = 2, 3,\ x_*^1(t) = y_*(t) = a^1$,
(when $a^2 = a^3 = 0$).

For the facets $F_{1,2}$:

$$y^1 = x^1 - x^2 x^3, \qquad y^2 = x^3,$$

$$\dot{y}^1 = (x^2)^2 + (y^2)^2,\ \dot{y}^2 = x^2 + u^1,$$

$$u^1 = \pm 1, \qquad I = y^1(t_F);$$

$$x_*^2(t) = 0, \qquad u_*^1 = -\mathrm{sign} x_*^3,$$

$$|x_*^3| = \begin{cases} |a^3| - t, & t \le |a^3|, \\ 0 & t > |a^3|. \end{cases}$$

For the facets $F_{3,4}$:

$$y^1 = x^1 - x^2 x^3, \qquad y^2 = x^2,$$

$$\dot{y}^1 = (x^3)^2 + (y^2)^2,\ \dot{y}^2 = x^3 + u^2,$$

$$u^2 = \pm 1, \qquad I = y^1(t_F);$$

$$x_*^3(t) = 0, \qquad u_*^2 = -\mathrm{sign} x_*^2,$$

$$x_*^2 = \begin{cases} |a^2| - t, & t \le |a^2|, \\ 0, & t > |a^2|. \end{cases}$$

It is easily seen that the condition (9) of theorem 1 is satisfied for the facets $F_{1,2}$ when $|a^3| < 1$ and for the facets $F_{3,4}$ when $|a^2| < 1$. As a result 5 corresponding families of sliding modes are received. Corresponding sliding modes trajectories

$$x^1 = a^1, \quad x^2 = 0, \qquad x^3 = 0;$$

$$x^1 = \frac{1}{3}((a^2)^3 - (x^2)^3),\ x^3 = 0;$$

$$x^1 = \frac{1}{3}((a^3)^3 - (x^3)^3),\ x^2 = 0$$

lay on special surfaces that are coordinate planes (x^1, x^2), (x^1, x^3) and their intersection — x^1 axis.

To formulate the necessary optimality conditions for sliding modes we introduce the following adjoint system:

$$\dot{\psi} = -\mathcal{H}_x(t, x_*(t), \psi(t), u_*(t)),$$

$$\psi(t_F) = -\Phi_x(x_*(t_F)),$$

$$\frac{\partial x}{\partial z} = h(t, x), \quad x = \xi(t, z), \qquad \xi(t, 0) = x_*(t), \qquad (10)$$

$$\frac{\partial p}{\partial z} = -(p^T h(t, x))_x, \qquad p(t, 0) = \psi(t),$$

$$\frac{\partial q}{\partial z} = -q h_t(t, x), \qquad q(t, 0) = 0,$$

where $x, \psi, p \in \mathbb{R}^n$, $q \in \mathbb{R}$, $z \in \mathbb{R}^k$,
$\mathcal{H}(t, x, \psi) = \max\limits_{v \in V_c(t,x)} \psi^T v = \max\limits_{u \in U(t,x)} \psi^T f(t, x, u)$.

Theorem 2. Let $x_*(t)$ be a solution for the problem ((2), (7)) with Lipshizian $\mathcal{H}(t,x,\psi)$, and $F(t,x)$ be a facet of $V_c(t,x)$, with and internal basis

$$h(t,x) = \{h_1(t,x), \dots, h_k(t,x),\ k \geq 1\},$$

$h_l(t,x)$ being continuous together with their derivatives in some open domain $C \subset \mathbb{R}^{n+1}$ that contains the $x_*(t)$ graph. Let

$$x_*(t) \in \operatorname{int} F(t, x_*(t)), \quad t \in T,$$

and this facet generate in C a corresponding F-derived system (5).

Then there exists a solution of (10),

$$(\psi(t), \xi(t,z), p(t,z), q(t,z)),$$

such that

1) $\psi^T \dot{x}_* = \mathcal{H}(t, x_*(t), \psi(t))$;

2) $\psi^T(t) h(t, x_*(t), \psi(t)) = 0$;

3) $\mathcal{H}(t, x_*(t), \psi(t)) =$

$= \operatorname{loc}\max_z (\mathcal{H}(t, \xi(t,z)), p(t,z) + q(t,z))$;

4) $\Phi(x_*(t_F)) = \operatorname{loc}\min_z \Phi(\xi(t_F, z))$.

Proof. The conditions 1) and 2) are known necessary conditions of the Pontryagin-type maximum principle (Clarke, 1983) for the F-sliding mode under consideration.

Consider the condition 3). Denote

$$K(t,z) = \mathcal{H}(t, \xi(t,z), p(t,z) + q(t,z). \qquad (11)$$

It is easily seen that $K(t,z)$ is continuous due to continuity of $\mathcal{H}$, ξ, p, q. Then

$$K(t,0) = \mathcal{H}(t, x_*(t), \psi(t)).$$

Suppose that 3) is not valid for some $t' \in T$, that is in any vicinity of $z_* = 0$ there exists a point $\bar{z}(t')$ such that

$$K(t',0) < K(t', \bar{z}(t')). \qquad (12)$$

Then, due to continuity of $K(t,z)$, there exists a vicinity of t', $T' = (t'_I, t'_F) \ni t'$ where (12) is valid. Let us show that $x_*(t)$ may be "improved" by some appropriate variation of $z(t)$ on T'.

Denote

$$Z(t) = \{z\colon K(t,0) < K(t,z)\},$$

that is some open domain in (z) space. Let us construct a solution $\bar{x}(t)$ of the inclusion

$$\dot{x} \in F(t,x), \qquad (13)$$

by construction a function $\bar{z}(t)$ such that

$$\bar{x}(t) = \xi(t, \bar{z}(t)), \quad \bar{z}(t) \in Z(t)$$

for any $t'_I < t < t'_F$, and

$$\bar{z}(t'_I) = \bar{z}(t'_F) = z_* = 0.$$

Consider (13) in the form

$$\dot{x} = v_0(t,x) + h(t,x)w, \quad w \in W(t,x), \qquad (14)$$

where $v_0(t,x) \in F(t,x)$, and $W(t,x)$ is some convex domain in $\mathbb{R}^k$. Transform (14) to the variables

$$y = \eta(t,x), \ z = \zeta(t,x)$$

($\zeta(t,x)$ is orthogonal complement to $\eta(t,x)$):

$$\dot{y} = \eta_x v_0(t,x) + \eta_t,$$

$$\dot{z} = \zeta_x v_0(t,x) + \zeta_t + \zeta_x h(t,x) w, \qquad (15)$$

$$w \in W(t,x), \quad x = \chi(t,y,z),$$

where $\chi(t,y,z)$ is the result of conversion of

$$(y = \eta(t,x),\ z = \zeta(t,x)).$$

Evidently

$$\bar{z}_*(t) = 0, \quad \bar{w}_*(t) \in \operatorname{int} W(t, x_*(t)). \qquad (16)$$

Take $\epsilon > 0$ and $\bar{z}(t')$ so that $|\bar{z}(t')| < \epsilon$, and let

$$\bar{z}(t) = \begin{cases} \bar{z}(t') \dfrac{(t - t'_I)}{(t' - t'_I)} & t'_I \leq t < t', \\[2ex] \bar{z}(t') \dfrac{(t - t'_F)}{(t' - t'_F)} & t' \leq t \leq t'_F. \end{cases}$$

Then

$$\bar{w}(t) = [\bar{\zeta}_x \bar{h}]^{-1} \left(\dot{\bar{z}}(t) - (\bar{\zeta}_x \bar{v}_0 + \bar{\zeta}_t)\right),$$

$$\dot{\bar{z}} = \begin{cases} \bar{z}(t')(t' - t'_I)^{-1}, & t'_I \leq t < t' \\ \bar{z}(t')(t' - t'_F)^{-1}, & t' \leq t \leq t'_F. \end{cases}$$

This means that

$$\delta w(t) = \bar{w}(t) - w_*(t) \to 0$$

uniformly on T' when $\bar{z}(t') \to 0$, and there exists $\epsilon > 0$ such that

$$w_*(t) + \delta w(t) \in W(t, \bar{x}(t)),$$

when (16) holds.

Consider the increment of the functional due to above variation of $z(t)$ with the use of the following representation of I:

$$I = F(x(t_F)) + \varphi(t_F, x_F) - \varphi(t_I, x_I) - \\ - \int_{t_I}^{t_F} R(t,x,u)dt.$$

where

$$R(t,x,u) = \varphi_x^T f(t,x,u) + \varphi_t,$$

$\varphi(t,x)$ is an arbitrary continuous and smooth scalar function. Evidently

$$R = \frac{d}{dt}\varphi(t,x(t)),\ \varphi_F - \varphi_I - \int_{t_I}^{t_F} R dt = 0,$$

when $\dot{x} = f(t,x,u)$. Let $\psi^{\mathrm{I}} \in \mathbb{R}^{n-k}$,

$$\varphi(t,x) = \varphi^{\mathrm{I}}(t,\eta(t,x)) = (\psi^{\mathrm{I}}(t))^T \eta(t,x), \quad t \in T.$$

Note that the partial equations in (10) describe in general the gradient of this function

$$(p(t,y,z), q(t,y,z)) = (\varphi_x(t,x), \varphi_t(t,x))$$

along with $x = \chi(t,y,z)$, and

$$\xi(t,z) = \chi(t,y_*(t),z).$$

Denote $m = x(t),\ t \in T$. Then

$$\delta I = I(\bar{m}) - I(m_*) = \delta\Phi + \delta\varphi_F - \int_{t'_F}^{t_F} \delta R^{\mathrm{I}} dt,$$

$$\delta\Phi + \delta\varphi_F = (\Phi_x(x_*(t_F)) +$$

$$+\varphi_x^T(t_F, x_*(t_F))\delta x(t_F) + o(\delta x(t_F)),$$

$$\delta R^{\mathrm{I}} = \delta_z R^{\mathrm{I}} + R^{\mathrm{I}}_{y_*}\delta y + o(\delta y),$$

$$\delta_z R^{\mathrm{I}} = R^{\mathrm{I}}(t, y_*(t) + \delta y, \delta z) -$$

$$-R^{\mathrm{I}}(t, y_*(t) + \delta y, 0), \tag{17}$$

$$R^{\mathrm{I}}(t,y,z) = (\varphi^{\mathrm{I}}_y)^T \eta_x v_0(t,x) +$$

$$+(\varphi^{\mathrm{I}}_y)^T \eta_t) + \varphi^{\mathrm{I}}_t = \tag{18}$$

$$= p^T(t,y,z)v_0(t,x) + q(t,y,z) + \varphi^{\mathrm{I}}_t,$$

$$x = \chi(t,y,z),$$

and

$$R^{\mathrm{I}}_y(t,y_*(t),0) = 0, \quad \delta\Phi + \delta\varphi_F = o(\delta x(t_F)),$$

due to condition 1) of the theorem.

Further

$$\delta z = \bar{z}(t) - z_*(t) = \bar{z}(t) = O(\Delta t),$$

and

$$\delta_z K = K(t,\bar{z}(t)) - K(t,0) = O(\Delta t),$$

where $\Delta t = t'_F - t'_I$.

It is seen from the expression (11) and (17) that

$$\delta_z K = R^{\mathrm{I}}(t, y_*(t), \bar{z}(t)) - R^{\mathrm{I}}(t, y_*(t), 0),$$

what implies together with (16) that

$$(\delta_z R^{\mathrm{I}} - \delta_z K) = O(\delta y) = o(\Delta t),$$

and

$$\delta I = o(\delta x(t_F)) - \int_{t'_F}^{t_F} o(\delta x) dt -$$

$$- \int_{t''_I}^{t'_F} \delta_z K + o(\Delta t)) dt.$$

Then

$$\delta x(t'_F) = O(\delta y(t'_F)),\ \delta x(t) = O(\delta y(t'_F)),\ t \in [t'_F, t_F]$$

because $\bar{z}(t'_F) = 0$. But

$$\delta y(t'_F) = o(\Delta t), \quad \int_{t'_I}^{t'_F} o(\Delta t) dt = o(\Delta t)^2),$$

hence

$$\delta I = o((\Delta t)^2) - \int_{t'_I}^{t'_F} \delta_z K dt = o((\Delta t)^2) -$$

$$-O((\Delta t))^2 = \left(\frac{o(\Delta t)^2}{\Delta t^2} - \frac{O((\Delta t)^2)}{\Delta t^2}\right)(\Delta t)^2,$$

where $O((\Delta t)^2) > 0$.

It is seen that $\delta I < 0$ for sufficiently small Δt, that is there exists such $\bar{m}$ that $I(\bar{m}) < I(m_*)$, when the condition 3) does not hold, what contradicts the optimality of m_*.

Similarly the necessity of condition 4) is proved. Suppose that 4) is not valid that is in any vicinity of $z_{F*} = 0$ there exists a point $\bar{z}_F$ such that $\Phi(\xi(t_F, \bar{z}_F) < \Phi(x_*(t_F))$. Then m_* also can be "improved".

Indeed, let us construct the solution $(\bar{y}(t), \bar{z}(t), \bar{w}(t))$ of the system (15) that correspond to the straight line

$$\bar{z}(t) = \frac{\bar{z}_F}{t_F - t'}(t - t')$$

for some $t' < t_F$ and sufficiently small $\bar{z}_F$. Take

$$x = x_*(t),\ u = u_*(t),\ t \in [t_I, t'].$$

Then

$$\delta I = \delta G^{\mathrm{I}} - \int_{t'}^{t_F} \delta R^{\mathrm{I}} dt, \quad \delta R^{\mathrm{I}} = \delta_z R^{\mathrm{I}} + R^{\mathrm{I}}_{y_*}\delta y + o(\delta y)$$

$$G^{\mathrm{I}} = \Phi^{\mathrm{I}}(y,z) + \varphi^{\mathrm{I}}(t_F, y),$$

$$\Phi^{\mathrm{I}}(y,z) = \Phi(\chi(t_F, y, z)),$$

$$\delta G^{\mathrm{I}} = \delta_z G^{\mathrm{I}} + G^{\mathrm{I}}_{y_*}\delta y + o(\delta y) =$$

$$= \delta_z \Phi^{\mathrm{I}} + G^{\mathrm{I}}_{y_*}\delta y + o(\delta y),$$

$$\delta_z \Phi^{\mathrm{I}} = \Phi^{\mathrm{I}}(y,z) - \Phi^{\mathrm{I}}(y,0).$$

Conditions 1) and 2) of the theorem, together with $\psi(t_F)$ from (10) imply

$$R^{\rm I}_{y*} = 0, \quad G^{\rm I}_{y*} = 0,$$

$$\delta I = \delta_z \Phi^{\rm I} + o(\delta y) - \int_{t'}^{t_F} (\delta_z R^{\rm I} + o(\delta y))dt,$$

$$\delta_z \Phi^{\rm I} = \delta_z \Phi(\xi(t_F, z)) + \delta_z \Phi^{\rm I} - \delta_z \Phi(\xi(t_F, z)) =$$

$$\delta_z \Phi(\xi(t_F, z)) + o(\delta y),$$

because

$$\Phi(\xi(t_F, z)) = \Phi^{\rm I}(\chi(t_F, y_*(t_F), z).$$

But $\bar{z} = O(\Delta t)$, $\Delta y = o(\delta t)$, $\Delta t = t_F - t'$, therefore

$$\delta I = O(\Delta t) + o(\Delta t) = \left(\frac{O(\Delta t)}{\Delta t} + \frac{o(\Delta t)}{\Delta t}\right)\Delta t,$$

where $O(\Delta t) < 0$, by supposition. Hence, $\delta I < 0$ for sufficiently small Δt, what is again a contradiction with optimality of m_*.

Example 2.

$$\dot{x}^1 = -u - \frac{1}{2}(x^2 - x^1)^2 - 1, \quad \dot{x}^2 = u, \quad 0 \le u \le 1,$$

$$T = [0, 1], \quad x(0) = x_I, \quad I = x^1(1) + x^2(1),$$

$$x_I^1 = x_I^2.$$

Here

$$F(t, x) = V(t, x) = V_c(t),$$

$$\mathcal{H}(t, x, \psi) = -\psi^1\left(\frac{1}{2}(x^2 - x^1)^2 + 1\right) + (\psi^2 - \psi^1)^+$$

Consider the following sliding mode

$$x_*^i = x_I^i + \frac{1}{2}t, \quad i = 1, 2, \quad \psi^1 = \psi^2 = -1,$$

corresponding adjoint system and its solution

$$\frac{dx}{dz} = \begin{pmatrix} -1 \\ 1 \end{pmatrix}, \quad \frac{dp}{dz} = 0, \quad , \frac{dq}{dz} = 0,$$

$$x^1 = -z + x_*^1(t), \quad x^2 = z + x_*^2(t),$$

$$p = \psi(t), \quad q = 0, \quad x_{1*}(t) = x_{2*}(t).$$

Here

$$\mathcal{H}(t, \xi(t, z), p(t, z)) + q(t, z) = \left(\frac{1}{2}(2z)^2 + 1\right).$$

This expression has is a minimum, not a maximum, at $z = 0$, so the regime under consideration is not optimal.

4. OPTIMALITY CONDITIONS: CONTINUOUS FAMILY OF FACETS

Theorems 1 and 2 can be applied to the cases, when the facet under consideration $F(t, x, p)$ does not depend on p on some set of p. In general case a family of facets is generated by p as some vector parameter, when it is possible to speak apriori of the correspondence the solution under investigation only to this family, not to some definite facet, what is revealed only aposteriori. Now above theorems will be extended on the case of a continuous family of facets.

Let us make preliminary one important notion. When solving optimal control problem 2 it is possible to replace equivalently the original differential restriction (2) by the following one

$$\dot{x} \in F(t, x, p), \quad p \in L_p(t, x) \tag{19}$$

where L_p is dual cone for the set V_c ($L_p = \mathbb{R}^n$ for the bounded V_c), when full class of facets is considered or its subcone, when some special family is considered.

The system (18) have the following property

$$p^T F(t, x, p) = p^T v_0(t, x, p) =$$
$$= \mathcal{H}(t, x, p) = \max_{v \in V_c} p^T v,$$

where $v_0(t, x, p) \in F(t, x, p)$ is an arbitrary fixed point.

Suppose that (19) generates for every p corresponding F-derived system so that the following family of derived systems

$$\dot{y} = \eta_x(t, x, p)F(t, x, p) + \eta_t, \quad y = \eta(t, x, p)$$

appears (where $\eta_x F = \eta_x v_0$, due to (4)).

Accordingly there appears corresponding family of adjoint systems that leads to the following nonlinear (with respect to p) system:

$$\dot{\psi} = -H_x(t, x_*(t), \psi(t), u_*(t)),$$

$$\psi(t_F) = -F_x(x_*(t_F)),$$

$$\frac{\partial x}{\partial z} = h(t, x, p), \quad x = \xi(t, z), \quad \xi(t, 0) = x_*(t), \tag{20}$$

$$\frac{\partial p}{\partial z} = -(p^T h(t, x, p))_x, \quad p(t, 0) = \psi(t),$$

$$\frac{\partial q}{\partial z} = -q h_t(t, x, p), \quad q(t, 0) = 0.$$

In this terms the desired generalizations of theorems 1 and 2 are formulated.

Theorem 3. Let $F(t, x, p)$ be a continuous family of facets of $V_c(t, x)$ and there exists $\tilde{p}(t)$, $t \in T =$

$[t, t_F]$ such that m_* is a solution for the following derived problem

$$\dot{y} = \tilde{\eta}_x \tilde{F}(t,x) + \tilde{\eta}_t, \quad y = \tilde{\eta}(t,x),$$

$$x(t_I) = x_I, \quad I = \Phi(x(t_F)),$$

where $\tilde{\eta}(t,x) = \eta(t,x,\tilde{p}(t))$, and

$$\dot{x}_* \in \tilde{F}(t,x) = F(t,x,p(t)).$$

Then m_* is a solution of the original problem ((2), (7)), and it is an optimal F-sliding mode for the problems ((1), (7)) if

$$\dot{x}_* \in \operatorname{int} \tilde{F}(t, x_*(t)).$$

Theorem 4. Let m_* be an optimal solution for the problem ((2), (7)), with Lipshizian $\mathcal{H}(t,x,\psi)$, and $F(t,x,p)$ be a continuous family of facets of $V_c(t,x)$, $p \in L_p(t,x)$,

$$h(t,x,p) = \{h_l(t,x,p),\ l = 1,2,\ldots,k > 0\}$$

be an internal basis for each facet. Let this family generates a corresponding family of F-derived systems in some open domain C of (t,x) space that contains the graph of $x_*(t)$, and $h_l(t,x,p)$ are continuous together with their derivatives in C for any p. Let there exist a piecewise smooth function $p_*(t)$, $t \in T$, such that $x_*(t)$ is F-sliding mode, that is

$$\dot{x}_* \in \operatorname{int} \tilde{F}(t, x_*(t)).$$

Then there exists a solution

$$(\psi(t), \xi(t,z), p(t,z), q(t,z))$$

of the system (20) such that

1) $\psi^T \dot{x}_* = \mathcal{H}(t, x_*(t), \psi(t))$;

2) $\psi^T(t) h(t, x_*(t), \psi(t)) = 0$;

3) $\mathcal{H}(t, x_*(t), \psi(t)) =$

$$= \operatorname{loc}\max_z (\mathcal{H}(t, \xi(t,z), p(t,z) + q(t,z));$$

4) $\Phi(x_*(t_F)) = \operatorname{loc}\min_z \Phi(\xi(t_F, z))$.

Proofs of these statements are quite similar to above ones for the theorems 1 and 2, and we will not repeat them.

Example 3. Let

$$\dot{x} = (x_1, x_2), \quad u = (u_1, u_2), \quad x_1, u_1 \in \mathbb{R}^2,$$

$$x_2, u_2 \in \mathbb{R}, \quad I = [0,1],$$

$$\dot{x}_1 = -u_1 u_2 - \left(\frac{1}{m}(x_2 - |x_1|)^m - 1\right) x_1 |x_1|^{-1},$$

$$\dot{x}_2 = u_2, \quad |u_1| = 1, \quad 0 \le u_2 \le 1, \quad m > 1, \tag{21}$$

$$x_1(0) = x_{1I}, \quad x_2(0) = x_{2I},$$

$$I = |x_1(1)| + x_2(1).$$

The velocity set of this system being a conic-type body

$$V_c = \{v\colon |v_1 + a(x)| = v_2, \quad 0 \le v_2 \le 1\},$$

$$v_1 \in \mathbb{R}^2, \quad v_2 \in \mathbb{R},$$

has a continuous family of 1-dimensional facets with

$$L_p = \{p\colon |p_1| + p_2 = 0\},$$

where $p_1 \in \mathbb{R}^2$, $p_2 \in \mathbb{R}$.

We have

$$H(t,x,p,u) = p_2 u_2 + p_1^T\Big(-u_1 u_2 -$$

$$-\left(\frac{1}{m}(x_2 - |x_1|)^m - 1\right) x_1 |x_1|^{-1}\Big).$$

It is seen that if $p \in L_p$ then H reaches its maximum at $u_1 = -p_1|p_1|^{-1}$ and at any u_2 that is at one of facets (defined by $p_1|p_1|^{-1}$) from above family, so that

$$\mathcal{H} = -p_1^T\left(\frac{1}{m}(x_2 - |x_1|)^m - 1\right) x_1|x_1|^{-1} +$$

$$(|p_1| + p_2)^+$$

$$\dot{\psi}_1 = -\mathcal{H}_{x_1} = -\psi_1^T (x_2 - |x_1|)^{m-1}\Big(1 +$$

$$+\left(\frac{1}{m}(x_2 - |x_1|)^m - 1\right)\left(|x_1|\psi_1 - \frac{\psi_1^T x_x}{|x_1|^3} x_1\right) \tag{22}$$

$$\dot{\psi}_2 = -\mathcal{H}_{x_2} = \psi_1^T (x_2 - |x_1|)^{m-1},$$

$$\psi_1(1) = -\frac{x(1)}{|x(1)|}, \quad \psi_2(1) = -1,$$

$$\psi \in L_p = \{\psi\colon |\psi_1| + \psi_2 = 0\}.$$

These are usual maximum principle equations for the case under, consideration. They are satisfied by various solutions of (21)–(22) for $u_1 = \psi_1|\psi_1|^{-1}$, that lay on the manifold

$$x_2 = |x_1|$$

and have the following properties:

$$u_2 = 1/2, \quad \psi_2 = \text{const},$$

$$\psi_1(t) = \text{const} = -x_1(1)|x_1(1)|^{-1},$$

$$\dot{x}_1 = \frac{1}{2}|x_1||x_1|^{-1}, \quad x_1|x_1|^{-1} = \rho = \text{const},$$

$$|x_1| = x_2 = |x_{1I}| + t/2.$$

Let us check the theorem 4 conditions. Consider the system (20) and its solution

$$\frac{dx}{dz} = \begin{pmatrix} p_1|p_1|^{-1} \\ 1 \end{pmatrix} \frac{1}{\sqrt{2}}, \quad \frac{dp}{dz} = 0, \quad \frac{dq}{dz} = 0,$$

$$p(z)\psi, \quad q(z) = 0, \quad x_1 = \psi|\psi_1|^{-1}\frac{z}{\sqrt{2}} + x_{1*}(t),$$

$$x_2 = \frac{z}{\sqrt{2}} + x_{2*}(t),$$

that is

$$x_1 = -\rho_* \frac{z}{\sqrt{2}} + \rho_*|x_{1*}| = \rho_* \left(|x_{1*}| - \frac{z}{\sqrt{2}}\right),$$

$$x_2 = \frac{z}{\sqrt{2}} + |x_{2*}|,$$

We use these expressions to receive

$$K(t,z) = \mathcal{H}(t, \xi(t,z,\psi), \psi) =$$

$$\rho_*^T \left(\frac{1}{m}(x_2 - |x_1|)^m - 1\right)\rho_* =$$

$$= \frac{1}{m}\left(\frac{2z}{\sqrt{2}}\right)^m = \frac{1}{m}2^{m/2}(z)^m.$$

for any solution of above class.

It is seen that $K(t,z)$ reaches a minimum, not a maximum, at $z_* = 0$ that is all sliding modes under consideration are not optimal.

5. DISCUSSION

Let us point out that on the contrary to known necessary conditions (Kelley, 1964; Kopp and Moyer, 1965; Gabasov and Kirillova, 1973) theorems 2 and 4 gives us more general local conditions in nondifferential form. Its advantage is visually demonstrated in the example 2. Here full (exact) dependence K on z whereas the known differential conditions of local extremum are not applicable for

$$2n + 1 < m < 2(n+1), \quad n = 0, 1, \ldots.$$

For example,

$$K_z = 2^{m/2}\, z^{m-1} = 0,$$

but

$$K_{zz} = 2^{m/2}(m-1)z^{m-2}$$

does not exist at $z_* = 0$ when $1 < m < 2$. Also

$$K_z = K_{zz} = K_z^{(3)} = 0,$$

but $K_z^{(4)}$ does not exist at $z_* = 0$ when $s < m < 4$, and so on.

Theorems 1 and 3 contain sufficient optimality conditions for sliding regimes in some general form: to be solutions for the appropriate derived problems independently of the methods we apply to the latter ones. One can refer to any available sufficient conditions for the derived problem as an optimal control problem, for instance (Krotov, 1997), to drew corresponding concrete sufficient conditions for sliding modes under consideration if required. Also it is possible to apply special methods of control improvement based on transformations to derived problems (Gurman, et al., 1987; 1988).

In the whole the pair of sufficient and necessary conditions are mutual compatible in the sense that they are based on the same notions (convexified velocity set, facet, facet-bound basis system, and its full integral or description of its elements by the adjoint system). This makes them as convenient tools in applications as well as in further theoretical studies, for the case of general dependence $F(t,x,p)$ that can be considered as a proper combination of above two cases.

6. REFERENCES

Clarke F.H. (1983). *Optimization and nonsmooth analysis.* John Wiley and Sons, New York.

Gabasov R. and F.M.Kirillova (1973). *Singular optimal controls.* Nauka, Moscow.

Goh B. (1966). Necessary conditions for singular Bolza problem. *SIAM J. Control,* **4**, 716-731.

Gurman V.I. (1965). On optimal processes of singular control. *Automatics and Remote Control,* **26**, 782-791.

Gurman V.I. (1972). On optimal processes with unbounded derivatives. *Automatics and Remote Control,* N 12, 14-21.

Gurman V.I. (1977). *Degenerate problems of optimal control.* Nauka, Moscow.

Gurman V.I. (1983). *Extension principle in control problems.* Nauka, Moscow.

Gurman V.I., et.al. (1987). *New methods of control processes improvement.* Nauka, Novosibirsk.

Gurman V.I., et.al. (1988). *Methods of improvement in computational experiments.* Nauka, Novosibirsk

Kelley H.J. (1964). A second variation test for singular extremals. *AIAA J.*, **2**, 26-29.

Kopp R.E., and H.G.Moyer (1965). Necessary optimality conditions for singular extremals.- *AIAA J.*, **3**, 84-91.

Krotov V.F. (1996). *Global methods in optimal control.* Marcel Dekker, New York.

Krotov V.F., V.Z.Bukreev and V.I.Gurman (1969). *New variational methods in flight dynamics* (NASA Tech. Transl. TTF-657, 1971), Mashinostroenie, Moscow.

Leichtweiss K. (1980). *Konvexe Mengen.* Berlin: VEB Deutsher-Verlag der Wissenschaften.

CONTRAST STRUCTURES IN SIMPLEST VARIATIONAL VECTOR PROBLEM AND ITS ASYMPTOTICS

Michael G. Dmitriev [1], **Ni Ming Kang**

Program Systems Institute of Russian Academy of Science, Pereslavl-Zalessky

Abstract: Contrast structures (trajectories with inner and boundary layers) in simplest vector variational problem are studied. Necessary optimality conditions lead to singularly perturbed vector nonlinear boundary value problems, for which contrast structures were not investigated before. Applying Krotov's functions to these solutions leads to new results: structures of spike type are connected with points of local maximum and structures with inner passages are connected with the points of global maximum of some functional.

Keywords: Optimal control, singular control, singular perturbations, regularization, cheap control, nonlinear control

,

1. PROBLEMS STATEMENT

Last years attention of many investigators was attracted to so-called contrast structures. Their solutions have zones of fast space-time deviation (inner and boundary layers). Large bibliography on this question is contained in (Butuzov and Vasil'eva, 1987; Butuzov and Vasil'eva, 1990; Vasil'eva, 1990). Let us consider next perturbed problem P_ϵ

$$J_\epsilon(u_\epsilon) = \int_0^T (a(x_\epsilon, t) + b'(x_\epsilon, t)u_\epsilon + \tfrac{1}{2}\epsilon^2 u'_\epsilon u_\epsilon)\, dt \longrightarrow \inf_{u_\epsilon} \qquad (1)$$

$$\dot{x}_\epsilon(t) = u_\epsilon(t), x_\epsilon(0,\epsilon) = 0, x_\epsilon(T,\epsilon) = 0, \quad 0 < \epsilon \ll 1 \qquad (2)$$

where $x_\epsilon(t) \in R^n$, $u_\epsilon(t) \in R^n$, $b(x_\epsilon, t) \in R^n$, $a(x_\epsilon, t) \in R$, T - are given positive number, the prime sign denotes the transposition, $(x_\epsilon, u_\epsilon) \in D = X \times U$, D - an admissible set of pairs (x_ϵ, u_ϵ) with $x_\epsilon(t)$ - an absolutely continuous vector-function.

[1]The work is particularly supported by Russian Foundation of Basic Investigations, project № 96-01-00804

Necessary optimality conditions lead to the special singularly perturbed boundary value problems:

$$\epsilon^2 \ddot{x}_\epsilon = g(x_\epsilon, t), \qquad (3)$$

$$x(0,\epsilon) = 0, \qquad x(T,t) = 0, \qquad (4)$$

where $g(x_\epsilon, t) = \dfrac{\partial a}{\partial x}(x_\epsilon, t) - \dfrac{\partial b}{\partial t}(x_\epsilon, t)$.

In Vasil'eva A. and Butuzov V. (1987) showed for the scalar case (n=1),, that singularly perturbed problem (3), (4) may have solutions with inner layers of spike type and solutions with inner passages. Contrast structures theory was rapidly developed in the last time. However the vector case, which has its own peculiarities, was practically not considered in the literature. Here only the algorithm of asymptotics construction, which is similar to the solution of spike type and the solution with passages from one root to another root (roots of degenerated algebraic equations), is described. The questions of the existence of a solution near constructed asymptotics for the problem (1),(2) is not studied.

It is well known that trajectories with spikes form the separatrix loop on the phase plane, and trajectories with passages from one root to other root form the nucleus. Here, due to the variational nature of boundary value problem (3), (4), these contrast structures may be connected with points of local and global maximum for some characteristic functional, and on the other hand, the hypothesis that definite variational sence have contrast structures in other singularly perturbed boundary value problems may be stated. See, for example, (Boglaev, 1970), where the analogous result for (3),(4) problem was obtained for contrast structures with inner passage.

2. ASYMPTOTICS OF SPIKE TYPE

Let us consider the so-called degenerated problem $\bar{P}$ which we get from (1), (2) for $\epsilon = 0$

$$\bar{J}(\bar{u}) = \int_0^T (a(\bar{x},t) + b'(\bar{x},t)\bar{u})\, dt \longrightarrow \inf_{\bar{u}}, \quad (5)$$

$$\dot{\bar{x}}(t) = \bar{u}(t),$$
$$\bar{x}(0) = 0, \qquad \bar{x}(T) = 0. \quad (6)$$

I. Let $\bar{J}^* = \inf_{\bar{u}} \bar{J} > -\infty$.

With the help of Krotov's function $\varphi(x,t)$ (Krotov and Gurman, 1973) we may rewrite the functional (5) in the form

$$\bar{J}(\bar{u}) = -\varphi(0,0) + \varphi(0,T) - \int_0^T P(\bar{x},t)\, dt,$$

where $P(x,t) = -a(x,t) + \varphi_t(x,t)$ and $\varphi(x,t)$ satisfies the next equation $\varphi_x(x,t) = b(x,t)$. For the asymptotics construction we use the direct scheme method (Belokopytov and Dmitriev, 1986).

Auxiliary lemma: Let function f has next expansion for any n

$$f(w_0,\ldots,w_n,\epsilon) = f_0(w_0) + \sum_{i=1}^n \epsilon^i f_i(w_i,\ldots,w_0) + O(\epsilon^{n+1}),$$

where all functions f, f_i, $i = \overline{0,n}$ have minimum points in some open domain. Then for sufficiently small $\epsilon > 0$

$$\min_{(w_0,\ldots,w_n)} f(w_0,\ldots,w_n,\epsilon) = \min_{w_0} f_0(w_0) + \sum_{i=1}^n \epsilon^i \min_{w_i} \tilde{f}_i(w_i) + O(\epsilon^{n+1}),$$

where $\tilde{f}_i(w_i) = f_i(w_i,\tilde{w}_{i-1},\ldots,\tilde{w}_0)$, $\tilde{w}_k = \arg\min_w f_k(w)$, $k = \overline{0,i-1}$.

Here the case, when there is only one point $t = t_*$ on [0,T] with a solution spike in its neighborhood, is considered. The case of several such points may be investigated by the same method.

Spike point t_* will be found in the form $t_* = t_0 + \epsilon t_1 + \ldots + \epsilon^n t_n + \ldots$, where $\|\dot{x}(t_*,\epsilon)\| = 0$.

By standard technique (Vasil'eva, 1972) the asymptotics is constructed in the form:

$$x(t,\epsilon) = \bar{x}(t,\epsilon) + \Pi x(\tau_0,\epsilon) + Qx(\tau,\epsilon) + Rx(\tau_1,\epsilon), \quad (7)$$
$$u(t,\epsilon) = \dot{x}(t,\epsilon),$$

where $\bar{x}(t,\epsilon) = \sum_{j=0}^{\infty} \bar{x}_j(t)\epsilon^j$ – is a regular series; $\Pi x(\tau_0,\epsilon) = \sum_{j=0}^{\infty} \epsilon^j \Pi x(\tau_0)$ – a boundary layer series in a neighborhood of point $t = 0$ ($\tau_0 = \frac{t}{\epsilon}$); $Qx(\tau,\epsilon) = \sum_{j=0}^{\infty} \epsilon^i Q_j x(\tau)$ – is a series, describing the solution's spike in a neighborhood of $t = t_*$ ($\tau = \frac{t-t_*}{\epsilon}$); $Rx(\tau_1,\epsilon) = \sum_{j=0}^{\infty} \epsilon^j R_j x(\tau_1)$ – is a boundary layer near $t = T$ ($\tau_1 = \frac{t-T}{\epsilon}$).

A1. Let the function $P(x,t)$ has a local maximum in $x = \alpha(t)$ (i.e. $P_x(\alpha(t),t) = 0$, $P_{xx}(\alpha(t),t) < 0$, $0 \le t \le T$) and there exists the function $x = \gamma(t)$ such that $P(\gamma(t),t) = P(\alpha(t),t)$ and $P_x(\gamma(t),t) \neq 0$, $0 \le t \le T$ (for definiteness $\alpha_i(t) < \gamma_i(t)$, $i = \overline{1,n}$).

From **A1** we obtain that $x = \alpha(t)$ is a stationary point of the saddle type for the adjacent to (3) equation (t is fixed).

A2. Let a separatrix, which go out from the saddle stationary point, forms the loop in the phase space , i.e. the separatrix contains only homoclinic points (Pilugin, 1988).

With the help of the standard technique for regular terms in approximation of the order i one gets

$$\dot{\bar{x}}_i(t) = \bar{u}_i(t), \quad \bar{x}_i(0) + \Pi_i x(0) = 0,$$
$$\bar{x}_i(T) + R_i x(0) = 0, \quad i = 0,1,\ldots$$

For boundary layers $Q_i x(\tau)$ one gets the next equations and boundary conditions

$$\begin{cases} \dfrac{dQ_i x(\tau)}{d\tau} = Q_i u(\tau), \\ \dot{Q}_i x(0) = 0, \qquad \lim_{\tau\to+\infty} Q_i x(\tau) = 0. \end{cases}$$

Functional expansion $J_\epsilon(u_\epsilon) = \sum_{j=0}^{\infty} \epsilon^j J_j$ gives the next

Lemma 1. If the conditions **I**, $\mathbf{A_1}$ take place, then the identity $J_0^* = \bar{J}^*$ holds, where the star is the optimum symbol.

Note that J_0 doesn't depend on boundary layers $\Pi_0 x$, $Q_0 x$, $R_0 x$. It can be shown, that after rather simple but long calculations one gets the next expression for $Q_1 J$

$$Q_1 J = 2\int_0^{+\infty} (P(\alpha(t_0), t_0) - P(\alpha(t_0) + Q_0 x(\tau), t_0) + \frac{1}{2}(Q_0 u(\tau))' Q_0 u(\tau)\Big) \, d\tau$$

From $\mathbf{A_1}$ follows the initial condition $Q_0 x(0) = \gamma(t_0) - \alpha(t_0)$.

Thus $Q_0 P$ can be rewritten in the form

$$Q_0 P : \begin{cases} Q_1 J = 2\int_0^{+\infty} (P(\alpha(t_0), t_0) \\ -P(\alpha(t_0) + Q_0 x(\tau), t_0) + \\ +\frac{1}{2}(Q_0 u(\tau))' Q_0 u(\tau)\Big) \, d\tau \longrightarrow \min\limits_{Q_0 u}, \\ \dfrac{dQ_0 x(\tau)}{d\tau} = Q_0 u(\tau), \\ Q_0 x(0) = \gamma(t_0) - \alpha(t_0), \\ \lim\limits_{\tau \to +\infty} Q_0 x(\tau) = 0. \end{cases}$$

Lemma 2 (Lukes, 1969). Provided that the conditions **I**, $\mathbf{A_1}$ hold, for any $Q_0 x(0) \in S$ there exists a unique optimal solution $Q_0^* x(\tau)$, $Q_0^* u(\tau)$ for $Q_0 P$, which satisfies the next estimates

$$\|Q_0^* x(\tau)\| \le C e^{-\beta\tau}, \; \|Q_0^* u(\tau)\| \le C e^{-\beta\tau}, \; \tau \ge 0,$$

where β and C – are some positive numbers, S – an attraction set for the problem $Q_0 P$.

t_0 may be found as a minimizing parameter in $Q_1^* J$, i.e. $t_0 = \arg\min\limits_{0<t_0<T} Q_1 J^*(\gamma(t_0) - \alpha(t_0))$. Then

$$\frac{d}{dt} Q_1 J^*(\gamma(t) - \alpha(t))|_{t=t_0} = 0. \tag{8}$$

$\mathbf{A_3}$. Let the equation (8) has a unique solution $t_0 \in (0, T)$, and $\dfrac{d^2}{dt^2} Q_1 J^*(\gamma(t) - \alpha(t))|_{t=t_0} > 0$.

Boundary layers functions $\Pi_0^* x(\tau_0)$ and $R_0^* x(\tau_1)$ may be determined analogously under the next condition

$\mathbf{A_4}$. Let initial values $\Pi_0 x(0) = -\alpha(0)$, $R_0 x(0) = -\alpha(T)$ belong to attraction sets in the problems $\Pi_0 P$, $R_0 P$ correspondingly.

The procedure described above may be repeated and leads to approximated decomposition problems for terms of higher approximation

$$\bar{P}_{n+1} : \begin{cases} \bar{J}_{2(n+1)} = -\int_0^T \Big(\frac{1}{2}(\bar{x}_{n+1}^0(t))' * \\ P_{xx}(\alpha(t), t)\bar{x}_{n+1}(t) + \bar{H}_{n+1}^1(t)\bar{x}_{n+1}(t) + \\ +\bar{H}_{n+1}^2(t)\bar{u}_{n+1}(t)\Big) \, dt \longrightarrow \inf\limits_{\bar{u}_{n+1}}, \\ \dot{\bar{x}}_{n+1}(t) = \bar{u}_{n+1}(t), \end{cases}$$

where $\bar{H}_{n+1}^1(t)$, $\bar{H}_{n+1}^2(t)$ depend on the already calculated terms

$\bar{w}_j(t)$, $0 \le j \le n$.

$$Q_{n+1}P : \begin{cases} Q_{2(n+1)+1} J = -2\int_0^{\infty} \Big(\frac{1}{2}(Q_{n+1}x)' * \\ P_{x^2}(\alpha(t_0) + Q_0 x(\tau), t_0) Q_{n+1} x(\tau) + \\ +\frac{1}{2}(Q_{n+1}u(\tau))' Q_{n+1} u(\tau) + \\ +H_{n+1}^1(\tau) Q_{n+1} x(\tau) + \\ +H_{n+1}^2(\tau) Q_{n+1} u(\tau)\Big) \, d\tau \longrightarrow \min\limits_{Q_{n+1}u}, \\ \dfrac{dQ_{n+1}x(\tau)}{d\tau} = Q_{n+1} u(\tau), \\ Q_{n+1}\dot{x}(0) = 0, \; \lim\limits_{\tau \to +\infty} Q_{n+1} x(\tau) = 0, \end{cases}$$

and $\bar{H}_{n+1}^1(\tau)$, $\bar{H}_{n+1}^2(\tau)$ depend on the known terms $Q_j w(\tau) = (Q_j x(\tau), Q_j u(\tau))'$, $\bar{w}_l(t) = (\bar{x}(t), \bar{u}(t))$, $0 \le j \le n$, $0 \le l \le n+1$.

For arbitrary n, $\bar{x}_n^*(t)$, $Q_n^* x(\tau)$, $\Pi_n^* x(\tau_0)$, $R_n^* x(\tau_1)$ can be determined from linear-quadratic problems $\bar{P}_{n+1}$, $Q_{n+1}P$ and Π_{n+1}, $R_{n+1}P$ (Krotov and Gurman, 1973).

After minimization procedure in J_j in functional expansion J_ϵ, we find t_n, $n \ge 1$. Let $\tilde{x}_n(t, \epsilon)$ and $\tilde{J}_n$ denote the particular sums of asymptotics (7) and the functional expansion correspondingly, where

$$\tilde{x}_n(t, \epsilon) = \sum_{j=0}^{n} \epsilon^j (\bar{x}_j(t) + \Pi_j x(\tau_0) + Q_j x(\tau) + R_j x(\tau_1)), \tag{9}$$

$$\tilde{u}_n(t, \epsilon) = \dot{\tilde{x}}_n(t, \epsilon), \; \tilde{J}_n = \sum_{j=0}^{2n+1} \epsilon^j J_j. \tag{10}$$

Note, that $(\tilde{x}_n, \tilde{u}_n)$ is not an admissible pair, since boundary conditions (2) are not valid, i.e. $\tilde{x}_n(0, \epsilon) = P_1(\epsilon, n)$, $\tilde{x}_n(T, \epsilon) = P_2(\epsilon, n)$, where $P_i(\epsilon, n) = O(\epsilon^{n+1})$ $(i = 1, 2)$.

To obtain an admissible pair (x, u) next functions are introduced:

$$\theta_n(t, \epsilon) = A e^{-\frac{t}{\epsilon}} + B e^{-\frac{T-t}{\epsilon}}, \quad 0 \le t \le T,$$

where coefficients A, B are chosen so, that $\theta_n(0, \epsilon) = -P_1(\epsilon, n)$, $\theta_n(T, \epsilon) = -P_2(\epsilon, n)$. For that we need take $A = (-P_1(\epsilon, n) + e^{-\frac{T}{\epsilon}} P_2(\epsilon, n))/(1 - e^{-\frac{2T}{\epsilon}})$, $B =$

$(-P_2(\epsilon,n)+e^{-\frac{T}{\epsilon}}P_1(\epsilon,n))/(1-e^{-\frac{2T}{\epsilon}})$. Evidently, A, B - are the quantities of the order ϵ^{n+1}, so $\theta_n(t,\epsilon)=O(\epsilon^{n+1})$.

Let $X_n(t,\epsilon)=\tilde{x}_n(t,\epsilon)+\theta_n(t,\epsilon)$, $U_n(t,\epsilon)=\dot{X}_n(t,\epsilon)$, then the pair $(X_n(t,\epsilon), U_n(t,\epsilon))$ is admissible. Denoting $J_n(U_n)=\tilde{J}_n(U_n)$, we have $\tilde{J}_n(\tilde{u}_n)=J_n(U_n)+O(\epsilon^{n+1})$.

Theorem 1. If there exists a solution $(x_\epsilon^*, u_\epsilon^*)$ of the problem (1), (2) and the conditions **I**, $\mathbf{A_1}-\mathbf{A_4}$ hold then for sufficiently small $\epsilon>0$ we have

$$\|x_\epsilon^*-X_n\|\le C\epsilon^{n+1},\quad \|u_\epsilon^*-U_n\|\le C\epsilon^{n+1},$$
$$\|J_\epsilon^*-J_n(U_n)\|\le C\epsilon^{2n+2}.$$

The proof of this theorem is given in the Appendix.

3. ASYMPTOTICS OF PASSAGE TYPE

$\mathbf{B_1}$. Let there exist vector functions $\alpha(t)\in X$, $\gamma(t)\in X$ such that $P(\alpha(t),t)=P(\gamma(t),t)=\max_x P(x(t),t)$ and $P_{xx}(\alpha(t),t)<0$, $P_{xx}(\gamma(t),t)<0$, $0\le t\le T$, (for simplicity the maximum points $\alpha(t)$, $\gamma(t)$ are supposed to be neighboring).

$\mathbf{B_2}$. Let separatrixs, connecting saddle points in the phase space $(\alpha,0)$, $(\gamma,0)$, form a nucleus.

The value of the passage point t_* can be founded in form:

$$t_*=t_0+\epsilon t_1+\ldots+\epsilon^k t_k+\ldots.$$

The value of $x(t_*,\epsilon)$ in the point t_* is equal to k, this can be represented in the form: $k=k_0+\epsilon k_1+\ldots+\epsilon^n k_n+\ldots$.

The asymptotics of the problem solution may be constructed in the form:

$$x(t,\epsilon)=\begin{cases}\bar{x}^{(1)}(t,\epsilon)+\Pi x(\tau_0,\epsilon)+Q^{(1)}x(\tau,\epsilon),\\ 0\le t\le t_*,\\ \bar{x}^{(2)}(t,\epsilon)+Rx(\tau_1,\epsilon)+Q^{(2)}x(\tau,\epsilon),\\ t_*\le t\le T,\end{cases}$$

where $\tau_0=\dfrac{t}{\epsilon}$, $\tau_1=\dfrac{t-T}{\epsilon}$, $\tau=\dfrac{t-t_*}{\epsilon}$.

Lemma 3. If conditions **I**, $\mathbf{B_1}$ take place, then $\bar{J}^*=J_0^*$.

For problems $Q_0^{(1)}P$, $Q_0^{(2)}P$ we have

$$Q_0^{(1)}P:\begin{cases}Q_1^{(1)}J=-\displaystyle\int_0^{-\infty}\Big(P(\alpha(t_0),t_0)-P(\alpha(t_0)+Q_0^{(1)}x(\tau),t_0)+\frac{1}{2}(Q_0^{(1)}u(\tau))'(Q_0^{(1)}u(\tau))\Big)\,d\tau\longrightarrow\inf_{Q_0^{(1)}u},\\ \dfrac{dQ_0^{(1)}x(\tau)}{d\tau_0}=Q_0^{(1)}u(\tau),\\ Q_0^{(1)}x(0)=k_0-\alpha(t_0),\\ \lim\limits_{\tau\to-\infty}Q_0^{(1)}x(\tau)=0\end{cases}$$

$$Q_0^{(2)}P:\begin{cases}Q_1^{(2)}J=-\displaystyle\int_0^{+\infty}\Big(P(\gamma(t_0),t_0)-P(\gamma(t_0)+Q_0^{(2)}x(\tau),t_0)++\frac{1}{2}(Q_0^{(2)}u(\tau))'\times\\ \times(Q_0^{(2)}u(\tau))\Big)\,d\tau\longrightarrow\inf_{Q_0^{(2)}u},\\ \dfrac{dQ_0^{(2)}x(\tau)}{d\tau}=Q_0^{(2)}u(\tau),\\ Q_0^{(2)}x(0)=k_0-\gamma(t_0),\\ \lim\limits_{\tau\to+\infty}Q_0^{(2)}x(\tau)=0.\end{cases}$$

Lemma 4. *If conditions* **I**, $\mathbf{B_1}$, $\mathbf{B_2}$ *take place, then for any $Q_0^{(j)}x(0)\in S^j$ $(j=1,2)$ there exist unique optimal solutions $Q_0^{*(j)}x(\tau)$, $Q_0^{*(j)}u(\tau)$ of problems $Q_0^{(j)}P$ $(j=1,2)$, satisfying the next estimates:*

$$\|Q_0^{*(1)}x(\tau)\|\le Ce^{\beta\tau},\quad \|Q_0^{*(2)}x(\tau)\|\le Ce^{-\beta\tau},$$
$$\|Q_0^{*(1)}u(\tau)\|\le Ce^{\beta\tau},\quad \|Q_0^{*(2)}u(\tau)\|\le Ce^{-\beta\tau},$$
$$\tau\le 0,\qquad\qquad \tau\ge 0,$$

where C and β - are some positive numbers, S^j - attraction sets for $Q_0^{(j)}P$ $(j=1,2)$.

(k_0,t_0) can be found as a parameter vector, minimizing

$$M_0(t_0,k_0)=Q_1^{(1)}J^*(t_0,Q_0^{(1)}x(0))++Q_1^{(2)}J^*(t_0,Q_0^{(2)}x(0)),$$

where $Q_1^{(j)}J^*$ - is the optimal value $Q_1^{(j)}J$ for concrete $Q_0^{(j)}x(0)$ $(j=1,2)$, i.e. $(t_0,k_0)=\arg\min\limits_{(t,k)}M_0(t,k)$ or the pair (t_0,k_0) satisfies the next equations: $\dfrac{\partial M_0}{\partial t}(t_0,k_0)=0$, $\dfrac{\partial M_0}{\partial k}(t_0,k_0)=0$.

$\mathbf{B_3}$. Let the last system of the equations have a solution (t_0,k_0), and $\dfrac{\partial^2 M}{\partial(t_0,k_0)^2}(t_0,k_0)$ - be a positive definite matrix.

$\mathbf{B_4}$. Let initial values $\Pi_0x(0)=-\alpha(0)$, $R_0x(0)=-\gamma(T)$ belong to attraction sets of problems Π_0P and R_0P correspondingly.

Linear quadratic problems for the next terms of asymptotic expansion can be obtained as usual.

After construction of admissible pairs on the base of partial sums (X_n, U_n) - of asymptotics of order n we come to

Theorem 2. *If an exact solution $(x_\epsilon^*, u_\epsilon^*)$ of (1), (2) exists and conditions* **I**, $\mathbf{B_1} - \mathbf{B_4}$ *are fulfilled, then for sufficiently small $\epsilon > 0$:*

$$||x_\epsilon^* - X_n|| \le C\epsilon^{n+1}, \qquad ||u_\epsilon^* - U_n|| \le C\epsilon^{n+1},$$
$$||J_\epsilon^* - J_n(U_n)|| \le C\epsilon^{2n+2}.$$

4. CONCLUSION

Applying Krotov's sufficient optimality conditions technique to the singularly perturbed problems (3), (4) in scalar and vector cases proved to be very useful. It leads to some interesting conclusions concerning a variational nature of contrast structures. It proved to be, that the spike type structures are connected with the local maximum points of some characteristic functional $P(x,t)$ and passage contrast structures are connected with the global maximum points of the same functional.

APPENDIX

Proof of Lemma 1. J_0 can be represented in the form:

$$J_0 = \int_0^T (a(\bar{x}_0(t),t) + b'(\bar{x}_0(t),t)\bar{u}_0(t))dt+$$
$$+\int_0^{+\infty} b'(\bar{x}_0(0) + \Pi_0 x(\tau_0), 0)\Pi_0 u(\tau_0)d\tau_0+$$
$$\int_{-\infty}^{+\infty} b'(\bar{x}_0(t_0) + +Q_0 x(\tau), t_0)Q_0 u(\tau)d\tau+$$
$$\int_{-\infty}^{0} b'(\bar{x}_0(T) + R_0 x(\tau_1), T)R_0 u(\tau_1)d\tau_1.$$

Since, $\dfrac{d\varphi}{dt}(\bar{x}_0, t) = \dfrac{\partial\varphi}{\partial t}(\bar{x}_0, t) + \dfrac{\partial\varphi}{\partial x}(\bar{x}_0, t)\bar{u}_0,$

$$\frac{d\Pi_0 x(\tau_0)}{d\tau_0} = \Pi_0 u(\tau_0), \qquad \frac{dQ_0 x(\tau)}{d\tau} = Q_0 u(\tau),$$
$$\frac{dRx_0(\tau_1)}{d\tau_1} = R_0 u(\tau_1)$$

we have

$$J_0 = \int_0^T (a(\bar{x}_0(t),t) - \varphi_t(x,t) + (b(\bar{x}_0(t),t) -$$
$$-\varphi_x(x,t))'\bar{u}_0 + \frac{d\varphi}{dt}(x,t)(\bar{x}_0(t),t))\,dt+$$
$$+\int_0^{+\infty} \frac{d\varphi}{d\tau_0}(\bar{x}_0(0) + \Pi_0 x(\tau_0), 0)\,d\tau_0+$$
$$+\int_{-\infty}^{0} \frac{d\varphi}{d\tau_1}(\bar{x}_0(T) + R_0 x(\tau_1), T)\,d\tau_1$$
$$= \varphi(\bar{x}_0(T) + R_0 x(0), T) - \varphi(\bar{x}_0(0)+$$
$$+\Pi_0 x(0), 0) + \int_0^T (a(\bar{x}_0(t),t) - \varphi_t(x,t))\,dt$$
$$= -\varphi(0,0) + \varphi(0,T) - \int_0^T P(\bar{x}_0(t),t)\,dt.$$

From last expression we get $J_0{}^* = \bar{J}^*$.

Proof of Theorem 1. Let us introduce the extended functional

$$I_\epsilon(x,u) = \int_0^T \left(a(x,t) + b'(x,t)u + \tfrac{1}{2}\epsilon^2 u'u\right)\,dt+$$
$$\int_0^T \tilde{\varphi}_n'(\dot{x} - u)dt,$$

where $\tilde{\varphi}_n = \sum_{i=1}^{n} \epsilon^i(\bar{\varphi}_i(t) + \Pi_i\varphi(\tau_0) + R_i\varphi(\tau_1))$. After the integration of second term I_ϵ by parts I_ϵ can be rewritten in the form

$$I_\epsilon(x,u) = -\int_0^T H(x,u,\tilde{\varphi}_n,t)\,dt$$
$$-\int_0^T \dot{\tilde{\varphi}}_n'(x,t)x(t)\,dt$$
$$+\tilde{\varphi}_n'(T)x(T) - \tilde{\varphi}_n'(0)x(0),$$

where $H(x,u,\tilde{\varphi}_n,t) = \tilde{\varphi}_n' u - \left(a(x,t) + b'(x,t)u + \dfrac{1}{2}\epsilon^2 u'u\right)$, and $\bar{\varphi}_i(t)$, $\Pi_i\varphi(\tau_0)$, $R_i\varphi(\tau_1)$ - terms , corresponding to costate variables in problems P_i, $\Pi_i P$, $R_i P$ correspondingly.

Substituting $\tilde{x}_n(t,\epsilon)$, $\tilde{u}_n(t,\epsilon)$ in $I_\epsilon(x,u)$, we get

$$I_\epsilon(\tilde{x}_n, \tilde{u}_n) = \tilde{J}_n + O(\epsilon^{2n+2}). \qquad (11)$$

At the same time we have

$$\dot{\tilde{\varphi}}_n = -H_x(\tilde{x}_n, \tilde{u}_n, \tilde{\varphi}_n, t)+$$
$$+O\left(\epsilon^{n+1} + \epsilon^n exp(-\frac{Ct}{\epsilon})+\right.$$
$$+\epsilon^n exp(-\frac{|t-t_*|}{\epsilon}C)$$
$$\left.+\epsilon^n exp(-\frac{T-t}{\epsilon}C)\right), \qquad (12)$$

$$0 = -H_u(\tilde{x}_n, \tilde{u}_n, \tilde{\varphi}_n, t) + O(\epsilon^{n+1}), \qquad (13)$$

Due to $\mathbf{A_3}$ functional I_ϵ is strictly convex, so I_ϵ has minimum in the point $(\tilde{x}^*, \tilde{u}^*)$. As it is known from (Vasil'ev, 1981), for $\tilde{w}^* = (\tilde{x}^*, \tilde{u}^*)$, $\tilde{w}_n = (\tilde{x}_n, \tilde{u}_n) \in V$ the next inequality holds

$$I_\epsilon^*(\tilde{x}^*, \tilde{u}^*) - I_\epsilon(\tilde{x}_n, \tilde{u}_n) + \langle \tfrac{\partial I_\epsilon}{\partial w}(\tilde{w}_n), \tilde{w}^*$$
$$-\tilde{w}_n\rangle \ge C||\tilde{w}^* - \tilde{w}_n||^2, \qquad (14)$$

where $C > 0$.

Since $I_\epsilon^*(\tilde{x}^*, \tilde{u}^*) - I_\epsilon(\tilde{x}_n, \tilde{u}_n) < 0$ and

$$\left\langle \frac{\partial I_\epsilon}{\partial w}(\tilde{w}_n),\ \tilde{w}^* - \tilde{w}_n \right\rangle \le \left\| \frac{\partial I_\epsilon}{\partial w}(\tilde{w}_n) \right\| \|\tilde{w}^* - \tilde{w}_n\|,$$

we have $\|\tilde{w}^* - \tilde{w}_n\| \le \left\| \frac{\partial I_\epsilon}{\partial w}(\tilde{w}_n) \right\|$.

From

$$\frac{\partial I_\epsilon}{\partial x}(\tilde{x}_n, \tilde{u}_n) = -\int_0^T (H_x(\tilde{x}_n, \tilde{u}_n, \tilde{\varphi}_n, t) + \dot{\tilde{\varphi}}_n)\, dt,$$

$$\frac{\partial I_\epsilon}{\partial u}(\tilde{x}_n, \tilde{u}_n) = -\int_0^T H_u(\tilde{x}_n, \tilde{u}_n, \tilde{\varphi}_n, t) dt,$$

we get

$$\left\| \frac{\partial I_\epsilon}{\partial w}(\tilde{w}_n) \right\| \le C\epsilon^{n+1}, \qquad (15)$$

and hence

$$\|\tilde{w}^* - \tilde{w}_n)\| \le C\epsilon^{n+1}, \qquad (16)$$

i.e. $\|\tilde{x}^* - \tilde{x}_n)\| \le C\epsilon^{n+1}, \quad \|\tilde{u}^* - \tilde{u}_n\| \le C\epsilon^{n+1}$.

From (14) we get the estimate

$$\|I_\epsilon(\tilde{x}_n, \tilde{u}_n) - I_\epsilon^*\| \le \left\| \frac{\partial I_\epsilon}{\partial w}(\tilde{w}_n) \right\| \|\tilde{w}_n - \tilde{w}^*\|,$$

and from (11), (15), (16) we obtain

$$\|\tilde{J}_n - I_\epsilon^*\| \le C\epsilon^{2n+2}. \qquad (17)$$

If exact solution $(x_\epsilon^*, u_\epsilon^*)$ of the initial problem (1), (2) exists, then substituting $(x_\epsilon^*, u_\epsilon^*)$ in $I_\epsilon(x, u)$, we get $I_\epsilon(x_\epsilon^*, u_\epsilon^*) = J_\epsilon^*$. Analogous estimates (16), (17) are valid for $(x_\epsilon^*, u_\epsilon^*)$ instead of $(\tilde{x}_n, \tilde{u}_n)$

$$\|\tilde{w}^* - w_\epsilon^*\| \le C\epsilon^{n+1}, \qquad \|J_\epsilon^* - I_\epsilon^*\| \le C\epsilon^{2n+2}.$$

Using the triangle inequality and relations (16), (17) we immediately get desired estimates. Thus, the theorem is proved.

The proof of other statements follows the scheme from (Belokopytov and Dmitriev, 1986).

REFERENCES

Belokopytov S.V. and M.G.Dmitriev (1986). Direct scheme in optimal control problems with fast and slow motions. *System and Control Letters*. **8**, № 7, pp.129-135.

Boglaev Yu.P. (1970). About two-point problem for one class of ordinary differential equations with small parameter under derivative. it Journal атем. и матем. физ., bf 10, № 4, pp. 958-968.

Butuzov V.F. and A.B.Vasil'eva (1987). About asymptotics contrast structures. *Matematicheskie zametki*, **42**, № 6, pp.831-84 (In Russian).

Butuzov V.F. and A.B.Vasil'eva (1990). *Asymptotic methods in singular perturbations theory.* Vyschaya skola, Moscow.

Krotov V.F. and V.I.Gurman (1973). *Metody i zadachi optimalnogo upravleniya.* Nauka, Moscow (in Russian).

Lukes D.L. (1969). Optimal regulation of nonlinear dynamical system. *SIAM J. Control.* **7**, № 1, pp. 75-100.

Pilugin S.Yu. (1988). *Vvedenie v grubye sistemy differencial'nyh uravnenij.* Izdatelstvo Leningradskogo Universiteta, Leningrad (in Russian).

Vasil'ev F.P. (1981). *Metody resheniya ekstremal'nyh zadach.* Nauka, Moscow (in Russian).

Vasil'eva A.B. (1972). K voprosu o blizkih k razryvnym resheniyam v sisteme s malym parametrom pri proizvodnyh uslovno ustojchivogo tipa. it Differencial'nye uravneniya, **8**, № .9, pp.1560-1568 (in Russian).

Vasil'eva A.B. (1990). K voprosu ob ustojchivosti reshenij tipa kontrastnyh struktur. it Matematicheskoe modelirovanie, **2**, № .1, pp.119-125 (in Russian).

FAST PERIODIC OSCILLATIONS IN SINGULARLY PERTURBED RELAY CONTROL SYSTEMS

Fridman L. *

Dept. of Math., Samara State Architecture and Civil Engineering Academy, 13-64 Gagarina str., Samara 443079, Russia. e-mail: fridman@icc.ssaba.samara.ru

Abstract. The singularly perturbed relay control systems (SPRCS) are examined. The mathematical apparatus for investigation of the fast periodic oscillations of SPRCS is developed. The theorem about existence of fast periodic solution of SPRCS is proved. The theorem about averaging is given. The algorithm of asymptotic representation for the fast periodic solution of SPRCS is suggested. The algorithm of correction of the averaged equation is given. Their stability of the fast periodic solution is investigated.

Key Words. Variable Structure Systems, Singular perturbations, Periodic motion, Stability.

1.INTRODUCTION

In the paper we will consider the existence and stability of the fast periodic solutions for the singularly perturbed relay control system of the form

$$\mu dz/dt = g(z,\xi,x,u(\xi)), \qquad (1)$$

$$\mu d\xi/dt = h_1(z,\xi,x,u(\xi)),\ dx/dt = h_2(z,\xi,x,u(\xi)),$$

where $z \in \mathbf{R}^m, \xi \in \mathbf{R}, x \in \mathbf{R}^n, u(\xi) = sign\,\xi, g, h_1, h_2$ are smooth functions of their arguments.

Introducing the "fast time" $\tau = t/\mu$ into (1), we will obtain

$$dz/d\tau = g(z,\xi,x,u(\xi)), \qquad (2)$$

$$d\xi/d\tau = h_1(z,\xi,x,u(\xi)),\ dx/d\tau = \mu h_2(z,\xi,x,u(\xi)).$$

For the smoothly singularly perturbed system the existence and stability in the first approximation of the fast periodic solution was investigated by Pontriagin and Rodygin (1962). The existence and stability in the first approximation of fast periodic solution of (1) was investigated in Fridman (1986), Fridman (1990) . It turns out that for the investigation of the fast periodic solutions of singularly perturbed system (2) it's impossible to use standard methods of small parameter (see for example Coddington and Levinson (1955)) for autonomous systems because setting $\mu = 0$ in (2) we will obtain degenerate equation for the slow variables x.

In this paper we develop the mathematical apparatus for investigation of the fast periodic oscillations (1),(2) is developed. For this end we employ. the point mapping method (see Neimark (1963),Neimark (1972)). In section 1 the specific features of the point mapping which generated by system (2) is investigated. In section 3 the theorem about existence of fast periodic solution of system (1) is proved. A proof of the theorem about investigation of stability in the first approximation is given in section 4. In section 5 the auxiliary theorems about decomposition of two speed point mapping are formulated. The theorem about averaging is given in section 6. Section 7 is devoted to the algorithm of asymptotic representation for the fast periodic solution of system (1). The algorithm of correction of the averaged equation is suggested in section 8. The reduction principle theorem is given in section 9.

2 SOME PROPERTIES OF THE POINT MAPPING WHICH MADE BY SPRCS

Let us mark the variation domain as Z, X variables (z,s,x) and x.

Definition. *We shall call the surface $\xi = 0$ the surface without stable sliding towards trajectories of the system*

$$dz/d\tau = g(z,\xi,x,u(\xi)), \qquad (3)$$

$$d\xi/d\tau = h_1(z,\xi,x,u(\xi))$$

if all the trajectories of (1) which start outside the surface $\xi = 0$ cross it at the point $(z,0,x)$ where the conditions (0.3) are fulfilled.

Suppose that the following conditions are true:

1^0 $\quad h_1, h_2, g \in \mathbf{C}^2[\bar{Z} \times [-1,1]]$;

2^0 surface $\xi = 0$ under all $x \in \bar{X}$ is a surface without stable sliding towards trajectories of system (3);

3^0 system (3) for all $x \in \bar{X}$ has an isolated orbitally asymptotically stable solution $(z_0(\tau, x), \xi_0(\tau, x))$ with the period $T(x)$;

4^0 let $R(z, x)$ be a point mapping of the set $V = \{(z, x) : h_1(z, 0, x, 1) > 0\}$ on the surface $\xi = 0$ into itself, performed by system (3) which has a fixed point $z^*(x)$ corresponding to $(z_0(\tau, x), \xi_0(\tau, x))$;

5^0 suppose that for $\lambda_i(x_0)$ $(i = 1, ..., m)$ the eigenvalues of the matrix $\frac{\partial R}{\partial z}(z^*(x_0), x_0)$ the inequalities $|\lambda_i(x_0)| \neq 1$ are true;

6^0 the averaged system $dx/dt = p(x)$, where $p(x) =$

$$= \frac{1}{T(x)} \int_0^{T(x)} h_2(z_0(\tau, x), \xi_0(\tau, x), x, u(\xi_0(\tau, x))d\tau, \quad (4)$$

has an isolated equilibrium point x_0 such that

$$p(x_0) = 0, \quad det|\frac{dp}{dx}(x_0)| \neq 0.$$

Let us denote as $z^{\pm}(\tau, z, x, \mu), \xi^{\pm}(\tau, z, x, \mu)$ the solutions of system (3) with the initial conditions $z^{\pm}(0, z, x, \mu) = z, \xi^{\pm}(0, z, x, \mu) = 0$ for $\xi > 0$ and $\xi < 0$.

The point mapping of domain V of the surface $\xi = 0$ has the following form

$$\Phi(z, x, \mu) = (\Phi_1(z, x, \mu), \Phi_2(z, x, \mu)) =$$

$$(z^-(\Theta, z^+(\theta, z, x, \mu), x^+(\theta, z, x, \mu), \mu),$$

$$x^-(\Theta, z^+(\theta, z, x, \mu), x^+(\theta, z, x, \mu), \mu)),$$

where functions $\theta(z, x, \mu), \Theta(z, x, \mu)$ are determined by equations

$$\xi^+(\theta, z, x, \mu) = 0,$$

$$\xi^-(\Theta, z^+(\theta, z, x, \mu), x^+(\theta, z, x, \mu), \mu)) = 0.$$

This means that $\Phi_1(z, x, 0) = R(z, x)$.

The surface $\xi = 0$ is the surface without stable sliding for system (3). This means that there exists a neighbourhood of the point $(z^*(x_0), x_0)$ on the surface $\xi = 0$ for which

$$max\{|d\xi^+/d\theta|, |d\xi^-/d\Theta|\} > 0.$$

It follows from condition 1^0 and implicit function theorem that for some small μ_0 functions Φ, θ, Θ have the continuous derivatives into the some set $U \times [0, \mu_0]$ on the surface $\xi = 0$. This means that we can consider the function Φ as the point mapping of the set $U \times [0, \mu_0]$ on the surface $\xi = 0$ Moreover we can rewrite $\Phi(z, x, \mu)$ in the form

$$\Phi(z, x, \mu) = (\bar{R}(z, x, \mu), x + \mu\bar{Q}(z, x, \mu)),$$

where $\bar{R}(z, x, \mu), \bar{Q}(z, x, \mu)$ are the sufficiently smooth functions and $\bar{Q}(z^*(x_0), x_0, 0) = 0, \bar{R}(z^*(x), x, 0) = z^*(x)$.

Let's make in Φ the substitution of variables using the formula $\eta = z - z^*(x)$. Then the point mapping (4) takes the form

$$\Psi(\eta, x, \mu) = (\Psi_1(\eta, x, \mu), \Psi_2(\eta, x, \mu)) =$$

$$= (\bar{R}(\eta + z^*(x), x, \mu) - z^*(x), x + \mu\bar{Q}(\eta + z^*(x), x, \mu)), \quad (5)$$

and consequently $\Psi(0, x, 0) = (0, x)$.

3 EXISTENCE OF THE FAST PERIODIC SOLUTION

Theorem 1. *Under conditions $1^0 - 6^0$ system (1) has the isolated periodic solution with the period $\mu(T(x_0) + O(\mu))$ near to the circle $(z_0(t/\mu, x_0), \xi_0(t/\mu, x_0), x_0)$.*

Proof. We will prove the existence of the periodic solution as the existence of the fixed point $(\eta^*(\mu), x^*(\mu))$ of the point mapping Ψ. Let's rewrite the conditions of existence of this fixed point in the form

$$G(\eta^*, x^*, \mu) = \begin{pmatrix} G_1(\eta^*, x^*, \mu) \\ G_2(\eta^*, x^*, \mu) \end{pmatrix} =$$

$$= \begin{pmatrix} \eta^* - \Psi_1(\eta^*, x^*, \mu) \\ \frac{1}{\mu}[x^* - \Psi_2(\eta^*, x^*, \mu)] \end{pmatrix} = 0. \quad (6)$$

It is necessary to take into account that for $\mu = 0$ $\eta^*(0) = 0, x^*(0) = x_0$ and $G_2(0, x_0, 0) = -T(x_0)p(x_0) = 0$ and consequently for $\mu = 0$ conditions (6) are fulfilled. Moreover, taking into account that for all $x \in \bar{X}$ $G_1(0, x, 0) = 0$ we can conclude $\frac{\partial G_1}{\partial x}(0, x_0, 0) = 0$. Let us compute the Jacobian of function G with respect by variables η, x at $\mu = 0$.

$$\frac{\partial G}{\partial(\eta, x)}(0, x_0, 0) =$$

$$= \begin{vmatrix} I_m - \frac{\partial R}{\partial z}(z^*(x_0), x_0) & 0 \\ \frac{\partial G_2}{\partial \eta}(0, x_0, 0) & -T(x_0)\frac{\partial p}{\partial x}(x_0) \end{vmatrix} \neq 0.$$

This means that there exists an isolated fixed point $(z^*(\mu), x^*(\mu))$ of point mapping G which corresponds to the periodic solution of systems (1) and (3) and in this case $z^*(\mu) = z^*(x_0) + O(\mu), x^*(\mu) = x_0 + O(\mu)$.

4 STABILITY ON THE FIRST APPROXIMATION

Assume that

7^0 the eigenvalues $\lambda_i(x_0)$ $(i = \overline{1,m})$ of the matrix $\frac{\partial \bar{R}}{\partial z}(z(x_0), x_0)$ satisfy the inequalities $|\lambda_i(x_0)| < 1$ $(i = \overline{1,m})$;

8^0 the eigenvalues $\nu_j(x_0)$, $j = 1,...,n$ of matrix $\frac{dp}{dx}(x_0)$ satisfy the inequalities

$$\mathbf{Re}\,\nu_j(x_0) < 0.$$

Theorem 2. *Under conditions* $1^0 - 8^0$ *the periodic solution of* (1),(2) *is orbitally asymptotically stable.*

Proof. Let's find the derivatives Ψ by variables η, x

$$\frac{\partial \Psi}{\partial(\eta, x, \mu)} = \Gamma(\eta, x, \mu) =$$

$$\begin{bmatrix} I_m - \frac{\partial \bar{R}}{\partial z}(x_0) + O(\mu) & O(\mu) \\ \frac{\partial \Psi_2}{\partial \eta}(0, x_0, 0) + O(\mu) & I_m + \mu T(x_0)\frac{\partial p}{\partial x}(x_0) + O(\mu) \end{bmatrix}$$

Consequently the matrix $\Gamma(\eta, x, \mu)$ has at the small vicinity of $(0, x_0, 0)$ two groups of eigenvalues

$$\lambda_i(x_0) + O(\mu),\ i = 1,...,m,$$

$$1 + \mu T(x_0)\nu_j(x_0) + o(\mu),\ j = 1,...,n.$$

This means that under conditions of theorem 3 there exists some neighbourhood of $(0, x_0, 0)$ for which Ψ is contraction mapping.

5 SOME THEOREMS ABOUT DECOMPOSITION OF TWO - SPEED POINT MAPPINGS

It is obvious that the problems of stability of fast periodic solution of system (1) is equivalent to the problem of stability of fixed point $\eta^*(\mu), x^*(\mu)$ of $\Psi(\mu)$. Let's introduce into Ψ the new variables according the formulae $\kappa = \eta - \eta^*(\mu), \chi = x - x^*(\mu)$. Then taking into account that $\partial\Psi(0, x_0, 0)/\partial x = 0$, we have

$$\Lambda_1(\kappa, \chi, \mu) = P\kappa + Q(\kappa, \chi, \mu),$$

$$\Lambda_2(\kappa, \chi, \mu) = \chi + \mu R(\kappa, \chi, \mu), \qquad (D.1)$$

where Q, R - are smooth functions and under conditions $1^0 - 7^0$

$$P = \partial\bar{R}/\partial z(z(x_0), x_0),\ ||P|| < 1$$

$$Q(\kappa, \chi, \mu) = O(\mu)O(|\kappa| + |\chi|) + O(|\kappa|^2 + |\chi|^2),$$

$$R(\kappa, \chi, \mu) = O(|\kappa| + |\chi|).$$

Thus we can reduce Cauchy problem for system (1) with initial conditions

$$z(0, \mu) = z^0,\ s(0, \mu) = 0\ x(0, \mu) = x^0. \qquad (IC)$$

to the investigation of the two-speed discrete system

$$\kappa_{k+1} = P\kappa_k + Q(\kappa_k, \chi_k, \mu),\ \chi_{k+1} = \chi_k + \mu R(\kappa_k, \chi_k, \mu), \qquad (D.2)$$

$$\kappa_0 = z^0 - z^*(x^*(\mu)),\ \chi_0 = x^0 - x^*(\mu).$$

Below we will use the following theorems about decomposition of point mappings (D.1),(D.2) (see (?)).

Theorem D.1. Assume, that for system (D.1) conditions (D.2) are held. Then system (D.1) has the slow motions integral manifold in the form $\kappa = V(\chi, \mu)$ for the small μ. Then there exist C_1, C_2 such that

$$||V(\chi, \mu)|| < C_1,$$

$$||V(\chi, \mu) - V(\bar{\chi}, \mu)|| < C_2||\chi - \bar{\chi}||.$$

The motion on manifold $\kappa = V(\chi, \mu)$ described by the equation

$$\Lambda_1(V(\chi, \mu), \chi, \mu) = \chi + \mu R(V(\chi, \mu), \chi, \mu). \quad (D.3)$$

For the slow coordinate of solution (D.2) $\chi_k(\chi_0)$ and $\bar{\chi}_k(\tilde{\chi})$ the solution of system (D.3) with initial condition $\bar{\chi}_0 = \tilde{\chi}$ there exist $c > 0$, $0 < q < 1$ and $\tilde{\chi} \in \mathbf{R}$ for which the inequality

$$|\chi(\chi_0) - \bar{\chi}(\tilde{\chi})| < cq^k$$

is true.

Theorem D.2. (reduction principle). If

$$Q(0, 0, \mu) = 0; \quad R(0, 0, \mu) = 0,$$

then the problem of stability of zero solutions of systems(D.1) and (D.3) are equivalent. This means that the zero solution of (D.1) are stable (asymptotically stable, unstable) if and only if when the zero solution of (D.3) is stable (asymptotically stable,unstable).

The function $V(\chi, \mu)$ may be found from the equation with any level of precision

$$PV(\chi, \mu) + Q(V(\chi, \mu), \chi, \mu) = V(\chi + \mu R(V(\chi, \mu), \chi, \mu), \mu)$$

in form

$$V(\chi, \mu) = V_0(\chi) + \mu V_1(\chi) + \mu^2 V_2(\chi) + \cdots.$$

The function $V_0(\chi)$ is a solution of the equation

$$PV_0(\chi) + Q(V_0(\chi), \chi, 0) = V_0(\chi).$$

Function $V_1(\chi)$ can be found from equation

$$PV_1(\chi) + Q'_\mu(V_0(\chi), \chi, 0) = V_1(\chi).$$

An equation describing the flow on slow motions integral manifold have the form

$$\Lambda_2(V(\chi, \mu), \chi, \mu) = \chi + \mu R(V_0(\chi), \chi, \mu) + \quad (D.4)$$

$$+ \mu^2(R'_\kappa(V_0(\chi), \chi, 0)V_1(\chi) + R'_\mu(V_0(\chi), \chi, 0)) + O(\mu^3).$$

6 THEOREM ABOUT AVERAGING

Assume that

9^0 the solution $\bar{x}(t)$ of averaged system with initial conditions $\bar{x}(0) = x^0$ for $t \in [0, L]$ is situated into the closed subdomain $\bar{X} \in X$.

Theorem 3. *Under conditions* $1^0 - 7^0$ *and* 9^0 *the slow coordinate* $x(t, \mu)$ *of solution* (1),(2) *and* $\bar{x}(t)$ *satisfy the inequality*

$$\sup_{t\in[0,L]} |x(t,\mu) - \bar{x}(t)| = O(\mu).$$

7 FOUNDING OF THE PERIODIC SOLUTION

Assume now that

$1A^0 \quad h_1, h_2, g \in \mathbf{C}^{k+2}[\bar{Z} \times [-1, 1]].$

We will find the period of the desired periodic solution of (2) in form

$$T(\mu) = T_0 + \mu T_1 + \mu^2 T_2 + ... + \mu^3 T_3 + ..., \quad (7)$$

where $T_0 = T(x_0)$ and time interval for which $u = 1$ and $u = -1$ in form

$$\theta^{\pm}(\mu) = \theta_0^{\pm} + \mu\theta_1^{\pm} + \mu^2\theta_2^{\pm} + \cdots + \mu^k\theta_k^{\pm} + \cdots,$$

where $\theta_0 = \theta(x_0)$. Then the asymptotic representation of desired periodic solution on $[0, T(\mu)]$ takes the form

$$z(\tau,\mu) = z_0(\tau) + \mu z_1(\tau) + \mu^2 z_2(\tau) + ... + \mu^k z_k(\tau) + ...,$$

$$x(\tau,\mu) = \xi_0(\tau) + \mu\xi_1(\tau) + \mu^2\xi_2(\tau) + ... + \mu^k\xi_k(\tau) + ...,$$

$$x(\tau,\mu) = x_0 + \mu x_1(\tau) + \mu^2 x_2(\tau) + ... + \mu^k x_k(\tau) + ...$$

Denote

$$\tilde{T}_k(\mu) = T_0 + \mu T_1 + \mu^2 T_2 + ... + \mu^k T_k,$$

$$\tilde{\theta}_k^{\pm}(\mu) = \theta_0^{\pm} + \mu\theta_0^{\pm} + \mu^2\theta_2^{\pm} + ... + \mu^k\theta_k^{\pm}.$$

Let's find the $k-th$ approximation of asymptotic representation for desired periodic solution for $\tau \in [0, \tilde{T}_k(\mu)]$ in form

$$Z_k(\tau,\mu) = z_0(\tau) + \mu z_1(\tau) + \mu^2 z_2(\tau) + ... + \mu^k z_k(\tau),$$

$$\Xi_k(\tau,\mu) = \xi_0(\tau) + \mu\xi_1(\tau) + \mu^2\xi_2(\tau) + ... + \mu^k\xi_k(\tau),$$

$$X_k(\tau,\mu) = x_0 + \mu x_1(\tau) + \mu^2 x_2(\tau) + ... + \mu^k x_k(\tau),$$

where continuous functions z_i, ξ_i, x_i are smooth on $[0, \tilde{\theta}_k^+(\mu)) \cup (\theta_k^+(\mu), T_k(\mu)]$ but have the jumps in the derivatives at $\tau = \tilde{\theta}_k^+(\mu)$. Let's show that under conditions of theorem 1 the functions $z_i^{\pm}, \xi_i^{\pm}, x_i^{\pm}$ and constants θ_i, Θ_i for every $i = 1, ..., k$ can be uniquely found.

Let's introduce in system (2) two "new times" according the formulae

$$\tau^+ = \tau/\tilde{\theta}_k^+(\mu); \tau^- = (\tau - \tilde{\theta}_k^+(\mu))/\tilde{\theta}_k^-(\mu), \tau^{\pm} \in [0,1],$$

and the auxiliary functions $z_0^{\pm}(\tau^{\pm}), \xi_0^{\pm}(\tau^{\pm})$ as the solutions

$$dz_0^{\pm}/d\tau^{\pm} = \theta_0^{\pm} g(z_0^{\pm}, \xi_0^{\pm}, x_0, \pm 1), \quad (3.\pm)$$

$$d\xi_0^{\pm}/d\tau = \theta_0^{\pm} h_1(z_0^{\pm}, \xi_0^{\pm}, x_0, \pm 1)$$

with initial and periodicity conditions

$$z_0^+(0) = z^*(x_0) = z_0^*, \quad \xi_0^+(0) = 0; \quad (8)$$

$$z_0^-(0) = z_0^+(1), \quad \xi_0^-(0) = \xi_0^+(1) = 0;$$

$$z_0^-(1) = z^*(x_0), \quad \xi_0^-(1) = 0.$$

From the periodicity of functions $z_0(\tau), \xi_0(\tau)$ it follows that system (3±),(8) has the unique solution.

Functions $x_1^{\pm}(\tau)$ are described by the equations

$$dx_1^{\pm}/d\tau^{\pm} = \theta_0^{\pm} h_2(z_0^{\pm}(\tau^{\pm}, x_0), \xi_0^{\pm}(\tau^{\pm}, x_0), x_0, \pm 1), \quad (4.1)$$

with initial conditions and periodicity conditions given by

$$x_1^+(0) = x_1^*, \; x_0^-(0) = x_1^+(1), \; x_1^-(1) = x_1^*, \quad (9.1)$$

Moreover

$$[h_{20}](x_0) = \int_0^1 h_2(z_0(\tau^+, x_0), \xi_0^+(\tau^+, x_0), x_0, 1)d\tau^+ +$$

$$+ \int_0^1 h_2(z_0(\tau^-, x_0), \xi_0^-(\tau^-, x_0), x_0, -1)d\tau^- = 0 \quad (10)$$

$$det|\frac{d[h_{20}]}{dx}(x_0)| = T(x_0)\frac{dp}{dx}(x_0) \neq 0.$$

This means that for every x_1^* there exists the unique solution of (4.1) and (9.1) for which $\int_0^1 \tilde{x}_1^+(\tau^+)d\tau^+ + \int_0^1 \tilde{x}_1^-(\tau^-)d\tau^- = 0$ and we can define function $x_1(\tau)$ in form

$$x_1(\tau) = x_1^* + \tilde{x}_1(\tau) =$$

$$= \begin{cases} x_1^* + \tilde{x}_1^+(\tau/\tilde{\theta}_k^+(\mu)) & \text{for} \quad \tau \in [0, \tilde{\theta}_k^+(\mu)], \\ x_1^* + \tilde{x}_1^-((\tau - \tilde{\theta}_k^+(\mu))/\tilde{\theta}_k^-(\mu)) & \\ \text{for} \quad \tau \in [\tilde{\theta}_k^+(\mu), \tilde{T}_k(\mu)]. & \end{cases}$$

Functions $z_1^{\pm}(\tau^{\pm}, x_1^*), \xi_1^{\pm}(\tau^{\pm}, x_1^*)$ are defined by equations

$$dz_1^{\pm}/d\tau = \theta_0^{\pm}(g_z'^{\pm} z_1^{\pm} + g_\xi'^{\pm}\xi_1^{\pm} + g_x'^{\pm} x_1^{\pm}) + \theta_1^{\pm} g^{\pm}; \quad (3.1)$$

$$d\xi_1/d\tau = \theta_0^{\pm}(h_{1z}'^{\pm} z_1^{\pm} + h_{1\xi}'^{\pm}\xi_1^{\pm} + h_{1x}'^{\pm} x_1^{\pm}) + \theta_1^{\pm} h_1^{\pm},$$

where the values of functions $g^{\pm}, h_1^{\pm}$ and its derivatives are calculated at the points $(z_0^{\pm}(\tau^{\pm}, x_0), \xi_0^{\pm}(\tau^{\pm}, x_0), x_0, \pm 1)$,

Initial and periodicity conditions for system (3.1) are defined by equations

$$z_1^+(0,x_1^*) = z_1^-(1,x_1^*) = z_1^*, z_1^-(0,x_1^*) = z_1^+(1,x_1^*) \quad (8.1)$$

$$\xi_1^+(0,x_1^*) = \xi_1^+(1,x_1^*) = \xi_1^-(0,x_1^*) = \xi_1^-(1,x_1^*) = 0.$$

Equations (3.1) depend linearly on $z_1^\pm, \xi_1^\pm, \theta_1^\pm$ end consequently their solutions $z_1^\pm(\tau,x_1^*), \xi_1^\pm(\tau,x_1^*), \theta_1^\pm(x_1^*)$ are linearly dependent on the initial conditions $z_1^\pm(0,x_1^*)$. Expressing $z_1^\pm(\tau,x_1^*), \xi_1^\pm(\tau,x_1^*), \theta_1^\pm(x_1^*)$ through $z_1^+(0,x_1^*)$ and substitute the results in the first equation of (8.1) we have the linear on $z_i^+(0,x_1^*)$ system of algebraic equations which determinant coincides with $det|I_m - \partial R(z^*(x_0),x_0)/\partial z| \neq 0$.

Functions $x_2^\pm(\tau)$ are described by the equations

$$dx_2^\pm/d\tau = \theta_0^\pm(h'_{2\,z}z_1^\pm +$$

$$+h'_{2\,\xi}\xi_1^\pm + h'_{2\,x}x_1^\pm) + \theta_1^\pm h_2. \quad (4.2)$$

where the values of functions $h_2^\pm$ are calculated at the points $(z_0^\pm(\tau^\pm,x_0), \xi_0^\pm(\tau^\pm,x_0), x_0, \pm 1)$. Initial and periodicity conditions are

$$x_2^+(0) = x_2^*, \quad x_2^-(0) = x_2^+(1), \quad x_2^-(1) = x_2^*, \quad (9.2)$$

The condition under which system (9.2) for every x_2^* have the periodic solution with zero averaged value takes the form

$$\int_0^1 [\theta_0^+(h'^+_{2\,z}z_1^+(\tau^+,x_1^*) + h'^+_{2\,\xi}\xi_1^+(\tau^+,x_1^*)+$$

$$+h'^+_{2\,x}x_1^+(\tau^+,x_1^*)) + \theta_1^+(\tau^+,x_1^*)h_2^+]d\tau^+ +$$

$$+\int_0^1 [\theta_0^-(h'^-_{2\,z}z_1^-(\tau^-,x_1^*) + h'^-_{2\,\xi}\xi_1^-(\tau^-,x_1^*)+$$

$$+h'^-_{2\,x}x_1^-(\tau^-,x_1^*)) + +\theta_1^-(x_1^*)h_2^-]d\tau^- = 0 \quad (10.1)$$

Conditions (10.1) are a system of linear equations for obtaining of x_1^*, whose determinant coincides with $\frac{dp}{dx}(x_0) \neq 0$. This means that we can find uniquely the function $x_2(\tau)$ in form $x_2(\tau) = \tilde{x}_2^* + \bar{x}_2(\tau)$, where $\bar{x}_2(\tau)$ is the function with zero averaged value.

Suppose now that functions $z_{i-1}(\tau), \xi_{i-1}(\tau), x_i(\tau)$ and constants $x_{i-1}^*, \theta_{i-1}^\pm$ are found, moreover, for every x_i^* the function $x_i(\tau)$ can be represented in form of the sum of x_i^* and the function $\tilde{x}_i(\tau)$ with zero averaged value.

Then the functions $z_i^\pm(\tau^\pm,x_i^*), \xi_i^\pm(\tau^\pm,x_i^*), x_i^\pm(\tau^\pm,x_i^*)$ are defined by equations

$$dz_i/d\tau^\pm = \theta_0^\pm(g'^\pm_z z_i^\pm + g'^\pm_\xi \xi_i^\pm + g'^\pm_x x_i^\pm)+$$

$$+\theta_i^\pm(x_i^*)g^\pm + \Pi_{1i}^\pm(\tau^\pm); \quad (3.i)$$

$$d\xi_i/d\tau^\pm = \theta_0^\pm(h'^\pm_{1\,z}z_i^\pm + h'^\pm_{1\,\xi}\xi_i^\pm + h'^\pm_{1\,x}x_i^\pm)+$$

$$+\theta_i^\pm(x_i^*)h_1^\pm + \Pi_{2i}^\pm(\tau^\pm),$$

where the values of functions $g^\pm, h_1^\pm$ and its derivatives are calculated at the points $(z_0^\pm(\tau^\pm,x_0), \xi_0^\pm(\tau^\pm,x_0), x_0, \pm 1)$, and functions $\Pi_{ji}^\pm, j = 1,2$ are uniquely defined functions containing the terms of order μ^i in asymptotic representations of $g^\pm, h_1^\pm$ depending from $z_j^\pm, \xi^\pm, x_j^\pm, x_j^*, j \le i-1$. Initial and periodicity conditions for system (3.i) are defined by the equations

$$z_i^+(0,x_1^*) = z_i^-(1,x_i^*) = z_i^*, \; z_i^-(0,x_i^*) = z_i^+(1,x_i^*) \quad (8.i)$$

$$\xi_i^+(0,x_i^*) = \xi_i^+(1,x_i^*) = \xi_i^-(0,x_i^*) = \xi_i^-(1,x_i^*) = 0.$$

Equations (3.i) depend linearly on $z_i^\pm, \xi_i^\pm, \theta_i^\pm$ and consequently their solutions $z_i^\pm(\tau,x_i^*), \xi_i^\pm(\tau,x_i^*), \theta_i^\pm(x_i^*)$ are linearly depend on the initial conditions $z_i^\pm(0,x_i^*)$. Expressing $z_i^\pm(\tau,x_i^*), \xi_i^\pm(\tau,x_i^*), \theta_i^\pm(x_i^*)$ through $z_i^+(0,x_i^*)$ and substituting the results in the first equation of (8.i) we have linear in $z_i^+(0,x_i^*)$ system of algebraic equations whose determinant is coincide with $det|I_m - \partial R(z^*(x_0),x_0)/\partial z| \neq 0$.

Functions $x_{i+1}(\tau)$ are described by the equations

$$dx_{i+1}^\pm/d\tau = \theta_0^\pm(h'_{2\,z}z_1^\pm + h'_{2\,\xi}\xi_1^\pm +$$

$$+h'_{2\,x}x_1^\pm) + \theta_1^\pm h_{i+1} + \Pi_{3i}^\pm(\tau), \quad (4.i+1)$$

where the values of functions $h_2^\pm$ and its derivatives are calculated at the

$$(z_0^\pm(\tau^\pm,x_0), \xi_0^\pm(\tau^\pm,x_0), x_0, \pm 1)$$

and functions $\Pi_3^\pm, j = 1,2$ are uniquely defined functions containing the terms of order μ^i in asymptotic representations of $h_2^\pm$ depending from $z_j^\pm, \xi^\pm, x_j^\pm, x_j^*, j \le i-$ Initial and periodicity conditions are

$$x_{i+1}^+(0) = x_{i+1}^*, \; x_{i+1}^-(0) = x_{i+1}^+(1), \; x_{i+1}^-(1) = x_{i+1}^*. \quad (9.i+1)$$

The condition under which system (9.i+1) have a periodic solution with zero averaged value for every x_{i+1}^* takes the form

$$\int_0^1 [\theta_0^+(h'^+_{2\,z}z_i^+(\tau^+,x_i^*) + h'^+_{2\,\xi}\xi_i^+(\tau^+,x_i^*)+$$

$$+h'^+_{2\,x}x_i^+(\tau^+,x_i^*)) + \theta_i^+(\tau^+,x_i^*)h_2^+]d\tau^+ + \quad (10.i+1)$$

$$+\int_0^1 [\theta_0^-(h'^-_{2\,z}z_i^-(\tau^-,x_i^*) + h'^-_{2\,\xi}\xi_i^-(\tau^-,x_i^*)+$$

$$+h'^-_{2\,x}x_i^-(\tau^-,x_i^*)) + \theta_i^-(x_i^*)h_2^-]d\tau^- = 0$$

Conditions (10.i+1) are a system of linear equations for obtaining of x_1^*, whose determinant coincides with $\frac{dp}{dx}(x_0) \neq 0$. This means that we can uniquely find uniquely the function $x_{i+1}(\tau)$ in the form $x_{i+1}(\tau) = \tilde{x}_{i+1}^* + \bar{x}_{i+1}(\tau)$, where $\bar{x}_{i+1}(\tau)$ is a function with zero averaged value. To finish the algorithm for design of desired asymptotic representation it is necessary to define

$$(z_i(\tau), \xi_i(\tau)) =$$

$$= \begin{cases} (z_i^+(\tau/\tilde{\theta}_k^+(\mu), x_i^*), \xi_i^+(\tau/\tilde{\theta}_k^+(\mu), x_i^*)) \\ \text{for} \quad \tau \in [0, \tilde{\theta}_k^+(\mu)], \\ (z_i^-((\tau - \theta_k^+(\mu))/\tilde{\theta}_k^-(\mu), x_i^*), \\ \xi_i^+((\tau - \theta_k^+(\mu))/\tilde{\theta}_k^-(\mu), x_i^*)) \\ \text{for} \quad \tau \in [\tilde{\theta}_k^+(\mu), \tilde{T}_k(\mu)], j = 1, ..., k. \end{cases}$$

$$x_j(\tau) = \begin{cases} x_j^* + \tilde{x}_j^+(\tau/\tilde{\theta}_k^+(\mu)) \quad \text{for} \quad \tau \in [0, \tilde{\theta}_k^+(\mu)], \\ x_j^* + \tilde{x}_j^-((\tau - \tilde{\theta}_k^+(\mu))/\tilde{\theta}_k^-(\mu)) \\ \text{for} \quad \tau \in [\tilde{\theta}_k^+(\mu), \tilde{T}_k(\mu)], j = 1, ..., k. \end{cases}$$

8 SINGULAR CORRECTION OF AVERAGING EQUATIONS

Let us show how we can use the knowledge of the fast periodic solution for correction of averaged equations with any precision level according the small parameter degrees. The knowledge of such equations is necessary the case when the linear part of averaged equations (3) has the spectral points on the imaginary axis.

Assume that we have found the functions

$$\theta^\pm(x, \mu) = \theta_0^\pm(x) + \sum_{i=1}^{\infty} \mu^i \theta_i^\pm(x),$$

$$T(x, \mu) = \theta^+(x, \mu) + \theta^-(x, \mu)$$

and $z_i^\pm(\tau^\pm, x), \xi_i^\pm(\tau^\pm, x), x_j^\pm(\tau^\pm, x)$, then

$$z(\tau, x, \mu) = z_0(\tau, x) + \mu z_1(\tau, x) + ... + \mu^i z_i(\tau, x) + ...,$$

$$\xi(\tau, x, \mu) = \xi_0(\tau, x) + \mu \xi_1(\tau, x) + ... + \mu^i \xi_i(\tau, x) + ...,$$

$$x(\tau, x, \mu) = \mu x_1(\tau, x) + ... + \mu^i x_i(\tau, x) + ... \quad .$$

Then the precise averaged equation has the form

$$dx/dt = \frac{1}{T(x\mu)} \int_0^{T(x,\mu)} h_2(z(\tau, x, \mu), \xi(\tau, x, \mu), x + \tilde{x}(\tau, \mu), u(\xi(\tau, x, \mu)))d\tau. \qquad (PAE)$$

Equations (PAE) correspond to the system (D.3) which describes a flow on the slow motion manifold in system (D.1). In this case the first order approximation of (PAE) has the form

$$dx/dt = \frac{1}{T_0(x)} \Bigg\{ (1 - \mu T_1(x)) \int_0^{T_0(x)} h_2 d\tau +$$

$$+\mu \Bigg[\int_0^{T_0(x)} \Bigg(h'_{2z} z_1(\tau, x) + h'_{2\xi} \xi_1(\tau, x) + h'_{2x} \tilde{x}_1(\tau) \Bigg) d\tau +$$

$$+\theta_1^+(x) h_2(z_0(\theta_0(x), x), \xi_0(\theta_0(x), x), x, 1) +$$

$$+\theta_1^-(x) h_2(z_0(T_0(x), x), \xi_0(T_0(x), x), x, -1) \Bigg] \Bigg\} \qquad (FAAE)$$

where the values of functions h_2 and it's derivatives in integral terms are calculated at the points $(z_0(\tau, x), \xi_0(\tau, x), x, u(\xi_0(\tau, x)))$. Analogously we can obtain the averaged equations with any precision level expanding in powers of the small parameter.

9 INVESTIGATION OF STABILITY IN CRITICAL CASE

Theorem 4 (Reduction Principle). *Under conditions $1^0 - 7^0$ the periodic solution for original system* (1) *is stable (asymptotically stable, unstable) if and only if the equilibrium point of system* (PAE) *is stable (asymptotically stable, unstable).*

Corollary. *Assume that for system* (1) *conditions $1^0 - 7^0$ are true. If the equilibrium point of system* (FAAE) *is asymptotically stable (unstable) in the first approximation than the periodic solution for original system* (1) *is asymptotically stable (unstable).*

REFERENCES

Coddington E.A., N. Levinson (1955). *Theory of ordinary differential equations*, McGraw Hill, New-York.

Fridman L.M. (1986). Singular extension definition of discontinuous control systems, *Differential equation*, **22,8**, 1461-1463 (Russian).

Fridman L.M. (1990). Singular extension definition of discontinuous control systems and stability, *Differential equation*, **26,10**, 1759-1764.

Neimark Y.I. (1963). The method of averaging from the viewpoint of the point mapping method. *Izvestija Vusov, Radiophysika*, **VI,5**, 1023-1032 (Russian).

Neimark Y.I. (1972). *The point mapping method in the theory of nonlinear oscillations*, ,Nauka, Moscow, (Russian).

Pontriagin L.S., L.V. Rodygin (1963). Periodic solution of one system of differential equation with small parameter near the derivative. Doklady Academii Nauk, **132,3** ,537-540 (Russian).

AN APPROACH TO OPTIMIZATION OF NONLINEAR SYSTEMS VIA WEI-NORMAN REPRESENTATION

Andrew A. Galyaev *

* *Institute of Control Sciences, 65 Profsoyuznaya Str., Moscow, 117806, Russia, E-mail: galaev@ipu.rssi.ru*

Abstract: It is considered the method allows to reduce the path- finding problem of a differential equation $\dot{x}(t) = \sum_{i=1}^{m} u_i(t)f_i(x(t))$, (where $x \in X$, X is a Banach space, operators f_i act from X to X, $u_i(t) \in L^1[0,T]$,) to a problem of finding extremals or optimal controls (in some sense) of a finite-dimension system of differential equations.

Keywords: Control functions, Optimal control, Nonlinear system, Tradjectories, Operators.

1. INTRODUCTION

This paper studies the representation of the solution of an ordinary differential equation or a system of equations in a Banach space X

$$\dot{x}(t) = \sum_{i=1}^{m} u_i(t)f_i(x(t)), \qquad t \in [0,T] \quad (1)$$

with the initial condition $x(0) = x_0 \in X$. Here f_i are bounded, time-independent and continuous operators for all $i = 1, \dots, m$. It is further required that f_i generate an associative Lie algebra over the field of complex numbers with Lie brackets $[f_i, f_j] = f_i f_j - f_j f_i = (\mathrm{ad} f_i) f_j$. If this assumption failed, we suggest that there exist minimal $r \geq m$, so that Lie algebra is generated by f_i and by Lie brackets $[\dots[f_{i_1}, f_{i_2}]\dots f_{i_n}]$. In this case let us denote $[\dots[f_{i_1}, f_{i_2}]\dots f_{i_n}] = f_k$, for $k = m+1, \dots r$. r is chosen in such a manner that $[f_i, f_k] = \sum_{j=1}^{r} \alpha_{ikj} f_j$ with $\alpha_{ijk} = \text{const}$.

In terms of f_k, $k = 1, \dots, r$ it is considered the equation

$$\dot{x}(t) = \sum_{k=1}^{r} v_k(t)f_k(x(t)). \qquad (2)$$

If we put $v_k(t) \equiv u_k(t)$ for $k = 1, \dots, m$ and $v_k(t) \equiv 0$ for $k = m+1, \dots, r$ then the set of solutions of (1) coincides with the set of solutions of (2). In general case the set of solutions of the system (1) is dense in the set of solutions of (2), as shown in (Sussmann and Liu, to appear), (Sussmann and Liu, 1991). That is equations (1) and (2) are identical. So $f_{m+1}, \dots, f_r$ are the new admissible directions of motion in the agreement with (Sussmann and Liu, to appear) and (Sussmann and Liu, 1991).

Under the above assumptions (as will be shown in the main part of the paper) there exists a unique solution to (2), and hence to (1), given by the expression:

$$x(t) = V_1(V_2 \dots (V_r(x_0, g_r(t)) \dots g_2(t))g_1(t)), \qquad (3)$$

where V_i is a fundamental solution to the suitable homogeneous equation

$$\dot{x}(t) = f_i(x(t)), \qquad i = 1, \dots, r \quad (4)$$

and $g_i(t)$ are continuous scalar functions of time.

Let us consider that for $i = 1, \dots, r$

(i) operators f_i satisfy a linear growth condition,

$$\| f_i(x) \|_X \leq (\| x \|_X + 1)C_i, \quad C_i = \text{const},$$

(ii) functions v_i, which play the role of controls, are bounded in $L^1[0,T]$,

(iii) Lie algebra, generated by f_i, is solvable.

Definition 1. Lie algebra is called solvable if there exists a chain of ideals $0 \subset L_r \subset L_{r-1} \subset \ldots \subset L_1 = L$, where each L_i has a dimension $r-i+1$ for $i = 1, \ldots, r$.

This definition is equivalent to the requirement of existence the basis $f_1, \ldots, f_r$ of Lie algebra so that

$$[f_i, f_j] = \sum_{k=i}^{r} \alpha_{ijk} f_k \qquad i > j \qquad (5)$$

Note that any Abelian Lie algebra is solvable. Moreover any two-dimensional Lie algebra is solvable, because there exists basis $[f_1, f_2] = cf_1$ (see (Olver, 1986)).

The purpose of this paper is to give an explicit approach to the solution of the path finding problem for infinite-dimension system of differential equations (1) with finite-dimension vector of controls $u(t) = (u_1(t), \ldots, u_m(t)$ via the solution of the system (2). The technique of the representation of the scalar functions $g_i(t)$ in (3) was primarily shown by Wei and Norman in (Wei and Norman, 1964), but in the case when f_i are linear operators and Lie algebra generated by f_i is finite-dimension. We developed a similar approach, based on the representation of the solution of (2) via the solutions of the appropriate homogeneous equations. The methods of operating with Lie algebras are described by P.Olver in (Olver, 1986). The ability of a object motion on the trajectory in the direction of Lie bracket $[f_i, f_j]$ is discussed in (Olver, 1986), (Sussmann and Liu, to appear), (Sussmann and Liu, 1991).

2. MAIN RESULTS

A: global representation of $x(t)$

An explicit representation of solution of the equation (1) or (2) on time interval [0,T] according to (3) has the form

$$x(t) = \prod_{i=1}^{r} \exp\,(g_i(t) f_i)(x_0) =$$

$$= \exp(g_1(t) f_1) \ldots \exp(g_r(t) f_r)(x_0) \qquad (6)$$

where scalar functions $g_i(t)$ satisfy to the system of differential equations:

$$\sum_{i=1}^{r} v_i(t) f_i(x) = \sum_{i=1}^{r} \dot{g}_i(t) \prod_{j=1}^{i-1} \exp\,(g_j(t) \mathrm{ad} f_j) f_i(x) \qquad (7)$$

$$g_i(0) = 0, \qquad i = 1, \ldots, r.$$

The solution to (4) in terms of streams is $V_i = \exp(t f_i)(x_0)$, when $g_i(t) = t$. To obtain the equations for in (7) $g_i(t)$ we need to reduce the homogeneous terms staying before basic functions f_i (f_i generates a given Lie algebra.)

Theorem 1. Suppose that conditions (i), (ii) and (iii) are valid then the representation (6) of the solution to (2) is global.

Remark 1. This result was established by Wei and Norman (Wei and Norman, 1964), in the case, when f_i are linear operators.

Proof of theorem : According to (iii) there is a basis $f_1, \ldots, f_r$ for a given Lie algebra which can be arranged so that L_i is the ideal generated by $f_i, \ldots, f_r$. The multiplication table for Lie algebra generated by f_i takes the form

$$[f_i, f_k] = \sum_{j=i}^{r} \alpha_{ikj} f_j, \qquad i > k \qquad (see(5))$$

It means that $\prod_{j=1}^{i-1} \exp(g_j(t) \mathrm{ad} f_j) f_i = \sum_{j=i}^{r} a_{ji} f_j$. In [1] it was delivered that diagonal elements a_{ii} never vanish. It is easy to find that $i = 1, \ldots, r$

$$\exp(g_{i-1}(t) \mathrm{ad} f_{i-1}) f_i = f_i + g_{i-1} \alpha_{i-1ii} f_i + F_{i+1} +$$

$$+ \ldots = f_i \exp(g_{i-1} \alpha_{i-1ii}) + G_{i+1},$$

where terms $F_{i+1} \in L_{i+1}$ and $G_{i+1} \in L_{i+1}$, i.e. they are linear combination basic functions from L_{i+1}. Therefore $g_i(t)$ are bounded on segment [0,T]. Next will be shown that $\| x(t) \|_X$ is bounded :

$$\| x(t) \|_X = \| \prod_{i=1}^{r} \exp(g_i(t) f_i)(x_0) \|_X \leq$$

$$\leq (\prod_{i=1}^{r} \exp(g_i(t) C_i))(\| x_0 \|_X + C(C_1, \ldots, C_r)),$$

$C > 0$, where $C_1, \ldots, C_r$ from (i). Thus due to the continuity of $g_i(t)$ and $C > 0$,it follows that $x(t)$ is bounded and representation (5) is global. The proof is completed.

It follows from theorem 1 that

$$a_{ij}(g) \dot{g}_i(t) = v_j(t) \qquad (8)$$

with the initial conditions $g_i(0) = 0$ $(i = 1, \ldots, r)$.

Next let us discuss some properties of matrix $A(g)$ with elements $a_{ij}(g)$ from (8).

Corollary 1. In agreement with theorem 1 $A(g)$ depends only on the structural numbers α_{ijk} of Lie algebra.

Indeed, the assertion follows from (7) and from the fact that $\exp(g_i(t) \mathrm{ad} f_i) f_j$ is expressed in terms of α_{ikj} only.

Corollary 2. $A(g)$ is an upper-triangular matrix and moreover it is invertible, i.e. equation (8) is solvable with respect to $\dot{g}_i(t)$.

Next let us derive the differential equations for functions $g_i(t)$ in the case when a Lie algebra of functions f_i is two-dimensional. The following lemma shows that the production $f(x)\partial V/\partial x$ lies in the span of f_i $i = 1, 2$.

Lemma 1. Suppose that f_1 and f_2 are smooth vector fields, generating a Lie algebra with Lie bracket $[f_1, f_2] = \alpha f_1 + \beta f_2$, $V(g(t), x)$ is the solution of the equation $dV/dg = f_1(V)$ with the initial condition $V(0) = x$. Assume also that condition (i) is valid. Then

$$f_2(x)\partial V/\partial x = \exp(-g(t)\beta)f_2(V) + \\ +\alpha/\beta(1 - \exp(-g(t)\beta))f_1(V)$$

Proof : In terms of the streams $V = \exp(g(t)f_1))(x)$. If we consider the vector fields f_1 and f_2, then $f_2(x)dV/dx$ has the form $f_2 \exp(g(t)f_1)(x)$. It is easy to prove that

$$f_2 f_1^n = \alpha/\beta(f_1^{n+1} - (f_1 - \beta)^{n+1}) - \alpha(f_1 - \beta)^n + \\ +(f_1 - \beta)^n f_2.$$

It means :

$$f_2(x)\partial V/\partial x = \exp(-g(t)\beta)f_2(V) - \alpha \exp(-g(t)\beta)V \\ +\alpha/\beta(f_1(V) - \exp(-g(t)\beta)(f_1(V) - \beta V)) = \\ = \exp(-g(t)\beta)f_2(V) + \alpha/\beta(1 - \exp(-g(t)\beta))f_1(V).$$

The proof is completed.

With the help of this Lemma will be proved next theorem.

Theorem 2. Under the conditions of Lemma 1 and condition (ii) there exists a solution of the equation $\dot{x}(t) = u_1(t)f_1(x) + u_2(t)f_2(x)$ (*) expressed in the form $x(t, x_0) = V_1(g_1(t), V_2(g_2(t), x_0))$ (**), where V_i satisfies (4).

Proof : After differentiation of the equation (**) obtained

$$\dot{x}(t) = (\partial V_1/\partial V_2)\partial V_2/\partial g_2(t)\dot{g}_2(t) + \partial V_1/\partial g_1 \dot{g}_1(t).$$

Next is according to Lemma 1 and the conditions of the theorem

$$\dot{x}(t) = \dot{g}_1(t)f_1(x) + \dot{g}_2(\exp(-g_1(t)\beta)f_2(x) + \\ +\alpha/\beta(1 - \exp(-g_1(t)\beta))f_1(x).$$

Let us reduce the homogeneous terms before the functions f_1 and f_2 and compare them with the terms in the equation (*). So it is obtained a system of the differential equations for $g_1(t)$ and $g_2(t)$:

$$u_1(t) = \dot{g}_1(t) + \dot{g}_2(t)\alpha/\beta(1 - \exp(-g_1(t)\beta) \\ u_2(t) = \dot{g}_2(t)\exp(-g_1(t)\beta) \qquad (9)$$

This system is solvable with respect to $\dot{g}_1(t)$ and $\dot{g}_2(t)$. So we may find $g_i(t)$. This completes the proof of the theorem.

The similar result takes place and in a general case. Namely let us prove now the generalization of Lemma 1 and show that the production $\partial V_j/\partial x f_i(x)$ lies in the span of $f_k(x)$ for any x and depends only on the structural numbers of a Lie algebra.

Lemma 2. Let us suppose that $f_k(x)$ generate an associative Lie algebra, with Lie brackets $[f_i, f_j] = \sum_{k=1}^{r} \alpha_{ijk} f_k$ and $V_j(g_j, x)$ is the solution of the adequate homogeneous equation $dV_j/g_j = f_j(V)$ with the initial condition $V(0) = x$. Let the condition (i) is valid. Then

$$\partial V_j/\partial x f_i(x) = \sum_{k=1}^{r} f_k(V_j)(\exp(\alpha_j g_j(t)))_{ik}$$

where $\exp(\alpha_j g_j(t))_{ik}$ is a common matrix exponentials, when index j is fixed.

Proof : In terms of the vector fields it is easy to establish the result

$$f_i f_j^n = f_j f_i f_j^{n-1} + \sum_{k=1}^{r} \alpha_{ijk} f_k f_{j_{n-1}} =$$

$$= f_j \sum_{k=1}^{r} \delta_{ik} f_k f_j^{n-1} + \sum_{k=1}^{r} \alpha_{ijk} f_k f_j^{n-1} =$$

$$= \sum_{k_1=1}^{r} \cdots \sum_{k_n=1}^{r} (\delta_{ik_1} f_j + \alpha_{ijk_1})(\delta_{k_1 k_2} f_j + \alpha_{k_1 j k_2}) \cdots$$

$$\ldots (\delta_{k_{n-1}k_n} f_j + \alpha_{k_{n-1} j k_n}) f_{k_n} = f_j^n f_i +$$

$$f_j^{n-1} n \sum_{k=1}^{r} \alpha_{ijk} f_k + f_j^{n-2} n(n-1)/2 \sum_{k_1,k_2=1}^{r}$$

$$\sum_{k_1,k_2=1}^{r} \alpha_{ijk_1}\alpha_{k_1 j k_2} f_{k_2} +$$

$$+ \ldots + \sum_{k_1=1}^{r} \cdots \sum_{k_{n-1}=1}^{r} \sum_{k=1}^{r} (\alpha_{ijk_1}\alpha_{k_1 j k_2} \cdots \alpha_{k_{n-1} j k} f_k)$$

where δ_{ij} is an identity matrix. If V_j is considered as the flow of the vector field f_j then we obtain

$$\partial V_j/\partial x f_i(x) = \sum_{k=1}^{r} f_k(V_j)(\exp(\alpha_j g_j(t)))_{ik}.$$

Similar the proof of the Lemma 1 we used again the formula $F(\exp(gf)(x)) = \exp(gf)(F(x))$ (see (Olver, 1986)) and condition (i). This terminates the proof.

The previous Lemma gives us an explicit approach how to lead the proof in the general case, reducing all steps of the analysis of the given differential equation to a method described above. Now let us prove an important theorem.

Theorem 3. Let us suppose that f_j satisfy the conditions of Lemma 2 and let the condition (ii) is valid. Then there exists a solution of equation (1) expressed in a form (4).

Proof : Let us assume that the theorem is true. Then after the differentiation of expression (4) we shall gain the equation

$$\dot{x}(t) = \partial V_1/\partial g_1 \dot{g}_1(t) + \ldots +$$

$$+\partial V_1/\partial V_2 \ldots \partial V_{r-1}/\partial V_r \partial V_r/\partial g_r \dot{g}_r(t).$$

Using Lemma 2 the previous equation has the following form

$$\dot{x}(t) = f_1(V_1)\dot{g}_1(t) + \dot{g}_2(t)\sum_{k=1}^{r} f_k(V_1)(\exp(\alpha_1 g_1))_{2k} +$$

$$+ \ldots + \dot{g}_i(t)\sum_{k=1}^{r} f_k(V_1)(\exp(\alpha_1 g_1(t)))_{k_1 k} \ldots$$

$$\ldots \sum_{k_{i-2}=1}^{r} (\exp(\alpha_{i-1} g_{i-1}(t)))_{ik_{i-2}} +$$

$$+ \ldots + \dot{g}_r(t)\sum_{k=1}^{r} f_k(V_1)(\exp(\alpha_1 g_1(t)))_{k_1 k} \ldots$$

$$\ldots \sum_{k_{r-2}=1}^{r} (\exp(\alpha_{r-1} g_{r-1}(t)))_{rk_{r-2}}$$

here all indexes k_i, $i = 1, \ldots, r-2$ are summarized. To obtain the equations for the functions $g_i(t)$ we reduce the homogeneous terms in the equation above staying before f_k. We have

$$v_k = \delta_{1k}\dot{g}_1(t) + \sum_{i=1}^{r} \dot{g}_i(t)(\exp(\alpha_1 g_1(t)))_{k_1 k} \ldots$$

$$\ldots \sum_{k_{i-2}=1}^{r} (\exp(\alpha_{i-1} g_{i-1}(t)))_{ik_{i-2}} \qquad (10)$$

We know that at $t = 0$, $v_k = \dot{g}_k(t)$. Hence there must exist a neighborhood of $t = 0$ in which (10) is invertible. We may rewrite (10) in this neighborhood in the form

$$\dot{g}_i = \sum_{k=1}^{r} v_k b_{ik}, \qquad g(0) = 0 \qquad (11)$$

Since b_{ik} are analytic and because of the validity of the condition (ii), we may find the neighborhood of $t = 0$ in which solution of (11) exists and unique. The proof is completed.

Corollary 3. In the case when condition (iii) is valid representation (3) of the solution to (1) is global.

In equation (11) all matrix exponentials will have the triangular form. Since the production of a triangular matrix on a triangular one will rise again to a same one. Then the matrix with elements b_{ik} will be triangular. Hence (11) is solvable with respect to $g_i(t)$ on all time interval $t \in [0, T]$, and representation of the solution (1) is global.

B: consideration of the problem of control

For the auxiliary systems (8) and (11) the next statements are valid :

-there exist methods of proving the existence theorems to the problems of optimal control;

-there exist methods to research optimal controls in the widest class of restrictions;

-for the auxiliary systems (8) and (11) there exist methods to research the discontinuous solutions, which appears, when the restrictions $\int_{[0,T]} \| u(t) \| \, dt \leq M < \infty$ is considered;

-(as the main result) the nature of a given Banach space X doesn't influence on the auxiliary system.

-we can find extremals and optimal controls for the system (2), but as it is shown in (Sussmann and Liu, to appear) and (Sussmann and Liu, 1991) the problem of the "approximate tracking" may be solved, i.e. of approximating any given curve by admissible trajectories. Thus for any given trajectory γ of (2) is produced a sequence γ_i of trajectories of (1) that converges to γ.

Example 1. Consider the smooth vector fields f_1 and f_2 and corresponding equation

$$\dot{x}(t) = u_1(t)f_1(x(t)) + u_2(t)f_2(x(t)) \qquad (12)$$

with Lie bracket $[f_1, f_2] = f_1$.

It is easy to find, that $\exp(g_1(t)\mathrm{ad}f_1)f_2 = f_2 + g_1(t)f_1$. From the above one can find

$$A(g) = \begin{pmatrix} 1 & g_1 \\ 0 & 1 \end{pmatrix}$$

is an upper-triangular matrix.

One can find both $g_1(t)$ and $g_2(t)$ in quadratures under the condition $g_1(0) = 0, g_2(0) = 0$. The solution to (12) for any x_0 is :

$$x(t) = \exp(g_1(t)f_1)\exp(g_2(t)f_2)(x_0) \qquad (13)$$

3. CONCLUSION

We obtained that the path-finding problem of the infinite dimension system (1) reduces to the solving of path-finding problem of the finite dimension auxiliary system (8), as illustrated by example 1. Similar example 1 we may consider an auxiliary system (11). To find analytic decisions to the equation (11) we must be able to calculate matrix exponentials and invertible matrix. Using the above algorithms the exact formulas for functions $g_i(t)$ are gained. The solutions of systems (8) and (11) may be easily obtained by the methods of computer algebra.

4. REFERENCES

Olver, P. (1986). *Applications of Lie groups to the differential equations*. Springer-Verlag. New-York.

Sussmann, H. J. and W. Liu (1991). Limits of highly oscillatory controls and the approximation of general paths by admissible trajectories. In: *Proceedings of the 30th CDC IEEE*. Vol. I.

Sussmann, H. J. and W. Liu (to appear). A characterization of continuous dependence of trajectories with respect to the input for control-affine system.

Wei, J. and E. Norman (1964). On global representations of the solutions of linear differential equations as a product of exponentials. *Am. Math.Soc.* **15**, 327–334.

OPTIMAL CONTROL OF PARABOLIC SINGULAR PERTURBATED VARIATIONAL INEQUALITIES.

Vladimir Y. Kapustyan

Professor, Head of Department,
Dnepropetrovsk Transport University,
2, Acad. Lazaryan Street, 320010,
Dnepropetrovsk, Ukraine

Abstract. In this paper the author's investigations results in the optimal control singular perturbated problem's solution for parabolic variational inequality are reduced. The problem's solution is constructed on the basis of the optimality conditions in local form, which are represented by special form of the one-sided mathematical physics problems. They are characterized by the presence of sets, on which the solutions loose the smooth. Obtained and proved in the paper asymptotics consider indicated peculiarities of problem.

Keywords: optimal control, decomposition, asymptotics, singular pertubation method, partial differential equations.

1. INTRODUCTION.

In this paper author presents some investigations on the optimal control problem solution for parabolic singular perturbated variational inequalities. The presence of small parameter in the main part of differential operator allows to use asymptotic analysis methods and author's results in analogous problems for solution parabolic equations (Kapustyan, 1996). Asymptotic solutions are constructed on the basis of necessary conditions of first-order optimality in (Barby, 1984). Zeroth components of corresponding outer decompositions of optimality conditions break definition domain of problem on two classes of unintersectional sets: 1) the first one generated by solution's structure of variational inequality; 2) the second one is defined by control which assumes outermost meanings its limitations. The indicated sets boundaries as the solutions are specificated to arbitrary order of asymptotic exactness with inequalities optimality conditions preservation. Discovered asymptotic solutions are substantiated.

2. OPTIMALITY CONDITIONS.

Consider such optimal control problem about obstacle: find $u(t) \in U = \{v : v(t) \in L_2(0,T), \mid v(t) \mid \leq \xi$ for a.e. $t \in [0,T]\}$ such that

$$I(v) = \frac{1}{2}\int_0^T(\int_\Omega(y(x,t) - z(x))^2dx + \nu v^2(t))dt \rightarrow min, \quad (1)$$

where $y(x,t)$ is the solution of variational inequality of parabolic type in (Barby, 1984, 1993)

$$(y_t(x,t) - \epsilon^2\Delta y(x,t) - g(x)v(t))(y(x,t) - \psi(x)) = 0$$

a.e. in Q,

$$y_t(x,t) - \epsilon^2\Delta y(x,t) - g(x)v(t) \geq 0,$$

$$y(x,t) \geq \psi(x) \text{ a.e. in Q}, \quad (2)$$

$$y(x,o) = y_0(x), \text{ a.e. in } \Omega, \ y(x,t) = 0, \text{ a.e. in } \Sigma;$$

here $Q = \Omega \times (0,T)$, $\Sigma = \partial\Omega \times (0,T)$, $\Omega \in R^n$- has compact closure and smooth (from C^∞) (n−1)-measured boundary $\partial\Omega$, $z(x) \in L_2(\Omega)$, $g(x) \in L_q(\Omega)$, $y_0(x) \in W_0^{2-2/q,q}(\Omega)$, $\psi(x) \in H^2(\Omega)$, $\psi(x) \leq 0$ a.e. on $\partial\Omega$, $y_0 \geq \psi(x)$ a.e. in Ω, $q >$max(n,2), $0 < \epsilon \ll 1$, ν =const> 0, Δ is Laplase operator.
Problem (1)-(2) has at least one solution u. Let (y,u) be an arbitrary pair in problem (1)-(2). Then (Barby, 1984) exists a function $p \in L_2(0,T;H^1(\Omega)) \bigcap BV([0,T];Y^*)$, $Y = H^s(\Omega)\bigcap H^1(\Omega)$, $s > n/2$ which satisfies the equations:

$$-p_t - \epsilon^2\Delta p = y(x,t) - z(x)$$

a.e. in $\{(x,t) : y(x,t) > \psi(x)\}$,

$$p(x,t) = 0, \ a.e. \ in\Sigma;$$

$$p(x,t)(g(x)u(t) + \epsilon^2\Delta y) = 0 \quad (3)$$

a.e. in $\{y = \psi\}$,

$$p(x,t) = 0 \text{ a.e. in } \Omega,$$

$$u(t) = \begin{cases} -\xi, & (g, p(\cdot,t)) - \nu\xi > 0, \\ -\nu^{-1}(g, p(\cdot,t)), & \\ \xi, & (g, p(\cdot,t)) + \nu\xi < 0, \end{cases} \quad (4)$$

where $(g, p(\cdot,t)) = \int_\Omega g(x)p(x,t)dx$.

3. FORMAL ASYMPTOTICS.

Find the outer (Nazarov, 1990) decomposition of the solution of (2)-(4) in form

$$\bar{y}(x,t) = \sum_{i=1}^{\infty} \epsilon^i \bar{y}_i(x,t);$$
$$\bar{p}(x,t) = \sum_{i=1}^{\infty} \epsilon^i \bar{p}_i(x,t). \quad (5)$$

ASSUMPTION 1. Suppose $z(x)$, $y_0(x)$, $\psi(x)$, $0 \le g(x) \in C^\infty(\Omega)$•

Zero components of decomposition (5) are defined like the solution of problem

$$(\bar{y}_{0_t}(x,t) - g(x)\bar{u}_0(t))(\bar{y}_0(x,t) - \psi(x)) = 0 \text{ in } Q,$$

$$\bar{y}_{0_t}(x,t) - g(x)\bar{u}_0(t) \ge 0, \ \bar{y}_0(x,t) \ge \psi(x) \text{ in } Q,$$

$$\bar{y}_0(x,0) = y_0(x) \text{ in } \Omega; \quad (6)$$

$$-\bar{p}_{0_t} = \bar{y}_0(x,t) - z(x) \text{ in } \{(x,t) : \bar{y}_0(x,t) > \psi(x)\},$$

$$\bar{p}_0(x,t)g(x)\bar{u}_0(t) = 0 \text{ a.e. in } \{\bar{y}_0 = \psi\},$$

$$\bar{p}_0(x,t) = 0 \text{ in } \Omega,$$

$$\bar{u}_0(t) = \begin{cases} -\xi, & (g, \bar{p}_0(\cdot,t)) - \nu\xi > 0, \\ -\nu^{-1}(g, \bar{p}_0(\cdot,t)), & \\ \xi, & (g, \bar{p}_0(\cdot,t)) + \nu\xi < 0. \end{cases} \quad (7)$$

Intrude sets

$$Q_0 = \{(x,t) : \ y(x,t) = \psi(x) \text{ a.e.}\},$$

$$Q_+ = \{(x,t) : \ y(x,t) > \psi(x) \text{ a.e.}\}, \quad (8)$$

$$Q = Q_0 \bigcup Q_+, \ Q_0 \bigcap Q_+ = \emptyset,$$

where $y(x,t)$ is the solution of (2)-(4).
Then $\bar{Q}_0$, $\bar{Q}_+$ are their zeroth-order approximations which defined by conditions

$$\bar{Q}_0 = \{(x,t) : \ \bar{y}_0(x,t) = \psi(x)\},$$

$$\bar{Q}_+ = \{(x,t) : \ \bar{y}_0(x,t) > \psi(x)\}, \quad (9)$$

$$Q = \bar{Q}_0 \bigcup \bar{Q}_+, \ \bar{Q}_0 \bigcap \bar{Q}_+ = \emptyset,$$

where $\bar{y}_0(x,t)$ is the solution of (6)-(7).

ASSUMPTION 2. Suppose that $p(x,t) = 0$ in Q_0 and let Q_0 be a cylinder in R^{n+1}, its base be two-connected domain $(\Omega \backslash \Omega_0)$. Suppose that outer boundary is equivalent to $\partial\Omega$ and inner boundary $(\partial\Omega)$ possesses properties of outer boundary •

Then the correlations are fulfilled in $\bar{Q}_+$

$$(x \in \Omega_0, \ t \in T_1^0):$$

$$\begin{cases} \dot{\bar{y}}_0 = -g(x)\xi, \\ -\dot{\bar{p}}_0 = \bar{y}_0 - z, \ (g, \bar{p}_0)_0 - \nu\xi > 0; \end{cases} \quad (10)$$

$$(x \in \Omega_0, \ t \in T_2^0):$$

$$\begin{cases} \dot{\bar{y}}_0 = g(x)\xi, \\ -\dot{\bar{p}}_0 = \bar{y}_0 - z, \ (g, \bar{p}_0)_0 + \nu\xi < 0; \end{cases} \quad (11)$$

$$(x \in \Omega_0, \ t \in T_3^0):$$

$$\begin{cases} \dot{\bar{y}}_0 = -\nu^{-1}g(x)(g, \bar{p}_0)_0, \\ -\dot{\bar{p}}_0 = \bar{y}_0 - z, \end{cases} \quad (12)$$

$$T = \bigcup_{i=1}^{3} T_i^0, \ T_i^0 \bigcap T_j^0 = \emptyset, \ i \ne j,$$

where $(\cdot,\cdot)$ denotes the scalar products by $\bar{\Omega}_0$.
Let's go over to the problems for $(g, \bar{y}_0)_0$, $(g, \bar{p}_0)_0$ for definition of zero components of the control switching moments

$$t \in T_1^0:$$

$$\begin{cases} (g, \dot{\bar{y}}_0)_0 = - \| g \|_0^2 \xi, \\ -(g, \dot{\bar{p}}_0)_0 = (g, \bar{y}_0)_0, \ (g, \bar{p}_0)_0 - \nu\xi > 0; \end{cases} \quad (13)$$

$$t \in T_2^0:$$

$$\begin{cases} (g, \dot{\bar{y}}_0)_0 = \| g \|_0^2 \xi, \\ -(g, \dot{\bar{p}}_0)_0 = (g, \bar{y}_0)_0, \ (g, \bar{p}_0)_0 + \nu\xi < 0; \end{cases} \quad (14)$$

$$t \in T_3^0:$$

$$\begin{cases} (g, \dot{\bar{y}}_0)_0 = -\nu^{-1} \| g \|_0^2 (g, \bar{p}_0)_0, \\ -(g, \dot{\bar{p}}_0)_0 = (g, \bar{y}_0)_0, \ \bar{y}_0 = \bar{y}_0 - z. \end{cases} \quad (15)$$

The problems (13)-(15) may be solved by dint of phase picture (Boltyansky, 1969) which allows to define the control structure. Then two cases are possible: 1) phase point $((g, \bar{y}_0)_0, (g, \bar{p}_0))$ doesn't go on bounds, i.e. it belongs to set $P_1 = \{0 < (g, \bar{p}_0)_0 \le \nu\xi, \ \nu^{-1/2} \| g \|_0 (g, \bar{p}_0)_0 < (g, \bar{y}_0)_0 < \infty\} \bigcup \{-\nu\xi \le (g, \bar{p}_0)_0 < 0, \ -\infty < (g, \bar{y}_0)_0 < \nu^{-1/2} \times \| g \|_0 (g, \bar{p}_0)_0\}$; 2) phase point goes on bounds, i.e. it belongs to set $P_2 = \{(g, \bar{p}_0)_0 \ge \nu\xi, \ \nu^{-1/2} \| g \|_0 (g, \bar{p}_0)_0 \le (g, \bar{y}_0)_0 < \infty\} \bigcup \{-\nu\xi > (g, \bar{p}_0)_0, \ -\infty < (g, \bar{y}_0)_0 \le \nu^{-1/2} \| g \|_0 (g, \bar{p}_0)_0\}$.

In the case 1) the solution has an appearance:

$$\begin{cases} (g, \bar{y}_0)_0 = (y_0 - z, g)_0 ch((\nu^{-1/2} \| g \|_0 \times \\ \times (T - t))(ch(\nu^{-1/2} \| g \|_0 T))^{-1}, \\ (g, \bar{p}_0)_0 = \nu^{1/2} \| g \|_0^{-1} (y_0 - z, g)_0 \times \\ \times sh(\nu^{-1/2} \| g \|_0 (T - t)) \times \\ \times (ch(\nu^{-1/2} \| g \|_0 T))^{-1} \end{cases} \quad (16)$$

on condition that initial data satisfy the inclusion

$$\{(y_0 - z, g)_0, \nu^{1/2} \| g \|_0^{-1} (y_0 - z, g)_0 \times \\ \times th(\nu^{-1/2} \| g \|_0 T)\} \in P_1. \quad (17)$$

In the case 2) systems (13)-(14) have the solution in $0 \le t \le \tau_0$ (τ_0 is the moments of descent of control from limitation)

$$\begin{cases} (g, \bar{y}_0) = \mp\xi \| g \|_0^2 t + (g, y_0 - z)_0, \\ (g, \bar{p}_0) = \pm\xi(\nu - 1/2 \| g \|_0^2 \times \\ \times [\tau_0^2 - t^2]) + (g, y_0 - z_0)_0(\tau_0 - t). \end{cases} \quad (18)$$

On the segment $t \in (\tau_0, T]$ system (15) has the solution

$$\begin{cases} (g,\bar{y}_0)_0 = \pm\xi\nu^{1/2} \| g \|_0 \times \\ \times ch(\nu^{-1/2} \| g \|_0 (T-t))\times \\ \times(sh(\nu^{-1/2} \| g \|_0 (T-\tau_0)))^{-1}, \\ (g,\bar{p}_0)_0 = \pm\xi\nu sh(\nu^{-1/2} \| g \|_0 \times \\ \times(T-t))(sh(\nu^{-1/2} \| g \|_0 \times \\ \times(T-\tau_0)))^{-1}. \end{cases} \quad (19)$$

At that the inclusion takes place

$$\{(y_0-z,g)_0, \pm\xi(\nu - 1/2 \| g \|_0^2 \tau_0^2)+ \\ +(g,y_0-z)_0\tau_0\} \in P_2. \quad (20)$$

Since the function $(g,\bar{y}_0)$ has to be continuous by $t=\tau_0$, we obtain the equation for definition τ_0 from (18)-(19)

$$\mp\xi \| g \|_0 (\nu^{1/2}\mathrm{cth}(\nu^{-1/2} \| g \|_0 \times \\ \times(T-\tau_0))\pm \| g \|_0 \tau_0) = (g,y_0-z_0). \quad (21)$$

The equation (21) has the unique solution if the inclusion (20) is executed. Let's regard futher for definition that the condition

$$\{(y_0-z,g)_0, \xi(\nu - 1/2 \| g \|_0^2 \tau_0^2)+ \\ +(g,y_0-z)_0\tau_0\} \in P_2 \bigcap\{(g,\bar{p}_0)_0 > \nu\xi, \\ \nu^{-1/2} \| g \|_0 (g,\bar{p}_0)_0 \le (g,\bar{y}_0)_0 < \infty\}, \quad (22)$$

which guarantees uniqueness of the solution of equation

$$-\xi \| g \|_0 (\nu^{1/2}\mathrm{cth}(\nu^{-1/2} \| g \|_0 \times \\ \times(T-\tau_0))+ \| g \|_0 \tau_0) = (g,y_0-z)_0, \quad (23)$$

is fulfilled. Thus in case 1) the couple $(\bar{y}_0(x,t),\bar{p}_0(x,t))$ is fined from (12). In particular

$$\bar{y}_0(x,t) = y_0(x) - g(x) \| g \|_0^{-2} \times \\ \times(y_0-z,g)_0(ch(\nu^{-1/2} \| g \|_0 T)- \\ -ch(\nu^{-1/2} \| g \|_0 (T-t)))\times \\ \times(ch(\nu^{-1/2} \| g \|_0 T))^{-1}. \quad (24)$$

The question about choice of domain $\bar{\Omega}_0$ is solved this way. Let $\bar{\Omega}_0$ be a set from Ω, which satisfies the condition $y_0(x) > \psi(x)$ and suppose that for any $x \in \bar{\Omega}_0$ the inequality

$$\bar{y}_0(x,T) > \psi(x) \quad (25)$$

is fulfilled, at that the function $\mathrm{ch}(\nu^{-1/2} \| g \|_0 T) - \mathrm{ch}(\nu^{-1/2} \| g \|_0 (T-t))$ increases monotonically.

Systems (10)-(12) are the conditions of optimality in the optimal control problem: find $\bar{u}_0(t) \in U$ such that

$$I_0(v) = \frac{1}{2}\int_0^T(\int_{\bar{\Omega}_0}(\bar{y}_0(x,t)-z(x))^2dx+ \\ +\nu v^2(t))dt \to min$$

by bounds

$$\dot{\bar{y}}_0(x,t) = g(x)v(t), \bar{y}_0(x,0) = y_0(x).$$

Let $\bar{\bar{\Omega}}_0$ be a system of expanded sets which belong to Ω and contain $\bar{\Omega}_0$ (boundaries of the indicated sets have the properties of the boundary $\partial\Omega$). Then $\bar{\Omega}_0$ is the solution of the optimization problem

$$\tfrac{1}{2}\int_0^T(\int_{\bar{\bar{\Omega}}_0}(\bar{y}_0(x,t)-z(x))^2dx+ \\ + \int_{\Omega/\bar{\bar{\Omega}}_0}(\psi(x)-z(x))^2dx+ \\ +\nu(\int_{\bar{\bar{\Omega}}_0} g(x)\bar{p}_0(x,t))^2)dt \to \min \quad (26)$$

by bound (17),(25), $\bar{y}_0(x,t)$ is given by the representation (24) and scalar products $(\cdot,\cdot)_0$ are calculated by $\bar{\bar{\Omega}}_0$ in all terms.

REMARK 1. Suppose that $y_0(x) > \psi(x)\ \forall\ x \in \Omega$. Then $\bar{\Omega}_0 = \Omega$ and it is enough that the condition (25) is fulfilled. It fulfiles always if $(y_0-z,g) < 0$ •

q In case 2) the solution of (10),(12), continuous for $t \in [0,T]$ and smooth for $x \in \bar{\Omega}_0$, is given by the couple $(\bar{y}_0(x,t),\bar{p}_0(x,t))$. In particular, for $t \in [0,\tau_0]$

$$\bar{y}_0(x,t) = -\xi g(x)t + y_0(x),$$

for $t \in (\tau_0, T]$

$$\bar{y}_0(x,t) = -\xi g(x)\tau_0 + y_0(x) + \xi\nu^{1/2} \| g \|_0^{-1} \times \\ \times g(x)(sh(\nu^{-1/2} \| g \|_0 (T-\tau_0)))^{-1}(ch(\nu^{-1/2}\times \\ \times \| g \|_0 (T-t)) - ch(\nu^{-1/2} \| g \|_0 (T-\tau_0))). \quad (27)$$

The question about choice of domain $\bar{\Omega}_0$ is solved by analogy with preceding case with next changes: the function (26) is minimized by bounds (22),(23),(25) and $\bar{y}_0(x,t)$ is given by representation (27). Let's supplement the solutions $(\bar{y}_0(x,t),\bar{p}_0(x,t))$ on $\partial\bar{\Omega}_0$ by following boundary layer functions $\tilde{y}_0(\bar{t},s,t),\tilde{p}_0(\bar{t},s,t)$ (Vasil'eva and Butuzov 1990), defined from problems $(k=0)$

$$\begin{cases} \tilde{y}_{k_t} - \tilde{y}_{k_{\bar{t}\bar{t}}} = \\ = -\sum_{j=1}^{k} L_j(s,\bar{t},\partial/\partial s,\partial/\partial\bar{t})\times \\ \times\tilde{y}_{k-j}(\bar{t},s,t), \\ \tilde{p}_{k_t} + \tilde{p}_{k_{\bar{t}\bar{t}}} + \tilde{y}_k = \\ = \sum_{j=1}^{k} L_j(s,\bar{t},\partial/\partial s,\partial/\partial\bar{t})\times \\ \times\tilde{p}_{k-j}(\bar{t},s,t), \\ \tilde{y}_k(0,s,t) = -\bar{y}_k(0,s,t), \\ \tilde{p}_k(0,s,t) = -\bar{p}_k(0,s,t); \\ \tilde{y}_k(\bar{t},s,t_0) = \tilde{p}_k(\bar{t},s,T) = 0, \end{cases} \quad (28)$$

where $\bar{t} = -\epsilon\zeta$, ζ is a distance to $\partial\bar{\Omega}_0$ along outer normal, s is local coordinates on surface $\partial\bar{\Omega}_0$; L_j is differential operator not exceeded second order on variable s and not exceeded first order on variable $\bar{t}$; their coefficiens smoothly depend on coordinate s on $\partial\bar{\Omega}_0$ and

polymially depend on $\bar{t}$.
The problems (28) are solved by single way in parabolic boundary layer functional class (Vasil'eva and Butuzov 1990).
REMARK 2. Note that inequality $(g, \bar{p}_0(.,t)) - \nu\xi > 0, t \in [t_0, \tau_0)$ will be fulfilled with exactness $O(\epsilon)$, if $\bar{p}_0(x,t)$ is replaced by sum $\bar{p}_0(x,t) + \tilde{p}_0$. The inequality (25) is fulfilled by replacement $\bar{y}_0(x,t)$ on $\bar{y}_0(x,t) + \tilde{y}_0$ at the expense of initial data choice only•
Thereby the solution of (6)-(7) is constructed completely, i.e. zeroth components of decomposition (5) are fined.
ASSUMPTION 3. Suppose the problem's data such that the moment of control switching $\tau_0 \in (0,T)$ existes and $\partial\bar{\Omega}_0 = \partial\Omega$ •
Show further the algorithm of specification of the control switching moment. Let τ be a moment of control descent from bound in initial problem. Let's find it in the form of asymptotic series

$$\tau = \sum_{j=0}^{\infty} \epsilon^j \tau_j. \tag{29}$$

Find the first coefficients of decomposition (29). Assume $\tau^1 = \tau_0 + \epsilon\tau_1$, $\tau_1 > 0$. Then $\bar{y}_1(x,t) = 0$, $\bar{p}_1 = \bar{p}(x_1)$, $t \in [0, \tau^1)$, and if $t \in (\tau^1, T]$, then, according to (15),

$$\begin{cases} \dot{\bar{y}}_1 + \nu^{-1} g(x)(g, \bar{p}_1) = -\nu^{-1} g(x)\alpha_1(t), \\ \dot{\bar{p}}_1 = -\bar{y}_1, \end{cases} \tag{30}$$

where $\alpha_1(t) = \int_0^\infty \int_{\partial\bar{\Omega}_0} D(x)/D(0)(g(0,s))\bar{p}_0(\xi, s, t)ds d\xi$, $\bar{p}_0(\bar{t}, s, t)$ is the solution of (28) for $k = 0$; $D(x)/D(\zeta, s)$ is Jacobian of transformation $x \to (-\zeta, s)$ in environs $\partial\bar{\Omega}_0$. At that the inequality

$$(g, p(.,t) - \nu\xi > 0 \tag{31}$$

accepts the form

$$(g, \bar{p}_0(.,t) + \epsilon\bar{p}(.,t)) + \epsilon\alpha_1(t) - \nu\xi > 0, t \in [0, \tau^1). \tag{32}$$

Form (32) obtain the condition by $t = \tau^1$ with exactness $O(\epsilon^2)$

$$(g, \partial\bar{p}_0(., \tau_0+))/\partial t\tau_1 + \alpha_1(\tau_0) + (g, \bar{p}_1(., \tau_0)) = 0, \tag{33}$$

and according to (30) the function $(g, \bar{p}_1(.,t))$ is defined by

$$\begin{cases} (g, \ddot{\bar{y}}_1) + \nu^{-1}||g||^2(g, p_1) = \\ -\nu^{-1}||g||^2\alpha_1(t), \\ (g, \ddot{\bar{p}}_1) = -(g, \bar{y}_1). \end{cases} \tag{34}$$

The condition (33) shows that the functions $\bar{y}_1$, $\bar{p}_1$ will have only one break point $t = \tau_0$. System (34) with boundary conditions $(g, \bar{p}_1(., \tau_0)) = -[(g, \partial\bar{p}_0(., \tau_0+)/\partial t)\tau_1 + \alpha_1(\tau_0)]$, $(g, \bar{p}_1(., T)) = 0$ has the solution

$$\begin{aligned} (g, \bar{y}_1) = &-[(g, \partial\bar{p}_0(., \tau_0+)/\partial t)\times \\ &\times\tau_1 + \alpha_1(\tau_0)]\nu^{-1/2}||g||ch(\nu^{-1/2}||g||\times \\ &\times(T-t))(sh(\nu^{-1/2}||g||(T-\tau_0)))^{-1} + \nu^{-1}\times \\ &\times||g||^2 ch(\nu^{-1/2}||g||(t-\tau_0))(sh(\nu^{-1/2}||g||\times \\ &\times(T-\tau_0)))^{-1}\textstyle\int_{\tau_0}^T \alpha_1(t)\times \\ &\times sh(\nu^{-1/2}||g||(T-t))dt - \nu^{-1}||g||^2 \textstyle\int_{\tau_0}^t \alpha_1(\tau)\times \\ &\times ch(\nu^{-1/2}||g||(t-\tau))d\tau, \end{aligned} \tag{35}$$

$$\begin{aligned} (g, \bar{p}_1) = &-[(g, \partial\bar{p}_0(., \tau_0+)/\partial t)\times \\ &\times\tau_1 + \alpha_1(\tau_0)]sh(\nu^{-1/2}||g||(T-t))\times \\ &\times(sh(\nu^{-1/2}||g||(T-\tau_0)))^{-1} - \nu^{-1/2}||g||sh(\nu^{-1/2}\times \\ &\times||g||(t-\tau_0))(sh(\nu^{-1/2}||g||(T-\tau_0)))^{-1}\times \\ &\times\textstyle\int_{\tau_0}^T \alpha_1(t)sh(\nu^{-1/2}||g||(T-t))dt + \\ &+\nu^{-1/2}||g||\textstyle\int_{\tau_0}^t \alpha_1(t)sh(\nu^{-1/2}\times \\ &\times||g||(t-\tau))d\tau. \end{aligned} \tag{36}$$

Hence, $(g, \bar{p}_1(.,t) = -[(g.\partial\bar{p}_0(., \tau_0+)/\partial t)\tau_1 + \alpha_1(\tau_0)]$, $(g, \bar{y}_1(.,t)) = 0$, $t \in [0, \tau_0]$ and $t = \tau_0$ it can be obtained that

$$\begin{aligned} \tau_1 = &(g, \partial\bar{p}_0(., \tau_0+))^{-1}\times \\ &\times[\nu^{-1/2}||g||(ch(\nu^{-1/2}||g||(T-\tau_0)))^{-1}\times \\ &\times\textstyle\int_{\tau_0}^T \alpha_1(t)sh(\nu^{-1/2}||g||\times \\ &\times(T-t))dt - \alpha_1(\tau_0)] \end{aligned} \tag{37}$$

and at that the right side of (37) has to be positive. Return to the inequality (32). It accepts the form

$$\begin{aligned} &(g, \bar{p}_0(.,t) + \epsilon\bar{p}_1(.,t)) + \epsilon\alpha_1(t) - \nu\xi > 0, \\ &t \in [0, \tau^1). \end{aligned} \tag{38}$$

For $t \in [0, \tau_0)$ the inequality (38) is fulfilled by virtue of (22) and for $t \in [\tau_0, \tau^1)$ it accepts the form

$$(g, \partial\bar{p}_0(., \tau_0 + \Theta(t-\tau_0))/\partial t)(t-\tau_0) +$$
$$+\epsilon(g, \bar{p}_1(., \tau_0)) > 0, \Theta \in (0,1),$$

which can be written with exactness $O(\epsilon^2$ in the form

$$(g, \partial\bar{p}_0(., \tau_0+)/\partial t)(t - \tau_0 - \epsilon\tau_1),$$

that teake place by virtue of (19): $(g, \partial\bar{p}_0(., \tau_0+)/\partial t) < 0$; $t - \tau_0 - \epsilon\tau_1 < 0$. Assume $\tau_1 < 0$. At that case reasonings stay to be true for $\tau_1 > 0$ except such moments: in equality (33) change $(g, \partial\bar{p}_0(., \tau_0+)) \to (g, \partial\bar{p}_0(., \tau_0-))$; needs to be done; quantity τ_1 is defined by (37) with the same change. The inequality (38) doesn't fulfil on (τ^1, τ_0). Really it can be accemped by the form

$$(g, \partial\bar{p}_0(., \tau_0-)/\partial t)(t - \tau_0 - \epsilon\tau_1) > 0,$$

but $(g, \partial\bar{p}_0(., \tau_0-)/\partial t) = \xi||g||^2\tau_0 - (g, y_0) < 0$ by virtue of (18),(22) and $t - \tau_0 - \epsilon\tau_1 > 0$, hence inverse inequality takes place. Thereby the first step of algorithm for definition of coeficients of decomposition (29) is constructed completely: if

$$\begin{aligned} \Lambda_1 = &\nu^{-1/2}||g||(ch(\nu^{-1/2}||g||(T-\tau_0)))^{-1}\times \\ &\times\textstyle\int_{\tau_0}^T \alpha_1(t)sh(\nu^{-1/2}||g||(T-t))dt - \\ &-\alpha_1(\tau_0) < 0, \end{aligned} \tag{39}$$

then $\tau_1 > 0$ and it is defined by (37) otherwise $-\tau_1 < 0$ and it is defined by the same form with change $(g, \partial\bar{p}_0(., \tau_0+)) \to (g, \partial\bar{p}_0(., \tau_0-))$; if $\Lambda = 0$ then $\tau_1 = 0$. But functions $\bar{y}_1(x,t)$, $\bar{p}(x,t)$ are fined from the system (34) and completed by the boundary layer functions $\tilde{y}_1$, $\tilde{p}_1$, which are defined from (28). Easy to check by induction that next lemma is true.

LEMMA. Let's suppose that the assumptions 1-3 are true and (22) takes place. Then series (29) is constructed simply and if $\tau_1 > 0$, then

$$\begin{aligned}
&\tau_i = -(g, \partial\bar{p}_0(., \tau_0+)/\partial t)^{-1}\times\\
&\times[R_{i+}(\tau_0, ..., \tau_{i-1}) + \nu^{1/2}||g||^{-1}\times\\
&\times(th(\nu^{-1/2}||g||(T-\tau_0))\textstyle\int_0^{\tau_0}\times\\
&\times(g, \Delta\bar{y}_{i-2}(., t))dt + (ch(\nu^{-1/2}||g||\times\\
&\times(T-\tau_0)))^{-1}\textstyle\int_{\tau_0}^T[sh(\nu^{-1/2}\times\\
&\times||g||(T-t))((g, \Delta\bar{y}_{i-2}(., t)) - \nu^{-1}||g||^2\times\\
&\times\alpha_i(t)) + \nu^{-1/2}||g||(g, \Delta\bar{p}_{i-2}(., t))\times\\
&\times ch(\nu-1/2||g||(T-t))]dt)],\ i > 2,
\end{aligned} \tag{40}$$

where $R_{i+}(\tau_0, ..., \tau_{i-1})$, $\alpha_i(t)$ are known functions; for $i = 1$ they are puted above and if

$$\begin{aligned}
&\Lambda_i = R_{i+}(\tau_0, ..., \tau_{i-1}) + \nu^{1/2}||g||^{-1}\times\\
&\times(th(\nu^{-1/2}||g||(T-\tau_0))\textstyle\int_0^{\tau_0}(g,\\
&\Delta\bar{y}_{i-2}(., t))dt + (ch(\nu^{-1/2}||g||(T-\tau_0)))^{-1}\times\\
&\times\textstyle\int_{\tau_0}^T[sh(\nu^{-1/2}||g||(T-t))((g, \Delta\bar{y}_{i-2}(., t))-\\
&-\nu^{-1}||g||^2\alpha_i-(t)) + \nu^{-1/2}||g||(g,\\
&\Delta\bar{p}_{i-2}(., t))ch(\nu^{-1/2}||g||(T-t))]dt) > 0,
\end{aligned} \tag{41}$$

then $\tau_i > 0$, otherwise - $\tau_i < 0$; if $\tau_1 < 0$, then the form (40) together with inequality (41) preserve their sense with change $(g, \partial\bar{p}_0(., \tau_0+)/\partial t) \to (g, \partial\bar{p}_0(., \tau_0-)/\partial t)$, $R_{i+}(\tau_0, ..., \tau_{i-1}) \to R_{i-}(\tau_0, ..., \tau_{i-1})$ •

Decompositions (5) are completed by the boundary layer functions defined by (28).

4. SUBSTATIATION OF ASYMPTOTIC DECOMPOSITIONS.

Assume that the segment of asymptotic series (29) is constructed, i. e.

$$\tau^N = \sum_{i=0}^N \epsilon^i \tau_i. \tag{42}$$

Consider the asymptotic decompositions

$$y^{(N)}(x,t) = \sum_{j=0}^N (\bar{y}_j(x,t) + \tilde{y}_j(\bar{t}, s, t)), \tag{43}$$

$$p^{(N)}(x,t) = \sum_{j=0}^N (\bar{p}_j(x,t) + \tilde{p}_j(\bar{t}, s, t)), \tag{44}$$

$$u^{(N)}(t) = \begin{cases} -\xi,\ 0 \le t \le \tau^N, \\ -\nu^{-1}\,(g, p^{(N)}(., t)),\ \tau^N \le t \le T. \end{cases} \tag{45}$$

Futher using the results of papers (Kapustyan 1993, 1996; Nazarov 1990) obtain that next theorem is true.

THEOREM. Suppose that the conditions of lemma are fulfilled. Then the functions (43)-(45) are the asymptotics of initial problem solution and next valuations take place

$$\begin{gathered}
||grad(y - y^{(N)})||_{L_2(Q)} +\\
+ ||grad(p - p^{(N)})||_{L_2(Q)} \le C\epsilon^N,\\
||y - y^{(N)}||_{L_2(Q)} +\\
+ ||p - p^{(N)}||_{L_2(Q)} \le C\epsilon^{N+1},\\
||u - u^{(N)}||_{L_2(0,T)} \le C\epsilon^{N+1},\\
|I(u) - I(u^{(N)})| \le C\epsilon^{2(N+1)}.
\end{gathered}$$

REFERENCES

Barby V. (1984). *Optimal control Of variation inequalities.* Pitman, London.

Barby V. (1993). *Analysis and control of nonlinear infinite dimensional systems.* Academic Press, Ins.

Boltyansky V. G. (1969). *Mathematical methods of optimal control.* Nauka, Moscow.

Kapustyan V. Y. (1996). Asymptotics of locally boundedcontrol in optimal parabolic problems. *Ukr. math. J.* 48, N1, 50-56.

Kapustyan V. Y. (1993). Asymptotics of control in optimal singular pertyrbated parabolic problems. Global bounds on control. *Docl. AN (Russia),* 333, N4, 428-431.

Nazarov S. A. (1990). Asymptotic solution of variation inequalittes for linear operator with small parameter by senior derivatives. *Izv. AN USSR. Series of math.*, 54, N4, 754-773.

Vasil'eva A. B. and Butuzov V. F. (1990). *Asymptotic methods in singular perturbations theory.* Vysshaya Shkola, Moscow.

ASYMPTOTIC SOLUTION OF PERIODIC CONTROL PROBLEM PERTURBED BY MATRIX

G.A.Kurina, S.S.Shabanova

Voronezh State Forest Technical Academy, Timirjazeva ul. 8, Voronezh, 394613, Russia

Abstract: By the direct substitution of the postulated asymptotic expansion of the solution into the problem conditions the asymptotic expansion with respect to powers of the small parameter ε is constructed under some conditions for the solution of the classic nonlinear periodic optimal control problem with fixed period and with the linear operator $A + \varepsilon B$ before the derivative in the state equation, where $detA = 0$ and $det(A + \varepsilon B) \not\equiv 0$. The closeness estimates between approximate and exact solution are given; the lack of increase of the performance index is established when the new terms of the control asymptotic are used.

Keywords: optimal control, periodic motion, discriptor systems, singular perturbations, asymptotic approximation.

1. INTRODUCTION

Periodic optimal control problems occur in mechanics, in theory of regulation, in chemical technology (the optimal control by chemical reactors), in cardiology and in many other applications. The survey of the publications, devoted to the research of the conditions of the control optimality in such problems, is given in (Tonkov, 1977).

In the present paper, the following classic optimal control problem is being considered

$$P_\varepsilon : J_\varepsilon(u) = \int_0^T F(x(t), u(t), t, \varepsilon)dt \to \min_u, \quad (1)$$

$$(A + \varepsilon B)\frac{dx(t)}{dt} = f(x(t), u(t), t, \varepsilon), \quad (2)$$

$$x(0) = x(T). \quad (3)$$

Here $x(t) \in X, u(t) \in U; X$ and U are real finite-dimensional Euclidean spaces, $T > 0$ is fixed; $t \in [0, T]; \varepsilon > 0$ is a small parameter; the functions F, f are sufficiently smooth with respect to their arguments, this functions take the values in R and X respectively and are T - periodic with respect to t; $A, B \in L(X)$; the operator A is singular ; $dim\ kerA < dimX$; $det(A + \varepsilon B) \not\equiv 0$; admissible controls $u = u(t)$ are continuous T-periodic functions for which the solution of the problem (2),(3) exists.

For construction of the asymptotics of the problem (1)-(3) solution in the given paper the so-called direct scheme is being used, it consists of substituting the postulated asymptotic expansion of solution into the problem conditions and of constructing the series of the optimal control problems the solutions of which are the terms of the asymptotic expansion. The most complete development of the direct scheme of solution asymptotics construction is presented in (Belokopytov and Dmitriev, 1986, 1989) for singularly perturbed optimal control problems.

Using the direct scheme for linear - quadratic problems singularly perturbed by matrix (the operator $A + \varepsilon B$ stands before the derivative in the state equation) [p/2] of the regular terms (p is the length of the B - Jordan chains of the operator A) and zero boundary layer terms of the solution asymptotics were constructed in case of the fixed left-hand and free right-hand terminal points in (Kurina and Ovezov, 1996).

2. DIRECT SCHEME FORMALISM

The solution of P_ε is being sought in the form

$$x(t) = \sum_{j \geq 0} \varepsilon^j x_j(t), \quad u(t) = \sum_{j \geq 0} \varepsilon^j u_j(t). \quad (4)$$

After substituting (4) into (2),(3) with series expansion of (2) with respect to powers of ε and equating the coefficients in resulting expressions at equal powers of ε the relations for the terms of (4)

$$A\frac{dx_o(t)}{dt} = f(x_o(t), \quad u_o(t), t, 0) = \overline{f},$$

$$A\frac{dx_1(t)}{dt} = -B\frac{dx_o(t)}{dt} + \overline{f}_x x_1(t) + \overline{f}_u u_1(t) + \overline{f}_\varepsilon, \tag{5}$$

$$\ldots,$$

$$x_j(0) = x_j(T), \quad j \geq 0 \tag{6}$$

are obtained. In this case and further the bar denotes that the values of the functions f, F and of their derivatives are calculated at $x = x_0(t), u = u_0(t), \varepsilon = 0$.

After substituting (4) into (1) and performing the series expansion of the function F with respect to powers of ε the series expansion of the functional $J_\varepsilon(u)$ with respect to powers of ε

$$J_\varepsilon(u) = \sum_{j \geq o} \varepsilon^j J_j \tag{7}$$

is obtained.

For the determination of the functions pair $x_o(t), u_o(t)$ the singular problem

$$P_o : J_o(u_o) = \int_0^T F(x_0(t), u_o(t), t, 0)dt =$$

$$= \int_0^T \overline{F}dt \to \min_{u_o},$$

$$A\frac{dx_o(t)}{dt} = f(x_o(t), u_o(t), t, 0) = \overline{f}, \tag{8}$$

$$x_o(0) = x_o(T) \tag{9}$$

is considered.

Let P, Q denote the orthogonal projectors of the state X onto Ker A, Ker A' respectively, corresponding to decompositions into orthogonal sums

$$X = KerA \oplus ImA' = KerA' \oplus ImA.$$

Let A_+ denotes the operator inverse to the operator

$$(I-Q)A(I-P) : ImA' \to ImA.$$

Here A' denotes the conjugate operator for the operator A.

The equation (8) is equivalent to the system of two equations

$$\frac{d(I-P)x_o(t)}{dt} = A_+\overline{f},$$

$$0 = Q\overline{f}.$$

Let's assume that the following condition is fulfilled.

1. For all $t \in [0,T]$ the equation

$$Qf(Px_o + (I-P)x_o, u_o, t, 0) = 0 \tag{10}$$

is uniquely resolvable with respect to Px_o :

$$Px_o = \Phi((I-P)x_o, u_o, t).$$

Under the condition 1 the reduced problem $\overline{P}$ in the states space JmA' of the smaller dimension

$$\overline{P} : \overline{J}(u) = \int_0^T F(\Phi(y(t), u(t), t) +$$

$$+ y(t), u(t), t, 0)dt \to \min_u,$$

$$\frac{dy(t)}{dt} = A_+(I-Q)f(\Phi(y(t), u(t), t) +$$

$$+ y(t), u(t), t, 0), \tag{11}$$

$$y(0) = y(T), \tag{12}$$

$y(t) \in ImA', \quad u(t) \in U$, follows from the problem P_0.

Let's assume that the following condition 2 is fulfilled.

2. The problem $\overline{P}$ has the unique solution determined from the condition

$$\frac{\partial\overline{H}}{\partial u} = 0, \tag{13}$$

where $\overline{H}$ is the Hamiltonian for the problem $\overline{P}$ and the conjugate variable $\overline{\psi} = \overline{\psi}(t)$ is the solution of the problem

$$\frac{d\overline{\psi}(t)}{dt} = -\left(\frac{\partial\overline{H}}{\partial y}\right)', \quad \overline{\psi}(0) = \overline{\psi}(T). \tag{14}$$

Proposition 1. Under the conditions 1,2 the problem P_o has the unique solution.

It is easy to see that the latter statement is valid, if we set $u_o(t) = u(t), x_o(t) = \Phi(y(t), u(t), t) + y(t)$, where $u(t), y(t)$ is the solution of the problem $\overline{P}$.

In addition to tne conditions 1,2 let's assume, that the following condition 3 is fulfilled.

3. For all $t \in [0,T]$ the operator $Q\overline{f}_x P : KerA \to KerA'$ has the inverse $(Q\overline{f}_x P)^{-1} : KerA' \to KerA$.

Proposition 2. Under the conditions 1-3 the optimal control for the problem P_o satisfies the relation

$$\frac{\partial H_o}{\partial u} = \psi_o'(t)\overline{f}_u - \overline{F}_u = 0, \tag{15}$$

where

$$H_o = \psi_o'(t)\overline{f} - \overline{F},$$

and $\psi_o = \psi_o(t)$ is the solution of the problem

$$A'\frac{d\psi_o(t)}{dt} = -\left(\frac{\partial H_o}{\partial x_o}\right)' = -\overline{f}_x'\psi_o(t) + \overline{F}_x', \tag{16}$$

$$\psi_o(0) = \psi_o(T). \tag{17}$$

Here $v'w$ denotes the scalar product of the elements v, w.

It is easy to see, that if $u_o(t), x_o(t)$ is the T-periodic solution of the problem P_o, then $u(t) = u_o(t), y(t) = (I-P)x_o(t)$ is the T-periodic solution of the problem $\overline{P}$. In addition the relations (11)-(14) are satisfied. In order to prove the proposition 2 the notation

$$\psi_o = \psi_o(t) = (P\overline{f}'_x Q)^{-1} P(\overline{F}'_x - \overline{f}'_x A'_+ \overline{\psi}(t)) + $$
$$+ A'_+ \overline{\psi}(t)$$

is introduced and the relations (10), (13), (14) are used.

Let's transform the coefficient J_1. After using the expression for $\overline{F}_x$, obtained from (16), the expression for $\overline{F}_u$ from (15), the formula of integration by parts and the relations (6), (17), (5) J_1 has the form

$$J_1 = \int_0^T (\psi'_o(t)(B\frac{dx_o(t)}{dt} - \overline{f}_\varepsilon) + \overline{F}_\varepsilon)dt.$$

Therefore the coefficient J_1 in the expansion (7) depends on the solution of the problem P_o only.

After performing the transformations of the coefficient J_2, analogous to previous, it will be written in the form

$$J_2 = \int_0^T (\frac{1}{2}(\overline{F}_{x^2} - \psi'_o(t)\overline{f}_{x^2})x_1^2(t) + (\overline{F}_{xu} - \psi'_o(t)\overline{f}_{xu}) \times$$
$$\times u_1(t)x_1(t) + \frac{1}{2}(\overline{F}_{u^2} - \psi'_o(t)\overline{f}_{u^2})u_1^2(t) +$$
$$+ (\overline{F}_{x\varepsilon} - \psi'_o(t)\overline{f}_{x\varepsilon} - \frac{d\psi'_o(t)}{dt}B)x_1(t) +$$
$$+ (\overline{F}_{u\varepsilon} - \psi'_o(t)\overline{f}_{u\varepsilon})u_1(t) + \frac{1}{2}(\overline{F}_{\varepsilon^2} - \psi'_o(t)\overline{f}_{\varepsilon^2}))dt.$$

So, the coefficient J_2 depends on the solution of the problem P_o and on $u_1(t), x_1(t)$, in addition J_2 is quadratic with respect to u_1, x_1.

The equality

$$\frac{\partial H}{\partial u} = 0$$

follows from the Pontrjagin maximum principle for the problem P_ε, where H is the Hamiltonian.

If the change $\psi(t) = (A' + \varepsilon B')^{-1}\varphi(t)$, where $\varphi = \varphi(t)$is the conjugate variable, is made, then from the previous relation and from (2),(3) it follows that the solution of the problem P_ε satisfies the equations

$$(A + \varepsilon B)\frac{dx(t)}{dt} = f(x(t), u(t), t, \varepsilon),$$
$$x(0) = x(T), \qquad (18)$$

$$(A' + \varepsilon B')\frac{d\psi(t)}{dt} = -f'_x(x(t), u(t), t, \varepsilon)\psi(t) +$$
$$+ F'_x(x(t), u(t), t, \varepsilon), \quad \psi(0) = \psi(T), \qquad (19)$$
$$0 = \psi'(t)f_u(x(t), u(t), t, \varepsilon) -$$
$$- F_u(x(t), u(t), t, \varepsilon). \qquad (20)$$

The equations (8), (9), (16), (17), (15) are obtained after substituting $x(t)$ and $u(t)$ by the expansions (4) and $\psi(t)$ by the expansion

$$\psi(t) = \sum_{j\geq 0} \varepsilon^j \psi_j(t), \qquad (21)$$

in (18) - (20), then decomposing the right-hand parts of equations into series with respect to powers of ε and equating the terms without ε.

Therefore the zero approximation of the problem (18)-(20), following from the control optimality condition for the problem P_ε, represents the condition of the control optimality for the problem P_o.

Further the notation

$$h(\varepsilon) = \sum_{j\geq 0} \varepsilon^j h_j = \{h\}_{n-1} + \varepsilon^n [h]_n + \alpha(\varepsilon^{n+1}),$$

will be used for the expansion of a function with respect to powers of ε, where $\{h\}_{n-1} = \sum_{j=0}^{n-1} \varepsilon^j h_j$, $[h]_n = h_n$, and $\alpha(\varepsilon^{n+1})$ denotes the sum of the expansion terms of the order ε^{n+1} and higher ones.

Let us consider the problems $P_j, j \geq 0$. When $j = 0$, P_o is the singular problem, which was determined earlier, when $j \geq 1$, the problem P_j is the linear-quadratic problem of the form

$$P_j : \tilde{J}_j(u_j) = \int_0^T (\frac{1}{2}(\overline{F}_{x^2} - \psi'_o(t)\overline{f}_{x^2})x_j^2(t) + (\overline{F}_{xu} -$$
$$- \psi'_o(t)\overline{f}_{xu})u_j(t)x_j(t) + \frac{1}{2}(\overline{F}_{u^2} - \psi'_o(t)\overline{f}_{u^2})u_j^2(t) +$$
$$+ \left(-\frac{d\psi'_{j-1}(t)}{dt}B + [\tilde{F}_x - \tilde{\psi}'_{j-1}(t)\tilde{f}_x]_j\right) x_j(t) +$$
$$[\tilde{F}_u - \tilde{\psi}'_{j-1}(t)\tilde{f}_u]_j u_j(t))dt \to \min_{u_j},$$

$$A\frac{dx_j(t)}{dt} = \overline{f}_x x_j(t) + \overline{f}_u u_j(t) + [\tilde{f}]_j - B\frac{dx_{j-1}(t)}{dt},$$
$$x_j(0) = x_j(T).$$

The tilde denotes that the functions values and their derivatives are calculated at $x = \tilde{x}_{j-1}(t), u = \tilde{u}_{j-1}(t)$, where

$$\tilde{x}_{j-1}(t) = \sum_{i=0}^{j-1} \varepsilon^i x_i(t), \; \tilde{u}_{j-1}(t) = \sum_{i=0}^{j-1} \varepsilon^i u_i(t), \qquad (22)$$

the pair of the functions $x_i(t)$ and $u_i(t)$ is the solution of the problem P_i, and the function $\tilde{\psi}_{j-1}(t)$ is determined by the following formula

$$\tilde{\psi}_{j-1}(t) = \sum_{i=0}^{j-1} \varepsilon^i \psi_i(t), \tag{23}$$

where $\psi_i(t)$ $(i \geq 1)$ is determined from the solution of the problem

$$A\frac{dx_i(t)}{dt} = \overline{f}_x x_i(t) + \overline{f}_u u_i(t) + [\tilde{f}]_i -$$

$$-B\frac{dx_{i-1}(t)}{dt}, x_i(0) = x_i(T), \tag{24}$$

$$A'\frac{d\psi_i(t)}{dt} = -\overline{f}'_x \psi_i(t) + x'_i(t)(\overline{F}'_{x^2} - \overline{f}'_{x^2}\psi_o(t)) +$$

$$+u'_i(t)(\overline{F}'_{xu} - \overline{f}'_{xu}\psi_o(t)) - B'\frac{d\psi_{i-1}(t)}{dt} + [\tilde{F}'_x -$$

$$\tilde{f}'_x \tilde{\psi}_{i-1}(t)]_i, \psi_i(0) = \psi_i(T), \tag{25}$$

$$0 = \psi'_i(t)\overline{f}_u - (\overline{F}_{u^2} - \psi'_o(t)\overline{f}_{u^2})u_i(t) -$$

$$-(\overline{F}_{ux} - \psi'_o(t)\overline{f}_{ux})x_i(t) - [\tilde{F}_u - \tilde{\psi}'_{i-1}(t)\tilde{f}_u]_i = 0 \tag{26}$$

following from the condition of the control optimality in the problem P_j (see proposition 2).

It should be noted that the state equation in the problem P_j is the coefficient at ε^j in the equality, which follows from (2) after substituting the expansions (4) in it and expanding the resulting expressions into series with respect to powers of ε.

Proposition 3. The problem for the determination of the coefficients at $\varepsilon^m, m \geq 0$, in the asymptotic expansion (4),(21) of the solution of the problem (18)-(20) obtained from the condition of the control optimality in the problem P_ε coincides with the problem obtained from the condition of the control optimality in the problem P_m.

When $m = 0$ this statement has been proved. It is supposed that it is valid when $m < j$. For $j \geq 1$ the notations

$$\Delta x(t) = x(t) - \tilde{x}_{j-1}(t) = \varepsilon^j x_j(t) + \alpha(\varepsilon^{j+1}),$$

$$\Delta u(t) = u(t) - \tilde{u}_{j-1}(t) = \varepsilon^j u_j(t) + \alpha(\varepsilon^{j+1}), \tag{27}$$

$$\Delta \psi(t) = \psi(t) - \tilde{\psi}_{j-1}(t) = \varepsilon^j \psi_j(t) + \alpha(\varepsilon^{j+1})$$

are introduced, where $\tilde{x}_{j-1}(t)$, $\tilde{u}_{j-1}(t)$, $\tilde{\psi}_{j-1}(t)$ are given by the formulas (22), (23) respectively.

The relations (24)-(26) with $i = j$, which follow from the condition of the control optimality in the problem P_j, are obtained after substituting x, u and ψ by the representations from (27)in (18)-(20) and equating the coefficients at ε^j on both sides of resulting equations. The proposition 3 is proved for $m = j$.

Further, let us note, that in the problem P_1 the performance criterion $\tilde{J}_1(u_1)$ is obtained from the coefficient J_2, which was transformed earlier, if the terms, which are known after the solution of the problem P_o, have been rejected.

Theorem 1. If in the expansion (4) $x_j(t)$ and $u_j(t)$ are obtained as the solution of the problem P_j for $j = \overline{0,m}$, then in (7) the coefficient J_{2m+1} depends on the solutions of the problems $P_j, j = \overline{0,m}$, and the transformed expression for the coefficient J_{2m+2} coincides with the performance criterion $\tilde{J}_{m+1}$ in the problem P_{m+1} if the terms, which are known after the solution of the problems $P_j, j = \overline{0,m}$, have been rejected.

If $m = 0$, the theorem has been proved.

Let's prove the theorem for any $m = n > 0$ assuming it is true for $0 \leq m < n$.

After substituting x and u by the representations from (27) with $j = n+1$ $J_\varepsilon(u)$ can be written in the form

$$J_\varepsilon(u) = \int_0^T (\{\tilde{F}\}_{2n+2} + \{\tilde{F}_x\}_n \Delta x(t) + \{\tilde{F}_u\}_n \Delta u(t) +$$

$$+\varepsilon^{2n+2}(\frac{1}{2}(\overline{F}_{x^2}x^2_{n+1}(t) + \overline{F}_{u^2}u^2_{n+1}(t)) +$$

$$+ \overline{F}_{xu}u_{n+1}(t)x_{n+1}(t) + [\tilde{F}_x]_{n+1}x_{n+1}(t) +$$

$$+[\tilde{F}_u]_{n+1}u_{n+1}(t))dt + \alpha(\varepsilon^{2n+3}). \tag{28}$$

In the proof of the theorem the tilde denotes that the values of the functions and their derivatives are calculated at $x = \tilde{x}_n(t), u = \tilde{u}_n(t)$.

By using the expressions for $\{\tilde{F}'_x\}_n, \{\tilde{F}_u\}_n$, $\{\tilde{\psi}'_n(t)\tilde{f}_x\}_n \Delta x(t) + \{\tilde{\psi}'_n(t)\tilde{f}_u\}_n \Delta u(t)$ and also the formula of integration by parts and the condition of T - periodicity for the functions $x_j(t)$, $j \geq 0$, $\tilde{\psi}_n(t)$ the relation

$$\int_0^T (\{\tilde{F}_x\}_n \Delta x(t) + \{\tilde{F}_u\}_n \Delta u(t))dt =$$

$$= \int_0^T (-\varepsilon^{2n+2}(([\tilde{\psi}'_n(t)\tilde{f}_x]_{n+1} + \frac{d\psi'_n(t)}{dt}B)x_{n+1}(t) +$$

$$+[\tilde{\psi}'_n(t)\tilde{f}_u]_{n+1}u_{n+1}(t) + \psi'_o(t)(\frac{1}{2}(\overline{f}_{x^2}x^2_{n+1}(t) +$$

$$+\overline{f}_{u^2}u^2_{n+1}(t)) + \overline{f}_{xu}u_{n+1}(t)x_{n+1}(t))) -$$

$$-\frac{d\tilde{\psi}'_n(t)}{dt}(A + \varepsilon B)\tilde{x}_n(t) +$$

$$+\tilde{\psi}'_n(t)\tilde{f}\}_{2n+2})dt + \alpha(\varepsilon^{2n+3})$$

is obtained.

Due to the executed transformations the formula (28) for $J_\varepsilon(u)$ can be written as

$$J_\varepsilon(u) = \int_0^T (\{\tilde{F} - \frac{d\tilde{\psi}'_n(t)}{dt}(A + \varepsilon B)\tilde{x}_n(t) -$$

$$- \tilde{\psi}_n'(t)\tilde{f}\}_{2n+2}+\varepsilon^{2n+2}(\frac{1}{2}(\overline{F}_{x^2}-\psi_o'(t)\overline{f}_{x^2})x_{n+1}^2(t)+$$

$$+ (\overline{F}_{xu} - \psi_o'(t)\overline{f}_{xu})u_{n+1}(t)x_{n+1}(t)+$$

$$+ \frac{1}{2}(\overline{F}_{u^2} - \psi_o'(t)\overline{f}_{u^2})u_{n+1}^2(t) + (-\frac{d\psi_n'(t)}{dt}B +$$

$$+ [\tilde{F}_x - \tilde{\psi}_n'(t)\tilde{f}_x]_{n+1})x_{n+1}(t)+$$

$$+ [\tilde{F}_u - \tilde{\psi}_n'(t)\tilde{f}_u]_{n+1}u_{n+1}(t))dt \ + \alpha(\varepsilon^{2n+3}).$$

From here it is seen, that if the problems P_j, $j = \overline{0,n}$, have been solved the coefficient at ε^{2n+1} in the expansion of $J_\varepsilon(u)$ with respect to powers of ε is known, and if in the coefficient at ε^{2n+2} from the above given expression, the terms, which are known after the solution of the problems $P_j, j = \overline{0,n}$, are rejected, we obtain the performance criterion $\tilde{J}_{n+1}(u_{n+1})$ in the problem P_{n+1}. The theorem is proved.

Let us give the sufficient condition of the solvability of the linear-quadratic periodic problem.

Proposition 4. If the equation

$$\frac{dy}{dt} = A_+(I-Q)(I-C(t)\tilde{C}^{-1}(t)Q)C(t)(I-P)y, \tag{29}$$

where $C(t) = G(t) - D(t)R^{-1}(t)S'(t)$, has no non-trivial T-periodic solutions then the periodic linear-quadratic optimal control problem

$$J(u) = \int_0^T (\frac{1}{2}x'(t)W(t)x(t) + x'(t)S(t)u(t) +$$

$$+ \frac{1}{2}u'(t)R(t)u(t) + d'(t)x(t) +$$

$$+ \ c'(t)u(t))dt \rightarrow \min_u, \tag{30}$$

$$A\frac{dx}{dt} = G(t)x(t) + D(t)u(t)+$$

$$+ g(t), \quad x(0) = x(T) \tag{31}$$

is solvable. Here $A \in L(X)$, the functions $W(t) \in L(X), S(t) \in L(U,X), R(t) \in L(U), d(t) \in X$, $c(t) \in U, G(t) \in L(X), D(t) \in L(U,X), g(t) \in X$ are continuous and T-periodic with respect to t, for all $t \in [0,T]$ the operator $QC(t)P : KerA \rightarrow KerA'$ has the inverse $\tilde{C}^{-1}(t)$ and the relations $W(t) = W'(t), R(t) = R'(t)$, $\begin{pmatrix} W(t) & S(t) \\ S'(t) & R(t) \end{pmatrix} > 0$ are fulfilled.

In order to prove this proposition the unique solvability of the problem

$$A\frac{dx}{dt} = C(t)x + D(t)R^{-1}(t)D'(t)\psi, \quad x(0) = x(T),$$

$$A'\frac{d\psi}{dt} = (W(t) - S(t)R^{-1}(t)S'(t))x - C'(t)\psi,$$

$$\psi(0) = \psi(T)$$

is established.

The problems $P_j (j \geq 1)$ have the form (30),(31). Therefore in order to provide the solvability of the problems P_j the following conditions 4-6 are assumed.

4.
$$\begin{pmatrix} \bar{H}_{x^2} & \bar{H}_{xu} \\ \bar{H}_{xu}' & \bar{H}_{u^2} \end{pmatrix} < 0,$$

where the bar denotes that the derivatives of the function $H = \psi' f(x,u,t,\varepsilon) - F(x,u,t,\varepsilon)$ are calculated at $x = x_0(t), u = u_0(t), \psi = \psi_0(t)$.

5. The operator $QC(t)P : KerA \rightarrow KerA'$, where $C(t) = \bar{f}_x - \bar{f}_u\bar{H}_{u^2}^{-1}\bar{H}_{xu}'$, has the inverse operator $\tilde{C}^{-1}(t)$.

6. The equation (29), where $C(t)$ is determined from the condition 5, has no non-trivial T-periodic solutions.

3. ESTIMATES OF APPROXIMATE SOLUTIONS

Let $\tilde{x}(t)$ denotes the solution of the problem (2), (3) when $u(t) = \tilde{u}_n(t) = \sum_{i=0}^{n} \varepsilon^i u_i(t)$ i.e.

$$(A+\varepsilon B)\frac{d\tilde{x}}{dt} = \hat{f}, \quad \tilde{x}(0) = \tilde{x}(T), \tag{32}$$

where the hat denotes that the value of the function is calculated at $x = \tilde{x}, u = \tilde{u}_n$. Let $\tilde{\psi}(t)$ denotes the solution of the problem (19) when $x = \tilde{x}, u = \tilde{u}_n$, i.e.

$$(A'+\varepsilon B')\frac{d\tilde{\psi}}{dt} = -\hat{f}_x'\tilde{\psi} + \hat{F}_x', \quad \tilde{\psi}(0) = \tilde{\psi}(T). \tag{33}$$

Let us write the solution of the problem P_ε in the form $x = \tilde{x} + \delta x, u = \tilde{u}_n + \Delta u$. Then from (2),(3),(32) it follows that δx is the solution of the problem

$$(A+\varepsilon B)\frac{d\delta x}{dt} = \hat{f}_x\delta x + \hat{f}_u\Delta u+$$

$$+G_1(\delta x, \Delta u, t, \varepsilon), \quad \delta x(0) = \delta x(T). \tag{34}$$

Using the relations (33),(34) and the formula of integration by parts the expression

$$\Delta J_\varepsilon(\Delta u) = J_\varepsilon(u) - J_\varepsilon(\tilde{u}_n) = \int_0^T \Big(-\frac{1}{2}\hat{H}_{x^2}(\delta x(t))^2-$$

$$-\hat{H}_{xu}\Delta u(t)\delta x(t) - \frac{1}{2}\hat{H}_{u^2}(\Delta u(t))^2-$$

$$-\hat{H}_u\Delta u(t) + G_2(\delta x, \Delta u, \tilde{\psi}, t, \varepsilon)\Big)dt$$

is obtained, where the hat denotes that the derivatives of the function H are calculated at $x = \tilde{x}, u = \tilde{u}_n, \psi = \tilde{\psi}$.

Let us consider the problem ΔP_ε which consists of the minimization of the functional $\Delta J_\varepsilon(\Delta u)$ on the trajectories δx of the problem (34). It is easy to see that the problem P_ε has the unique optimal control if and only if the problem ΔP_ε has the unique optimal control.

For the investigation of the problem ΔP_ε the following conditions 7-9 supposed.

7. All the B-Jordan chains of the operator A are of the same length p.

8.The equation

$$\frac{dy}{dt} = A_+(I-Q)(I-\bar{f}_x(Q\bar{f}_xP)^{-1}Q)\bar{f}_x(I-P)y$$

has no non-trivial T-periodic solutions.

It should be noted that the latter condition is equivalent to the lack of non-trivial T-periodic solutions of the equation $A\frac{dx}{dt} = \bar{f}_x x$.

9. There exists the T-periodic continuously differentiable invertible operator $S(t) : KerA \longrightarrow KerA$ such that

$$S^{-1}(t)A_p^{-1}Q\bar{f}_xPS(t) = diag(D^-(t), D^+(t)),$$

where $A_p = (-1)^{p-1}QB(A_+(I-Q)B)^{p-1}P$, the operators $D^-(t), D^+(t)$ act in the subspaces X_1, X_2, the direct sum of which is $KerA$, for all $t \in [0,T]$ the operators $D^-(t)$ and $-D^+(t)$ are stable.

The condition 3 follows from the latter condition.

Proposition 5. Under sufficiently small $\varepsilon > 0$ there exist the unique solutions of the problems (32), (33) in the neighborhoods of $\tilde{x}_n, \tilde{\psi}_n$ respectively and the estimates

$$\tilde{x} - \tilde{x}_n = O(\varepsilon^{n+1}), \quad \tilde{\psi} - \tilde{\psi}_n = O(\varepsilon^{n+1})$$

are valid.

When the statement of this proposition is proved some ideas from (Flatto and Levinson, 1955) are used.

Proposition 6. There exist $\Delta_0 > 0, \varepsilon_0 > 0$ such that for every continuous function $\Delta u = \Delta u(t)$ satisfying to the inequality $||\Delta u||_{C[0,T]} \leq \Delta_0$ and for $\varepsilon \in (0, \varepsilon_0]$ the solution of the problem (34) exists and the inequality

$$||\delta x||_{C[0,T]} \leq c||\Delta u||_{C[0,T]}$$

is fulfilled.

All previons results make it possible to establish the following statements.

Theorem 2. Under sufficiently small $\varepsilon > 0$ the solution x^*, u^* of the problem P_ε exists and the next estimates are valid

$$x^* - \tilde{x}_n = O(\varepsilon^{n+1}), \quad u^* - \tilde{u}_n = O(\varepsilon^{n+1}),$$

$$J_\varepsilon(u^*) - J_\varepsilon(\tilde{u}_n) = O(\varepsilon^{2(n+1)}).$$

Theorem 3. Under sufficiently small $\varepsilon > 0$ and $u_i \neq 0$ the following inequalities are realized

$$J_\varepsilon(\tilde{u}_i) < J_\varepsilon(\tilde{u}_{i-1}), i = \overline{1, n-1},$$

where $\tilde{u}_i = \sum\limits_{j=0}^{i} \varepsilon^j u_j$.

4. CONCLUSION

By means of the direct scheme the asymptotic expansion is constructed for the solution of the nonlinear periodic optimal control problem perturbed by the matrix. The statements of the theorems 2, 3 are analogues to the results obtained by Belokopytov and Dmitriev (1986, 1989) for the singularly perturbed non-periodic optimal control problems.

It should be noted that the behaviour of the solution of the periodic linear–quadratic optimal control problem with singularly perturbed state equation was investigated by Dmitriev and Muradova (1980) when the small parameter tends to zero by means of studying the behaviour of the periodic solution of the corresponding singularly perturbed matrix differential Riccati equation, the optimal control was considered in the feedback form.

5. ACKNOWLEDGEMENT

This research was supported in part by the grant from the Russian Fundamental Investigations Foundation No 96–01–01639.

REFERENCES

Belokopytov, S.V. and M.G.Dmitriev (1986). Direct scheme in optimal control problems with fast and slow motions. *Systems and Control Letters.* **8**, pp. 129-135.

Belokopytov, S.V. and M.G.Dmitriev (1989). Solution of classic optimal control problems with boundary layer. *Avtomatika i Telemehanika.* №7, pp. 71-82(in Russian).

Dmitriev, M.G. and N.D.Muradova (1980). Substantiation of mathematical model idealization for one problem of periodic optimization. *Izvestiya Akademii Nauk Turkmenskoj SSR. Serija Fiziko–tehnicheskih, Himicheskih i Geologicheskih Nauk.* №3, pp .6-12 (in Russian).

Flatto, L. and N. Levinson (1955). Periodic solutions of singularly perturbed systems. *J. Rat. Mech. and Analysis,* **4**, pp.943-950.

Kurina, G.A. and H.A. Ovezov (1996). Asymptotic analysis of linear - quadratic optimal control problems singularly perturbed by matrix. *Izvestija Vysshih Uchebnyh Zavedenij.Matematika.* №12, pp. 63-74.(in Russian).

Tonkov, E.L.(1977). Optimal control of periodic motions. In: *Matematicheskaja Fizika Respublikanskij Mezhvedomstvennyj Sbornik. Akademija Nauk Ukrainskoj SSR.* Vyp. 22, pp. 54-64. Institut Matematiki, Kiev, "Naukova Dumka" (in Russian).

REPRESENTATION OF GENERALIZED SOLUTIONS FOR DIFFERENTIAL EQUATIONS WITH AFFINE DEPENDENCE ON UNBOUNDED CONTROLS

Boris M. Miller

Institute for Information Transmission Problems
GSP-4, B. Karetny Per. 19, 101447 Moscow, Russia
e-mail: bmiller@ippi.ac.msk.su

Abstract: The problem of jumps representation is considered for systems with affine dependence on unbounded controls. In the case of Lie algebra structure, i.e., when the system satisfies the involutivity condition, the explicite representation of jumps is obtained. It was proved that the jumps representation has the same form as in systems which satisfy the commutativity condition.

Keywords: impulsive control, generalized solutions, Lie algebra, optimal control.

1. INTRODUCTION

The general approach to representation of generalized solutions of differential equations with unbounded controls, based on the idea of replacing time by a new independent variable, was develloped in (Miller, 1995). The aim of this paper is to provide a detailed exposition of generalized solutions for differential equations with lenear dependence oh unbounded controls. Here we will be concerned with the class of differential equations which admit an explicit representation of jumps, in the terms of functions in the right-hand-side of the original system, instead of some auxiliary system as in general case. Moreover, here we will demonstrate how the general approach may be used directly for a class of systems which satisfies involutiviyt condition.

The main feature of this kind of differential equations is that *regularity assumptions* (see Miller, 1995) will satisfy almost automatically without any additional assumptions, and the appropriate representation of the auxiliary system can be obtained in explicit form. The examples of such explicit representation for jumps are well-known for systems satisfying commutativity condition (see for example: Miller, 1978; Bressan and Rampazzo, 1991).

In affine case the generalized solution of differential equation with unbounded controls can be represented by equation with a measure

$$X(t) = X(0-) + \int_0^t f(X(\tau),\tau)d\tau +$$

$$\int_0^t B(X(\tau),\tau)dV^c(\tau) + \sum_{\tau\le t} \Delta X(\tau),$$

where

$$\Delta X(\tau) = \int_0^{\|\Delta V(\tau)\|} B(y(s),\tau)e(s)ds,$$

and $\{y(s)\}$ is the solution of system

$$\dot{y}(s) = B(y(s),\tau)e(s)$$

with initial condition

$$y(0) = X(\tau-),$$

and some controls $e(s)$ satisfying constraints

$$||e(s)|| \leq 1.$$

By Frobeneus theorem in commutative case (Brockett, 1973) the general solution of the system

$$\dot{y}(s) = \sum_{i=1}^{N} B_i(y(s), \tau)e_i(s)$$

with initial condition $y(0) = X(\tau-)$, admits the representation

$$y(s) = \Phi_N^\tau(\xi_N(s), \Phi_{N-1}^\tau(\xi_{N-1}(s), ..., \Phi_1^\tau(\xi_1(s), y(0))...)), \quad (1)$$

where

$$\xi_k(s) = \int_0^t e_k(w)dw,$$

and $\Phi_k^\tau(t, y)$ is the solution of differential equation

$$\dot{y}(t) = B_k(y(t), \tau), \qquad y(0) = y. \quad (2)$$

The representation (1) is valid in some vicinity of $t = 0$, however it can be continued for all $t > 0$ if solutions of equations (2) are defined for any initial conditions for all t. Applying this result to the representation of jumps for generalized solution of equation

$$\dot{X}(t) = F(X(t), t) + B(X(t), t)u(t)$$

we conclude that $\Delta X(\tau) = y_\tau(||\Delta V(\tau)||) - X(\tau-)$, where $y_\tau(s)$ is the solution of the differential equation

$$\dot{y}_\tau(s) = B(y_\tau(s), \tau)e^\tau(s),$$

Therefore, the jumps $\Delta X(\tau)$ can be represented as follows

$$\Delta X(\tau) = \Phi_N^\tau(\Delta V_N(\tau), \Phi_{N-1}^\tau(\Delta V_{N-1}(\tau), ..., \Phi_1^\tau(\Delta V_1(\tau), X(\tau-))...)) - X(\tau-), \quad (3)$$

where

$$\Delta V_k(\tau) = \int_0^{||\Delta V(\tau)||} e_k(s)ds$$

The aim of this paper is to obtain the analogu-ous representation in involutive case.

2. REPRESENTATION OF JUMPS IN INVOLUTIVE CASE

2.1. Some definitions and auxiliary results

Let $f(x)$ and $g(x)$ are two smooth vector-fields in R^n.

Definition 1 The *product* or *Lie bracket* $[f, g]$ of two smooth vector-fields in R^n will be defined as

$$[f, g](x) = f_x'(x)g(x) - g_x'(x)f(x),$$

where $f_x'(x), g_x'(x)$ are $n \times n$ matrices of partial derivatives of vector-field components.

If $[f, g] = 0$ the pair of vector-fields f, g will be called *commutative*.

Definition 2

The set of vector-fields $\{f_i(x), i = 1, ..., N\}$ will be called *involutive* if the Lie brackets of any f_i, f_j are the linear combinations of $\{f_i(x), i = 1, ..., N\}$, i.e. there exists the set of constants $\{c_{i,j}^k, i, j, k = 1, ..., N\}$ such that

$$[f_i, f_j] = \sum_{k=1}^{N} c_{i,j}^k f_k. \quad (4)$$

The set $\{c_{i,j}^k, i, j, k = 1, ..., N\}$ will be called the *set of structural constants* of Lie algebra $\mathcal{L}\{f_1, f_2, ..., f_N\}$ generated by vector-fields $\{f_i(x), i = 1, ..., N\}$.

Relation (3) provides one of the possible analytical representations of jump function, which due to the semigroup property depends only on $\tau, X(\tau-)$, and the value of measure $\Delta V(\tau)$, localized at point τ.

To obtain the representation (3) in general case, when functions $\{f_i\}$ satisfy involutivity properties (see Def. 2), we need the following

Lemma 1 *Let $\Phi_i(\xi, y) \in R^N$, where $\xi \in R^1, y \in R^N$ is the solution of differential equation*

$$\frac{\partial}{\partial \xi}\Phi_i(\xi, y) = f_i(\Phi_i(\xi, y)) \quad (5)$$

with initial condition

$$\Phi_i(0, y) = y.$$

Suppose that functions $\{f_i, i = 1, \dots, N\}$ satisfy involutivity condition (4), then for any $i, j = 1, \dots, N$ we have the relation

$$\frac{\partial \Phi_i(\xi, y)}{\partial y} f_j(y) = \sum_{k=1}^{N} \gamma_{ij}^k(\xi) f_k(\Phi_i(\xi, y)), \quad (6)$$

where $\gamma_{ij}^k(\xi)$ are the analytical functions depending on the structural constants of the Lie algebra $\mathcal{L}\{f_1, \dots, f_N\}$ only.

These $\gamma_{ij}^k(\xi)$ are the solution of the system of linear differential equations

$$\frac{d\gamma_{ij}^k(\xi)}{d\xi} = \sum_{l=1}^{N} c_{il}^k \gamma_{ij}^l(\xi), \quad (7)$$

with initial conditions

$$\gamma_{ij}^{j} = 1, \quad \textit{and} \quad \gamma_{ij}^{k} = 0, \quad \textit{if} \quad k \neq j. \qquad (8)$$

Lemma 2 *Let the conditions of Lemma 2 hold. Let also given a set of differentiable functions* $\{\xi_i(t) \in R^1, i = 1, \dots, N\}$ *and*

$$Z(t) = \Phi_N(\xi_N(t), \Phi_{N-1}(\xi_{N-1}(t), \dots,$$
$$\Phi_1(\xi_1(t), y), \dots,)).$$

Then

$$\dot{Z}(t) = \sum_{i=1}^{N} f_i(Z(t)) \sum_{j=1}^{N} L^{ij}(\xi_N(t), \dots, \xi_1(t))\dot{\xi}_j(t), \qquad (9)$$

where functions $L^{ij}(\xi_N, \dots, \xi_1)$ *are analitical ones and depend on the structural constants of Lie algebra* $\mathcal{L}\{f_1, \dots, f_N\}$ *only.*

Here $L^{ij} = L_N^{ij}$, *where* $\{L_k^{ij}, i = 1, \dots, N; j = 1, \dots, k\}$ *can be found from the recurrent equations*

$$L_k^{ij}(\xi_k, \dots, \xi_1) = \sum_{\nu=1}^{N} \gamma_{k\nu}^{i}(\xi_k) L_{k-1}^{\nu j}(\xi_{k-1}, \dots, \xi_1), \quad \textit{if} \quad 1 \leq j \leq k-1, \qquad (10)$$

$$L_k^{ik}(\xi_k, \dots, \xi_1) = \begin{cases} 1, & \textit{if} \quad i = k, \\ 0, & \textit{if} \quad i \neq k. \end{cases}$$

with

$$L_1^{i1} = \begin{cases} 1, & \textit{if} \quad i = 1, \\ 0, & \textit{if} \quad i \neq 1 \end{cases}$$

and with $\gamma_k^{ij}(\xi)$ *being the solution of equation (7), (8).*

Using this two lemmas we are in the state to prove representation (3) in the case when functions $\{f_i, i = 1, \dots N\}$ are in involution.

2.2. Jump representation in involutive case

Theorem 1 *Let* $L(\vec{\xi})$ *be* $N \times N$ *matrix-valued function with the elements* $L^{ij}(\vec{\xi}) = L^{ij}(\xi_N, \dots, \xi_1)$, *where functions in right-hand-side are the same as in representation (9). Denote by*

$$\vec{\xi}(t) = \{\xi_N(t), \dots, \xi_1(t)\},$$

$$\vec{u}(t) = \{u_N(t), \dots, u_1(t)\}$$

and consider a system of differential equations

$$L(\vec{\xi}(t))\dot{\vec{\xi}}(t) = \vec{u}(t), \qquad (11)$$

with initial condition $\vec{\xi}(0) = 0$.

1. *Suppose system (11) has a some bounded solution* $\vec{\xi}(t)$ *on some interval* $[0, T]$, *and suppose also that the equation*

$$\dot{x}(t) = \sum_{i=1}^{N} f_i(x(t))u_i(t), \quad x(0) = x_0 \in R^N \qquad (12)$$

has on $[0, T]$ *the unique solution, then this solutions admits the representation*

$$x(t) = \Phi_N(\xi_N(t), \Phi_{N-1}(\xi_{N-1}(t), \dots,$$
$$\Phi_1(\xi_1(t), x_0), \dots,)); \qquad (13)$$

2. *Let controls* $\{u_N(t), \dots, u_1(t)\}$ *are integrable on* $[0, T]$, *then there exists some interval* $[0, t^*]$, $t^* > 0$, *where system (11) has the unique solution, and therefore representation (13) for the solution of (12) also takes place on this interval.*

As a corollary of Theorem 1 we can state the representation of jumps for involutive systems.

Proof. Statement (1) is the obvious consequence of Lemma 2 and can be obtained by a substitution of relation (11) into (9) which gives

$$\dot{Z}(t) = \sum_{i=1}^{N} f_i(Z(t)) \sum_{j=1}^{N} L^{ij}(\xi_N(t), \dots, \xi_1(t))\dot{\xi}_j(t) = \sum_{i=1}^{N} f_i(Z(t))u_i(t).$$

Hence, $Z(t)$ satisfies equation (12) and by virtue of the uniqieness of its solution coincides with $x(t)$ defined by formula (13).

To prove the statement (2) we will show that matrix-valued function $L(\vec{\xi})$ is non-singular in some vicinity of point $\vec{\xi} = 0$. First we can show that $L(0) = I_{N \times N}$, where $I_{N \times N}$ is the identity $N \times N$ matrix.

Indeed, every matrix $L_k(0)$ has a following blok-form

$$L_k(0) = \begin{pmatrix} I_{k \times k} \\ \dots\dots\dots \\ 0_{(N-k) \times k} \end{pmatrix}, \qquad (14)$$

where $I_{k \times k}$ is identity $k \times k$ matrix, and $0_{(N-k) \times k}$ is null matrix of appropriate dimension.

The representation for $L_k(0)$ follows from recurrent relations for L_k, because by (10) we have

$$L_{k+1}^{ij}(0) = \sum_{\nu=1}^{N} \gamma_{k+1\nu}^{i}(0) L_k^{\nu j}(0) \quad \text{if} \quad j \leq k$$

and

$$L_{k+1}^{ik+1} = \begin{cases} 1 & \text{if} \quad i = k+1, \\ 0 & \text{if} \quad i \neq k+1. \end{cases}$$

Assume for a moment that $L_k(0)$ satisfies (14) and notice that by (8)

$$\gamma^i_{k+1\nu}(0) = \begin{cases} 1 & \text{if} \quad i = \nu, \\ 0 & \text{if} \quad i \neq \nu \end{cases} \qquad (15)$$

for $i, \nu = 1, \ldots, N$. Substitution of (15) into above relation for $L^{ij}_{k+1}(0)$, $j \leq k$ and taking into account the relation for $L^{ik+1}_{k+1}(0)$ gives that under this assumption matrix $L^{ij}_{k+1}(0)$ also has form (14). Since L^{ij}_1 has form (14) it follows that all $L_k(0)$ have form (14) and $L(0) = L_N(0) = I_{N\times N}$. Therefore, matrix-valued function $L(\vec{\xi})$ with analytical coefficients has an inverse one in some vicinity of point $\vec{\xi} = 0$. Hence, the equation (11) can be resolved with respect to $\dot{\vec{\xi}}(t)$ and can be transformed into equavalent equation

$$\dot{\vec{\xi}}(t) = L^{-1}(\vec{\xi}(t))\vec{u}(t), \qquad (16)$$

with initial condition $\vec{\xi}(0) = 0$. By virtue of analytical properties of $L(\vec{\xi})$ the inverse matrix $L^{-1}(\vec{\xi})$ is also analytical in some vicinity of point $\vec{\xi} = 0$ and equation (16) has the unique solution on some interval $[0, t^*]$, $t^* > 0$. This solution satisfies also equation (11), moreover equation (12) has the unique solution due to the integrability of $u_i(t)$. Therefore, according to the first statement of this theorem solution of (12) concides on $[0, t^*]$ with function given by representation (13).

Corollary Let the columns of matrix-valued function $B(x,t)$ are in involution for any $t \in [0,T]$, i.e., satisfy the relations

$$[B_i(x,t), B_j(x,t)] = \sum_{k=1}^{m} c^k_{ij}(t) B_k(x,t)$$

and equation

$$L(\vec{\xi}(s), t)\frac{d\vec{\xi}(s)}{ds} = \vec{e}(s) \qquad (17)$$

with initial condition $\vec{\xi}(0) = 0$ has the unique solution for any functions $\{\vec{e}(s) : \|\vec{e}(s)\| \leq 1\}$. Here $L(\vec{\xi}, t)$ is the matrix from representation (9) corresponding to the set of constants $\{c^k_{ij}(t)\}$ at point $t \in [0,T]$.

Then the jump of generalized solution at point τ admits the representation

$$\Delta X(\tau) =$$

$$\Phi_N(\xi_N(\|\Delta V(\tau)\|, \tau, \Phi_{N-1}(\xi_{N-1}(\|\Delta V(\tau\|), \tau, \ldots,$$

$$\Phi_1(\xi_1(\|\Delta V(\tau)\|), \tau, X(\tau-))\ldots)) - X(\tau-), \qquad (18)$$

where $\|\Delta V(\tau)\|$ is a total value of vector measure $V(dt)$ localized at point τ, $\Phi_k(s, \tau, y)$ is the solution of differential equation

$$\dot{y}(s) = B_k(y(s), \tau), \qquad y(0) = y.$$

with $B_k(x,\tau)$ equals to the k-th column of matrix-valued function $B(x,t)$, and $\vec{\xi}(s)$ as a solution of equation (17) on the interval $[0, |\mathcal{V}|\{\tau\}]$ with some functions $\{\vec{e}(s) : \|\vec{e}(s)\| \leq 1\}$.

This result gives the explicite representation of possible jumps for affine involutive systems. It also gives the opportunity to direct synthesis of jumps by using equation (17). Indeed, any jump of generalized solution at point τ can be expressed in terms of the increment of vector-valued variable $\xi(s)$ on the interval $[0, \|\Delta V(\tau)\|]$. Therefore, if we put $\dot{\xi}(s) = \dfrac{\xi(\|\Delta V(\tau)\|)}{\|\Delta V(\tau)\|}$, then by using the appropriate choice of value $\|\Delta V(\tau)\| > 0$ we can obtain vector of controls $\vec{e}(s) \in \{\|\vec{e}\| \leq 1\}$, which gives us the generalized solution with the jump value equal to (18).

3. EXAMPLE

Consider a system with 3×2 - dimensional matrix $B(x)$, where $x = (x_1, x_2, x_3) \in R^3$ and

$$B(x) = (B_1(x), B_2(x)) = \begin{pmatrix} f_1(x_1) & 0 \\ 0 & f_2(x_2) \\ 1 & x_3 \end{pmatrix}.$$

It is easy to check that

$$[B_1(x), B_2(x)] = \begin{pmatrix} 0 \\ 0 \\ -1 \end{pmatrix} = B_3(x),$$

and

$$[B_1, B_2] = B_3, \quad [B_1, B_3] = 0, \quad [B_2, B_3] = B_3.$$

The set of structural constants

$$\{c^k_{ij}, \; i, j, k = 1, \ldots, 3\}$$

will be

$$c^1_{12} = 0 \quad c^2_{12} = 0 \quad c^3_{12} = 1$$

$$c^1_{13} = 0 \quad c^2_{13} = 0 \quad c^3_{13} = 0$$

$$c^1_{23} = 0 \quad c^2_{23} = 0 \quad c^3_{23} = 1$$

and the jump representation is as follows

$$\Delta X(\tau) =$$

$$\Phi_3(\xi_3(\|\Delta V(\tau)\|), \Phi_2(\xi_2(\|\Delta V(\tau)\|),$$

$$\Phi_1(\xi_1(\|\Delta V(\tau)\|), \tau, X(\tau-)))) - X(\tau-).$$

where $||\Delta V(\tau)||$ is a total value of vector measure $V(dt)$ localized at point τ, $\Phi_k(s,y)$ are the solutions of differential equations

$$\dot{y}(s) = B_k(y(s)), \qquad y(0) = y, \quad k = 1, ..., 3.$$

Functions $\{\xi_1(s), \xi_2(s), \xi_3(s)\}$ satisfy the equations

$$\dot{\xi}_1(s) = e_1(s),$$

$$\dot{\xi}_2(s) = e_2(s),$$

$$\dot{\xi}_3(s) = (exp\{\xi_2(s)\} - 1)e_1(s) + \xi_3(s)e_2(s),$$

where functions $\vec{e}(s) = \{e_1(s), e_2(s)\}$ satisfy on the interval $[0, ||\Delta V(\tau)||]$ the following constraints

$$||\vec{e}(s)|| \le 1,$$

$$\int_0^{||\Delta V(\tau)||} e_1(s) = \Delta V_1(\tau),$$

$$\int_0^{||\Delta V(\tau)||} e_2(s) = \Delta V_2(\tau).$$

This system has the solution

$$\xi_1(s) = \int_0^s e_1(v)dv$$

$$\xi_2(s) = \int_0^s e_2(v)dv$$

$$\xi_3(s) =$$

$$exp\{\xi_2(s)\}[\xi_1(s) - \int_0^s exp\{-\xi_2(v)\}e_1(v)dv]$$

the jump of generalized solution admits the representation

$$\Delta X_1(\tau) = \Phi_{11}(\Delta V_1(\tau), x_1(\tau-)) - x_1(\tau-),$$

$$\Delta X_2(\tau) = \Phi_{22}(\Delta V_2(\tau), x_2(\tau-)) - x_2(\tau-),$$

$$\Delta X_3(\tau) = (exp\{\Delta V_2(\tau) - 1)x_3(\tau-)+$$

$$exp\{\Delta V_2(\tau)\} \int_0^{||\Delta V(\tau)||} exp\{-\xi_2(s)\}e_1(s)ds.$$

It follows that jumps of first and second components do not depend on the way of approximation of impulsive input, i.e. choice of functions e_1, e_2, meanwhile the jump of third component depends on the way of approximation and can take various values. For example, if

$$0 \le e_1, e_2 \le 1,$$

which corresponds to the case of positive impulses, one can obtain various values of jump, i.e.

$$\Delta X_3(\tau) \in$$

$$[(exp\{\Delta V_2(\tau)\} - 1)x_3(\tau-)+ \Delta V_1(\tau),$$

$$(exp\{\Delta V_2(\tau)\} - 1)x_3(\tau-) + exp\{\Delta V_2(\tau)\}\Delta V_1(\tau)].$$

In the case of bipolar impulses, i.e., when $-1 \le e_1, e_2 \le 1$ one can obtain nonzero jump of third component even if $\Delta V_1(\tau) = \Delta V_2(\tau) = 0$. Suppose that

$$\int_0^{||\Delta V(\tau)||} e_1(s)ds = \int_0^{||\Delta V(\tau)||} e_2(s)ds = 0$$

$$\int_0^{||\Delta V(\tau)||} |e_1(s)|ds = |\Delta V_1(\tau)|,$$

$$\int_0^{||\Delta V(\tau)||} |e_2(s)|ds = |\Delta V_2(\tau)|$$

$$|\Delta V_1(\tau)| + |\Delta V_2(\tau)| = ||\Delta V(\tau)||,$$

then

$$\Delta X_1(\tau) = 0, \quad \Delta X_2(\tau) = 0,$$

however

$$\Delta X_3(\tau) = \int_0^{||\Delta V(\tau)||} exp\{-\xi_2(s)\}e_1(s)ds.$$

The estimation of ΔX_3 gives

$$|\Delta X_3(\tau)| \le \left[exp\left\{\frac{|\Delta V_2(\tau)|}{2}\right\} - 1\right]\frac{|\Delta V_1(\tau)|}{2}.$$

It is easy to check that lower and upper bounds can be achived with some $\{e_1(s), e_2(s)\}$. For example, to achive the upper bound the following $\{e_1(s), e_2(s)\}$ can be used

$$e_1(s) = \begin{cases} 1, \text{ if} \\ s \in \left(\frac{|\Delta V_2(\tau)|}{2}, \frac{|\Delta V_2(\tau)|}{2} + \frac{|\Delta V_1(\tau)|}{2}\right) \\ -1, \text{ if} \\ s \in \left(\frac{|\Delta V_1(\tau)|}{2} + |\Delta V_1(\tau)| + |\Delta V_2(\tau)|\right) \\ 0 \text{ otherwise,} \end{cases}$$

$$e_2(s) = \begin{cases} -1, \text{ if } s \in \left[0, \frac{|\Delta V_2(\tau)|}{2}\right], \\ 1, \\ \text{if } s \in \left[\frac{|\Delta V_2(\tau)|}{2} + \frac{|\Delta V_2(\tau)|}{2}, \right. \\ \left. \frac{|\Delta V_1(\tau)|}{2} + |\Delta V_2(\tau)|\right], \\ 0 \text{ otherwise.} \end{cases}$$

Acknowledgements This work was supported in part by NSF of USA Grant No CMS 94-1447s, INTAS Grants 94-697 and 93-2622, Australian Research Council Grant A 8913 2609, and Russian Basic Research Foundation Grant No 95-01-00573.

REFERENCES

Bressan A. and Rampazzo F. (1991). Impulsive control systems with commutative vector fields, *J. Optim. Theory Appl.*, **71**, No. 1, 67–83.

Brockett R. W. (1973). Lie algebras and Lie groups in control theory. In: *Geometric methods in system theory* (Mayne D. Q., Brockett R. W., (Eds.)). 43–82. Proceedings of the NATO Advanced study institute held at London, Aug. 27 - Sept. 7, 1973, Dordrecht - Boston, D. Reidel Publishinh Company.

Miller B. M. (1978). The nonlinear sampled-data control problem for systems described by ordinary measure differential equations I, II *Automat. Remote Control*, **39**, No. 1, 57–67, No. 3, 338–344.

Miller B. M. (1995). Generalized solutions of nonlinear optimization problems with impulse control I, II *Automat. Remote Control*, **55**, No. 4, 62–76, No. 5, 56–70.

AVERAGED EXTENSIONS OF EXTREMAL PROBLEMS

Anatoly M. Tsirlin * **Vladimir A. Kazakov** **

* *Programm Systems Instiute,Russian Academy of Science*
** *University of Sydney*

Abstract: The whole number of real statements of extremal problems contains both vector of desired variables $X \in V_x$ and either averaged values of these variables or averaged values of functions depending on these variables. In this case solution includes a distribution $P(x)$ and nonlinear programming problems turn into variational ones with one — sidedly bounded controls. It is shown that for wide type of the problems the optimal distribution $P^*(x)$ is lumped in finite number of point at V_x. The algorythm for calculation of these "base" points and weight coefficients corresponding to these points is represented. In calculus of variations averaged extension allows us to obtain a proof of maximum principle for a wide type of problems with scalar argument.

Keywords: averaging, extremal problems, extension, optimality conditions, canonical form

1. INTRODUCTION

There are many forms of extremal problems that contain not only vector and functional variables but also their averaged values or averaged values of a function which depends on these variables (Fromovitz, 1965). As a rule, incorporation of the averaging operation into formulation of the problem extends the set of feasible solutions D to the set $\bar{D}$.

Extremal problem B is called an extension of the original problem A (or an extended problem A) (Tsirlin, 1974) is the following two conditions hold:
1. The sets of the feasible solutions of the problems A and B relay to each other as

$$D_B \supset D_A. \tag{1}$$

2. The optimality criterion I_B of the problem B coincides with optimality criterion of the problem A on the set D_A

$$I_B(x) = I_A(x), \quad x \in D_A. \tag{2}$$

It turns out that in a problem with averaging the unknown variables are distributions of the variables of the original problem. For example, if the unknown variable in the original problem A is vector $x \in V_x$ then the unknown variable in the averaged problem B is such a distribution $P(x)$ that

$$\int_{V_x} P(x)dx = 1; \quad P(x) \geq 0. \tag{3}$$

Thus, the nature of the sets D_B is different from the nature of the set D_A and (1) is meaningless.

Let us change the definition of an extension of extremal problem. First we introduce the definition of the isomorphism of two extremal problem.
1.1 Problems A_0 and A_1 are isomorphic (identical with respect to solutions) if it is possible to find such one-to-one mapping between their sets of feasible solutions that from the inequality

$$I_A(y) \geq I_A(z), \quad (y, z) \in D_A \tag{4}$$

follows that

[1] This work is supported by INTAS (grant No. 93 12-81) and RFBR (grant No. 97-01-00451).

$$I_{A_1}(y_1) \geq I_{A_1}(z_1), \quad (y_1, z_1) \in D_{A_1}, \tag{5}$$

where y_1 and z_1 correspond to y and z. The equivalent class $\bar{A}$ is defined as a set of all the problems which are isomorphic to the problem A_1.

Assume that the element y_0 can not be improved on some subset $\Delta A \subset D_A$ (y_0 obeys necessary conditions of optimality of the problem A). The element y_1^0 which corresponds to y_0 obeys the necessary conditions of optimality of the problem A_1. Therefore optimality conditions for all the problems that belong to $\bar{A}$ are obtained if they are formulated for any problem from this class.

It turns out that for the nonlinear programming problem A is isomorphic to the averaged problem A_1 where distribution $P(x)$ of the form

$$P(x) = \delta(x - x_0). \tag{6}$$

corresponds to 1 each vector $x_0 \in V_x$.
1.2 We will call any problem B an extension of the problem A if the conditions (1) and (2) hold for some problem from the class $\bar{A}$.

Since an averaged problem in which unknown variables are distributions $P(x)$ is an extension of the problem with condition of the type (6), it is also an extension of the nonlinear programming problem.

2. AVERAGED EXTENSIONS OF NONLINEAR PROGRAMMING PROBLEM. STRUCTURE OF THE OPTIMAL SOLUTION.

Let us consider the following nonlinear programming problem (NP)

$$f_0(x) \to \max \Big/ f(x) = 0, \quad x \in V_x, \tag{7}$$

where $x \in R^n$, f_0 is a scalar function and f is an m-dimensional vector function, $m < n$. This is our original problem A.

The averaging is frequently introduced by replacing functions f_0 and f with their averaged values $\overline{f_\nu}$. Here function $\overline{f_\nu(x)}$ is defined as

$$\overline{f_j(x)} = \frac{1}{T}\int_0^T f_j(x(t))dt =$$

$$= \int_{V_x} f_j(x)P(x)dx, \; j = \overline{0,m}. \tag{8}$$

In the latter case the distribution $P(x)$ obeys the condition (3). The will call the problem

$$\overline{f_0(x)} \to \max_{P(x)} \Big/ \overline{f(x)} = 0 \tag{9}$$

as $\overline{NP}$ problem.

The following statement is true 2.1: *The optimal solution of $\overline{NP}$ problem $P^*(x)$ has the following form*

$$P^*(x) = \sum_{\nu=0}^{m} \gamma_\nu \delta(x - x_\nu), \tag{10}$$

where

$$\gamma_\nu \geq 0, \quad \sum_{\nu=0}^{m} \gamma_\nu = 1. \tag{11}$$

Non-zero vector of Lagrange multipliers $\lambda = (\lambda_0, \ldots \lambda_m)$ can be found such that at points x_ν function

$$R = \sum_{\nu=0}^{m} \lambda_j f_j(x) \tag{12}$$

has its global maximum with respect to $x \in V_x$. x^ν are called the basic values of x. If the optimal solution of $\overline{NP}$ problem as a function of time $x(t)$ exists then it switches from one basic value to another, being equal to each of them during γ_ν fraction of the total duration of the process T.

The averaging in NP problem can be done not for all variables but for part of them only. Let us divide the variables of the problem (7) into two groups - deterministic x and randomized u. The averaging is done only with respect to u. The $\overline{NP}^u$ problem has the following form

$$\overline{f_0(x,u)}^u \to \max_{x,p(u)} \Big/ \overline{f(x,u)}^u = 0. \tag{13}$$

Here

$$\overline{f_j(x,u)}^u = \frac{1}{T}\int_0^T f_j(x, u(t))dt = \int_{V_u} f_j(x,u)P(u)du. \tag{14}$$

$$P(u) \geq 0; \qquad \int_{V_u} P(u)du = 1.$$

It is assumed that functions f_j are continues with respect to u and continuously differentiable on x. (Statement 2.2) Optimality conditions of the problem (13) have the form:

1. The optimal distribution of the randomized variable has the following form

$$P^*(u) = \sum_{\nu=0}^{m} \gamma_\nu \delta(u - u^\nu), \tag{15}$$

$$\gamma_\nu \geq 0, \qquad \sum_{\nu=0}^{m} \gamma_\nu = 1.$$

2. Non-zero vector $\lambda = (\lambda_0, \ldots \lambda_m)$ can be found such that the Lagrange function $R = \sum_{j=0}^{m} \lambda_j f_j(x,u)$, which is computed using this λ, can not be improved locally with respect to the deterministic variables and has global maximum on randomized variables on the set V_u at each of the basic points u^ν:

$$u^\nu = \arg\max_{u \in V_u} R(\lambda, x^*, u), \quad \nu = 0, m. \tag{16}$$

$$\frac{\delta \overline{R}}{\delta x}\delta x = \frac{\delta}{\delta x}\Big\{\sum_{\nu=0}^{m} \gamma_\nu R(x,u^\nu)\Big\}\delta x \le 0. \qquad (17)$$

here δx is a variation allowed by the constraints $x \in V_x$

Since it is possible that not all the constraints in NP problem depend on both deterministic and randomized variables, and it can include not only averaging of the functions but also functions of the averaged values of the time-dependent variables etc., many different averaged extensions of the NP problem can be found. It does not make sense to derive optimality conditions for each one of these versions. It is much more reasonable to write down the canonical form of the average extension of NP problem and to derive its necessary conditions of optimality. The statements 2.1 and 2.2 will follow from these conditions.

The canonical form of the averaged extension of the NP problem has the form

$$F_0\Big[\overline{f(x,u)}, x\Big] \to \max \qquad (18)$$

subject to constraints

$$F_j\Big[\overline{f(x,u)}, \phi(x,u)\Big] = 0, \quad j=\overline{1,r}; \quad x \in V_x, \qquad (19)$$

the overline f corresponds to the averaging on u over the closed and bounded set V_u. The dimensionality of the vector-function f is m, function F is continuously differentiable on all of its arguments, and f is continues on u and continuously differentiable on x.

Theorem 2.3 (The optimality conditions of the averaged extension of the NP problem):
1. The optimal distributions of the randomized variables have the following form

$$P^*(u) = \sum_{\nu=0}^{m} \gamma_\nu \delta(u-u^\nu) \qquad (20)$$

where γ_ν obey the conditions (11).
2. Non-zero vector $\lambda = (\lambda_0, \lambda_{j\nu})(j=\overline{1,r}; \quad \nu=\overline{0,m})$ can be found such that for each basic value u^ν of vector u function

$$L = \lambda_0 \frac{\delta F_0}{\delta \overline{f}} f(x,u) + \sum_{j=1}^{r}\sum_{\nu=0}^{m} \lambda_{j\nu}\frac{\delta F_{j\nu}}{\delta \overline{f}} f(x,u),$$

attains its maximum on V_u. Here

$$F_{j\nu} = F_j(\overline{f(x,u)}, \phi(x,u^\nu)); \quad \bar{f} = \sum_{\nu=0}^{m}\gamma_\nu f(x,u^\nu).$$

Hence

$$u^\nu = \arg\max_{U\in V_u} L(x^*, \lambda, u). \qquad (21)$$

3. *Function*

$$R = \lambda_0 F_0 + \sum_{j=1}^{r}\sum_{\nu=0}^{m}\lambda_{j\nu}F_{j\nu} \qquad (22)$$

can not be improved locally with respect to its deterministic arguments

$$\frac{\delta R}{\delta x}\delta x \le 0. \qquad (23)$$

It is easy to show that after reduction of the problems $\overline{NP}$ and $\overline{NP}^u$ to the form (18), (19) the optimality conditions for them follow from the theorem 2.3.

3. AVERAGING IN VARIATIONAL PROBLEMS

Introduction of averaging in variational problems where unknown variables depend on the scalar argument t allows to obtain a solution in a form of maximizing sequences and to formulate optimality conditions in a form of maximum principle for any arbitrary form of optimality criterion and constraints. We shall start by giving some auxiliary statements and definitions.

Definition 3.1. *Problem A: $I_A(y) \to \max \big/ y \in D_A$ is called correct with respect to its value if infinitesimal change of any of the constraints D_A leads to infinitesimal change of the value of the problem I_A^*.*

Naturally, this definition requires a definition of how to measure the closeness to each other of two conditions which define D_A. If this is done then it can be formulated in terms of $E \sim \delta$.

It is sufficient for correctness of nonlinear programming problem in terms of the definition 3.1. that the Slater's complementary slackness conditions are satisfied.

Definition 3.2. *Extension B: $I_B(y) \to \max \big/ y \in D_B$ of the problem A is equivalent if*

$$I_{\bar{A}}^* = \sup_{y\in D_{\bar{A}}} I_{\bar{A}}(y) = I_B^* = \sup_{y\in D_B} I_B(y). \qquad (24)$$

Lemma 3.3. *The sufficient condition for the extension to be equivalent is the possibility for any solution of the extended problem $y^0 \in D_B$ to find such a sequence of the solutions of the original problem $\{y_i\} \subset D_A$ that*

$$\lim_{i\to\infty} I_{\bar{A}}(y_i) = I_B(y^0). \qquad (25)$$

If the problem is correct with the respect to its value then it is possible that the sequence $\{y_i\}$ does not belong to $D_{\bar{A}}$. The only requirement is that any constraint of the original problem is satisfied with arbitrary accuracy in the limit $i \to \infty$ (in accordance with the definition 3.1).

Lemma 3.4. *If y_A^* is the optimal solution of the problem A, extension B is equivalent to A and $D_B \supset D_A$ then y_A^* obeys necessary conditions of optimality of the extension problem.*

We will call the following problem the canonical form of variational problem

$$I = \int_0^T \Big[f_{01}(t, x(t), u(t), a) + \tag{26}$$

$$+ \sum_l f_{02}(t, x(t), a)\delta(t - t_l)\Big] dt \to \max$$

subject to constraints

$$J_j(\tau) = \int_0^T \Big[f_{j1}(t, x(t), u(t), a, \tau) +$$

$$+ f_{j2}(t, x(t), a, \tau)\delta(t - \tau)\Big] dt = 0; \tag{27}$$

$$\forall \tau \in [0, T], \quad j = \overline{1, m}, \quad u \in V_u, \quad a \in V_a.$$

Here a is vector of parameters, which are constant on $[0, T]$, functions f_{j1} and f_{j2} are continuously differentiable on x, a and t and continuous on u.

Lemma 3.5. *Assume that the problem (26), (27) is correct with respect to its value (according to the definition 3.1, where a closeness of each initial and variated condition (27) should be understood in uniform metrics) then the averaged extension of this problem is*

$$\bar{I} = \int_0^T \Big[\overline{f_{01}(t, x, u, a)}^u +$$

$$+ \sum_l f_{02}(t, x, a)\delta(t - t_l)\Big] dt \to \max \tag{28}$$

subject to constrains

$$\bar{J}_j(\tau) = \int_0^T \Big[\overline{f_{j1}(t, x, u, a, \tau)}^u +$$

$$+ f_{j2}(t, x, a, \tau)\delta(t - \tau)\Big] dt = 0, \tag{29}$$

$$\forall \tau \in [0, T], j = \overline{1, m}, u \in V_\lambda, a \in V_a$$

is equivalent to (26), (27).

Here

$$\overline{f_{j1}}^u = \int_{V_u} f_{j1}(t, x, A, u, a, \tau) P(u, t) du. \tag{30}$$

Distribution $P(u, t)$ obeys the conditions

$$p(u, t) \geq 0; \quad \int_{V_u} P(u, t) du = 1 \quad \forall t \in [0, T]. \tag{31}$$

The proof of this statement is based on Lemma 3.3.

The solution of the problem (28), (29) consists of distribution $P^*(u, t)$, function $x(t)$ and vector a. It obeys the following conditions (**Theorem 3.5.**):
1. Optimal distribution has the form

$$P^*(u, t) = \sum_{\nu=0}^m \gamma_\nu(t)\delta(u - u^\nu(t)), \tag{32}$$

where piece-wise continues functions $\gamma_\nu(t) \geq 0$ $\forall t \in [0, T]$ and $\sum_{\nu=0}^m \gamma_\nu(t) = 1$.
2. Scalar $\lambda_0 \geq 0$ *vector function* $\lambda(\tau) = (\lambda_1(\tau), \ldots \lambda_m(\tau))$, *piece-wise continuous for almost everywhere on* $[0, T]$ *that is defined and non-zero simultaneously with* λ_0 *on the interval* $[0, T]$ *and equal zero outside of this interval can be found such that the functional*

$$S = \lambda_0 \bar{I} + \sum_{j=1}^m \int_0^T \lambda_j(\tau) \bar{J}_j(\tau) d\tau = \int_0^T R dt \tag{33}$$

and its integrand

$$R = \lambda_0 R_0 + \sum_{j=1}^m R_j^c, \tag{34}$$

$$R_0 = \sum_{\nu=0}^m \gamma_\nu(t) f_{01}(t, x(t), u^\nu(t), a) +$$

$$\sum_l f_{02}(t, x(t), a)\delta(t - t_l),$$

$$R_j^c = \int_0^T \lambda_j(\tau) \Big[\sum_{\nu=0}^m \gamma_\nu(t) f_{j1}(t, x(t), u^\nu, a, \tau) +$$

$$+ f_{j2}(t, x(t), a, \tau)\delta(\tau - t)\Big] d\tau \tag{35}$$

obey the following conditions

$$\frac{\delta S}{\delta a}\delta a \leq 0, \tag{36}$$

$$\frac{\delta \overline{R}}{\delta x} = 0, \tag{37}$$

$$u^\nu(t) = \arg\max_{u \in V_u} R(x, \lambda, a^*, u). \tag{38}$$

Since the extension (28), (29) is equivalent to the problem (26), (27), from the Lemma 3.4 it follows that if the optimal solution of the latter one $(u^*(t), x^*(t), a)$ exists then it obeys the optimality conditions (36) -(38).

Conditions for existence of the optimal solution of the problem (26), (27) assume that $\gamma_0(t) = 1$, and the other multipliers $\gamma_j(t)$ in (32) are equal zero.

Conditions (36) - (38) allow to derive necessary conditions of optimality in a form of maximum principle for a problem with arbitrarily combination of criterion type and constraints. This can be done simply by writing down items R_0 and R_j^c for each type of criterion and constraints, denoting $u(t)$ these variables which after reducing the problem to the canonical form are present in function f_{01} only (variables of the first group), writing down function R according to (34) and substituting it into (36), (38).

It is also important that this allows to trace easily how changes or addition of some condition effect optimality conditions - the changes it causes in one of the items in function R and in participation of some variables in the first group.

Example: Let us consider the following optimal control problem

$$I = \int_0^T f_{01}(x, u, t)dt \to \max \qquad (39)$$

$$\dot{x}_j = f_{j1}(x, u, t), \quad u \in V_u, \quad j = \overline{1, m}, \quad x(0) = x_0.$$

with the usual assumptions about the functions f_0 and f_j. From the comparison of the problems (39) and (26), (27) it is clear that $R_0 = f_{01}(x, u, t)$. Differential equations can be rewritten in (27) form as

$$J_j(\tau) = \int_0^T \Big[f_{j1}(x(t), u(t), t)h(\tau - t) -$$

$$- x(t)\delta(\tau - t)\Big] dt = 0.$$

Here $h(t)$ is Heaviside function and $\delta(t)$ is Dirac function. The term

$$R_j^c = \int_0^T \lambda_j(\tau)\Big[f_{j1}(x, u, t)h(\tau - t) - x_j(t)\delta(\tau - t)\Big] dt =$$

$$= f_{j1}(x, u, t) \int_t^T \lambda_j(\tau)d\tau - \lambda_j(t)x(t) =$$

$$= f_{j1}(x, u, t)\psi_j(t) + \dot{\psi}_j(t)x_j(t), \qquad (40)$$

where $\psi_j(t) = \int_t^T \lambda_j(\tau)d\tau$. Function R is

$$R = \lambda_0 f_{01}(x, u, t) + \sum_j \psi_j(t) f_{j1}(x, u, t) +$$

$$+ \sum_j \dot{\psi}_j(t)x_j(t).$$

Conditions (36), (37) yield equations of the Pontryagin's maximum principle. Note that inclusion of various constraints at the final instance of time yields transversality conditions directly, without any special derivations.

Different applications of this approach can be found in (Tsirlin, 1992; Tsirlin, 1985; Tsirlin, 1997).

REFERENCES

Fromovitz St. (1965) Non - Linear programming with randomization, *Management Sci*, **vol. 11**, N 9.

Tsirlin A. M (1974) Optimization in average and sliding regimes in optimal control problems, *Izv. AN SSSR, Technicheskaja kibernetika*, N 2, p.27 - 33.

Tsirlin A. M (1992) Optimality conditions of the averaged mathematical programming problems, *DAN*, **vol. 323**, N 1, p. 43 - 47.

Tsirlin A. M (1985) *Optimal cycles and cycle regimes*, Energoatomizdat, Moscow.

Tsirlin A. M (1997) *The methods of averaging optimization and their application*, Nauka, Moscow.

SINGULARLY PERTURBED BOUNDARY VALUE PROBLEMS MODELLING FAST BIMOLECULAR REACTIONS *

V.F.Butuzov[1], **N.N.Nefedov**[2], **K.R.Schneider**[3]

[1,2] *MSU, Faculty of Physics, Department of Mathematics,*
119899 Moscow, Russia
[3] *Weierstraß-Institut für Angewandte Analysis und Stochastik,*
Mohrenstraße 39, D–10117 Berlin, Germany

Abstract: Singularly perturbed boundary value problems for systems with fast and slow variables in the case of exchange of stability are considered. Such systems appear in particular by modeling of fast bimolecular reactions.

Keywords: singular perturbations, exchange of stability.

1. INTRODUCTION.

In this paper we study boundary value problems for singularly perturbed systems of the form

$$\begin{aligned} \varepsilon^2 \tfrac{d^2u}{dx^2} &= g(u,v,x,\varepsilon)\,, \\ \tfrac{d^2v}{dx^2} &= f(u,v,x,\varepsilon) \end{aligned} \tag{1}$$

where u and v are scalars, $0 < x < 1$. Systems of this type describe steady state solutions of the reaction diffusion system

$$\begin{aligned} \tfrac{\partial u}{\partial t} + \varepsilon^2 \tfrac{\partial^2 u}{\partial x^2} &= g(u,v,x,\varepsilon), \\ \tfrac{\partial v}{\partial t} + \tfrac{\partial^2 v}{\partial x^2} &= f(u,v,x,\varepsilon). \end{aligned} \tag{2}$$

We consider (1) under the assumption that the corresponding degenerate equation

$$g(u,v,x,0) = 0 \tag{3}$$

has two solutions $u = \varphi_1(v,x)$ and $u = \varphi_1(v,x)$ which intersect transversally. This can be interpreted as an exchange of stability of the two branches of equilibria $u = \varphi_1(v,x)$ and $u = \varphi_1(v,x)$ consisting of a saddle point or a center of the associated system

$$\frac{d^2u}{d\xi^2} = g(u,v,x,0)$$

where v and x are considered as parameters. Under the assumption that (3) has two solutions the standard theory for singularly perturbed systems can not be applied to (1). Our goal is to extend results of the (Nefedov,et.all, 1994) concerning the solution of initial value problems for singularly perturbed systems of the form

$$\begin{aligned} \varepsilon^2 \tfrac{du}{dt} &= g(u,v,t,\varepsilon)\,, \\ \tfrac{dv}{dt} &= f(u,v,t,\varepsilon) \end{aligned} \tag{4}$$

in case of exchange of stability to boundary value problems for (1).

To motivate our investigations we consider the following differential system modelling a bimolecular reaction with fast bimolecular reaction rate $r(\bar{u},\bar{v})/\varepsilon^2$, slow monomolecular reaction rates $g_1(\bar{u})$ and $g_2(\bar{v})$, and inputs $I_a(x)$ and $I_b(x)$ depending only on the space variable x.

$$\begin{aligned} \tfrac{\partial \bar{u}}{\partial t} &= \tfrac{\partial^2 \bar{u}}{\partial x^2} + I_a(x) - g_1(\bar{u}) - \tfrac{r(\bar{u},\bar{v})}{\varepsilon^2}, \\ \tfrac{\partial \bar{v}}{\partial t} &= \tfrac{\partial^2 \bar{v}}{\partial x^2} + I_b(x) - g_2(\bar{v}) - \tfrac{r(\bar{u},\bar{v})}{\varepsilon^2}. \end{aligned} \tag{5}$$

A stationary solution of (5) satisfies

$$\begin{aligned} \varepsilon^2 \tfrac{d^2\bar{u}}{dx^2} &= -\varepsilon^2(I_a(x) - g_1(\bar{u})) + r(\bar{u},\bar{v})\,, \\ \varepsilon^2 \tfrac{d^2\bar{v}}{dx^2} &= -\varepsilon^2(I_b(x) - g_2(\bar{v})) + r(\bar{u},\bar{v})\,. \end{aligned} \tag{6}$$

After the coordinate transformation $u = \bar{u}, v = \bar{u} - \bar{v}$ system (6) can be rewritten as

$$\begin{aligned} \varepsilon^2 \tfrac{d^2u}{dx^2} &= -\varepsilon^2(I_a(x) - g_1(u)) + r(u,u-v)\,, \\ \tfrac{d^2v}{dx^2} &= I_b(x) - I_a(x) - g_2(u-v) + g_1(u) \end{aligned} \tag{7}$$

which has the form (1).

*This work is partially supported by Grant № 96-01-00694 from the Russian Foundation for Fundamental Researches and partially supported by joint Grant № 96-01-000116 from the RFFR and German Scientific Sosiety.

The main results of this paper concern the existence and the asymptotic expansion of the solution of some boundary value problem related to system (1). As a consequence of these investigations we obtain that the reaction rate $r(u(x), v(x))/\varepsilon^2$ in (6) has an interior layer as a function of the space variable x in case of exchange of stability. This is an extension of the jumping behavior of the fast reaction rate of a bimolecular reaction in a homogeneous medium (Nefedov, *et al.*, 1994).

2. NOTATION. FORMULATION OF THE PROBLEM. ASSUMPTION.

Let $I_{\varepsilon_0}, I_{v_0}, I_{u_0}$ be the intervals defined by $I_{\varepsilon_0} := \{\varepsilon \in R : 0 < \varepsilon \leq \varepsilon_0\}, 0 < \varepsilon_0 << 1, I_{u_0} := \{u \in R : |u| < u_0\}, u_0 > 0, I_{v_0} := \{v \in R : |v| < v_0\}, v_0 > 0$, let the set G_0 be defined by $G_0 := I_{v_0} \times (0,1)$, let D_0 be defined by $D_0 := G_0 \times I_{u_0}$.

In what follows we study the singularly perturbed nonlinear boundary value problem

$$\begin{aligned} &\varepsilon^2 \tfrac{d^2u}{dx^2} = g(u,v,x,\varepsilon), \\ &\tfrac{d^2v}{dx^2} = f(u,v,x,\varepsilon), \quad x \in (0,1), \\ &u'(0) = u'(1) = 0, \; v(0) = v^0, \; v(1) = v^1 \end{aligned} \tag{8}$$

under the following assumptions.

(A$_1$) *$f, g \in C^2(D_0 \times I_{\varepsilon_0}, R)$ where all derivatives are continuous in closure $\overline{D_0}$ of D_0.*

The boundary value problem

$$\begin{aligned} &\tfrac{d^2v}{dx^2} = f(u,v,x,0), \\ &0 = g(u,v,x,0), \quad x \in (0,1), \\ &v(0) = v^0, \quad v(1) = v^1 \end{aligned} \tag{9}$$

is called the degenerate problem to (8).

In case that $g(u,v,x,0) = 0$ has a unique solution $u = \varphi(v,x)$, the degenerate problem (9) can be written in the form

$$\begin{aligned} &\tfrac{d^2v}{dx^2} = f(\varphi(v,x),v,x,0), \\ &v(0) = v^0, \quad v(1) = v^1, \quad x \in (0,1), \end{aligned}$$

and under some additional assumptions the standard theory can be applied to (8).

In the sequel we study (8) by assuming

(A$_2$) *In D_0, the equation (3) has two different solutions $u = \varphi_1(v,x)$ and $u = \varphi_2(v,x)$ defined on G_0 with $\varphi_1(v,x), \varphi_2(v,x) \in C^2(G_0, R)$ where all derivatives are continuous in $\overline{G_0}$.*

(A$_3$) *There exists a continuous function*

$$k : [0,1] \to R$$

such that $\varphi_1(k(x),x) \equiv \varphi_2(k(x),x)$ for $\forall x \in [0,1]$.

Assumption (A_3) says that the surfaces $u = \varphi_1(v,x)$ and $u = \varphi_2(v,x)$ intersect at the curve whose projection into the region G_0 is described by $v = k(x)$.

(A$_4$) *There is a point $x_0 \in (0,1)$ such that the boundary value problems*

$$\begin{aligned} &\tfrac{d^2v}{dx^2} = f(\varphi_1(v,x),v,x,0), \quad 0 < x < x_0, \\ &v(0) = v^0, \quad v(x_0) = k(x_0), \end{aligned}$$

and

$$\begin{aligned} &\tfrac{d^2v}{dx^2} = f(\varphi_2(v,x),v,x,0), \quad x_0 < x < 1, \\ &v(x_0) = k(x_0), \; v(1) = v^1, \end{aligned}$$

have solutions $\hat{v}_1(x)$ and $\hat{v}_2(x)$ defined on $[0,x_0]$ and $[x_0,1]$ respectively and satisfying

$$\hat{v}_1'(x_0) = \hat{v}_2'(x_0).$$

We define the function $\hat{v}(x)$ by

$$\hat{v}(x) = \begin{cases} \hat{v}_1(x) & \text{for} \quad 0 \leq x \leq x_0, \\ \hat{v}_2(x) & \text{for} \quad x_0 \leq x \leq 1. \end{cases}$$

It is easy to see that $\hat{v}(x)$ is twice continuously differentiable and represents a solution of the degenerate problem

$$\begin{aligned} &\tfrac{d^2v}{dx^2} = f(\hat{\varphi}(v,x),v,x,0), \\ &v(0) = v^0, \quad v(1) = v^1, \quad 0 < x < 1, \end{aligned} \tag{10}$$

where $\hat{\varphi}(v,x)$ is defined by

$$\hat{\varphi}(v,x) := \begin{cases} \varphi_1(v,x) & \text{for} \quad 0 \leq x \leq x_0, \\ \varphi_2(v,x) & \text{for} \quad x_0 \leq x \leq 1. \end{cases}$$

By means of $\hat{v}(x)$ we introduce the functions

$$\begin{aligned} \psi_1(x) &:= \varphi_1(\hat{v}(x),x), \\ \psi_2(x) &:= \varphi_2(\hat{v}(x),x), \end{aligned} \tag{11}$$

which are twice continuously differentiable and satisfy

$$\psi_1(x_0) = \psi_2(x_0). \tag{12}$$

Concerning the relative position of the curves $u = \psi_1(x)$, $u = \psi_2(x)$ we assume

(A$_5$) (i) $\psi_1(x) > \psi_2(x)$ *for* $0 \leq x < x_0$,
(ii) $\psi_1(x) < \psi_2(x)$ *for* $x_0 < x \leq 1$.

To motivate the following assumptions we introduce the associated equation to system (1)

$$\frac{d^2u}{dt^2} = g(u,v,x,0) \tag{13}$$

in which v, x have to be considered as parameters. From hypothesis (A_2) it follows that (13) has two intersecting families of equilibria the stability of which is determined by the sign of g_u at these families. The following asssumption describes the behavior of the sign of these expressions as a function of x and characterizes an exchange of stabilities of these families.

(A_6)

$$g_u(\psi_1(x), \hat{v}(x), x, 0) > 0 \quad \text{for} \quad 0 \le x < x_0,$$

$$g_u(\psi_1(x), \hat{v}(x), x, 0) < 0 \quad \text{for} \quad x_0 < x \le 1,$$

$$g_u(\psi_2(x), \hat{v}(x), x, 0) < 0 \quad \text{for} \quad 0 \le x < x_0,$$

$$g_u(\psi_2(x), \hat{v}(x), x, 0) > 0 \quad \text{for} \quad x_0 < x \le 1.$$

Under the assumptions (A_1) - (A_6) the vector function $(\hat{u}(x), \hat{v}(x))$ where $\hat{u}(x)$ is defined by

$$\hat{u}(x) = \begin{cases} \psi_1(x) & , 0 \le x \le x_0, \\ \psi_2(x) & , x_0 \le x \le 1, \end{cases} \tag{14}$$

is referred to as the composed stable solution of (9). We note that $\hat{u}(x)$ is not differentiable at $x = x_0$. The composed stable solution satisfies

$$g(\hat{u}, \hat{v}, x, 0) = 0. \tag{15}$$

Finally we suppose

(A_7) *There are positive constants μ, d_1, d_2 and ρ such that for $x \in [0, 1]$*
(i) $\mu - \hat{\varphi}_v(\hat{v}(x), x) \ge d_1$.
(ii) $\hat{g}_{uu}(x_0)\mu^2 + 2\hat{g}_{uv}(x_0) + \hat{g}_{vv}(x_0)\mu \ge d_2$ *where "^" indicates that the derivative has to be evaluated at $(\hat{u}(x), \hat{v}(x), x, 0)$.*
(iii) $\hat{f}_u\mu + \hat{f}_v \ge -\pi^2 + \rho$.

(A_8) *For $x \in \{(0,1) \setminus x_0\}$ and $\varepsilon \in I_{\varepsilon_0}$* it it holds

$$\varepsilon^2 \frac{d^2\hat{u}}{dx^2} \ge g(\hat{u}(x), \hat{v}(x), x, \varepsilon),$$
$$\frac{d^2\hat{v}}{dx^2} \ge f(\hat{u}(x), \hat{v}(x), x, \varepsilon).$$

(A_9) *The vector function*

$$\Big(g(u, v, x, \varepsilon), f(u, v, x, \varepsilon)\Big)$$

is quasimonotone in some region near the composed stable solution. The assumptions (A_7) - (A_7) are concerned with the method of the proof which is based on the concept of lower and upper solutions.

3. EXISTENCE AND ASYMPTOTIC BEHAVIOR OF THE SOLUTION

In this section we state that the boundary value problem (8) has a solution which is close to the composed stable solution.

Theorem. *Assume hypotheses (A_1) - (A_9) to be valid. Then there exists a sufficiently small ε_1 such that for $0 < \varepsilon \le \varepsilon_1$ the boundary value problem (8) has a solution $(u(x,\varepsilon), v(x,\varepsilon))$ satisfying for $x \in [0,1]$*

$$\lim_{\varepsilon \to 0} u(x, \varepsilon) = \hat{u}(x),$$
$$\lim_{\varepsilon \to 0} v(x, \varepsilon) = \hat{v}(x).$$

Moreover we have for $x \in [0,1]$

$$u(x, \varepsilon) = \hat{u}(x) + O(\varepsilon^{\frac{1}{2}}),$$
$$v(x, \varepsilon) = \hat{v}(x) + O(\varepsilon^{\frac{1}{2}}).$$

The concept of ordered lower and upper solutions of (8) plays a central role in our approach.

REFERENCES

Nefedov N.N., K.R. Schneider and A. Schuppert, Jumping behavior in singularly perturbed systems modelling bimolecular reactions, Weierstraß–Institut für Angewandte Analysis und Stochastik, Berlin, Preprint Nº 137, 1994.

Bifurcational phenomena of the singularly perturbed equation with delay

Kaschenko S.A.

1. Introduction. The scalar differential equation of the first order

$$\frac{dx}{dt} + x = F(x(t-T)) \quad (T > 0) \qquad (1)$$

is one of the simplest and at the same time the most often occurring in the applied problems (Ref.[1]) equations with delay. Another, more convenient form of this equation (which we obtain from (1) as a result of norming time $t \to Tt$) is

$$\varepsilon\frac{dx}{dt} + x = F(x(t-1)), \qquad (2)$$

here $\varepsilon = T^{-1}$. In a number of References (for example, Refs.[1,2]) the importance of studying dynamics of Eqs.(1) and (2) when

$$0 < \varepsilon \ll 1 \qquad (3)$$

was noted.

In this paper we study the dynamics of a more complicated object, i.e. the equation with two delays provided (3)

$$\frac{dx}{dt} + x(t-h) = F(x(t-T)),$$

where $h > 0$, or using the form corresponding to Eq.(2):

$$\varepsilon\frac{dx}{dt} + x(t-\varepsilon h) = F(x(t-1)). \qquad (4)$$

This equation is singularly perturbed. For $\varepsilon = 0$ we formally obtain the difference equation

$$x(t) = F(x(t-1)). \qquad (5)$$

The dynamical properties of Eq.(5) have been studied rather well (Ref.[1]). However, on the basis of the numerical analysis it is known that even provided (3) the dynamics of Eqs.(1) and (4) is not defined by the behaviour of the trajectories of Eq.(5). In Refs.[3,4] the author tried to explain this phenomenon for Eq.(2) on the basis of the special methods of local analysis if function $F(x)$ was quasi-linear, i.e.

$$F(x) = ax + \mu f(x), \qquad (6)$$

where $0 < \mu \ll 1$. It is comfortable to assume function $f(x)$ to be continuous and $f(0) = 0$. The results of this paper essentially supplement and develop the constructions given in Ref.[3]. As a fundamental result we will show that (complex) nonlinear parabolic Ginsburg-Landau equation will be a normal form of Eq.(4) provided (3) and (6) that will describe the dynamics of its solutions in cases close to critical ones in the problem of the equilibrium state stability.

The developed approach is used below for investigation of the dynamics of the neutral equation

$$\frac{dx}{dt} + x(t-h) = \frac{d}{dt}[ax(t-T) + \mu f(x(t-T))].$$

2. The problem statement. Linear analysis. At first we consider the linear (for $\mu = 0$) equation

$$\varepsilon\frac{dx}{dt} + x(t-\varepsilon h) = ax(t-1).$$

The question concerning stability of its solutions is connected with the behaviour of the roots of the characteristic quasi-polynomial

$$\varepsilon\lambda + \exp(-\varepsilon\lambda h) = a\exp(-\lambda). \qquad (7)$$

Introduce into consideration the complex function $P(\delta) = i\delta + \exp(-i\delta h)$ of real argument δ and let

$$a_0 = \min_{0\le\delta<\infty} |P(\delta)| = |P(\delta_0)|.$$

Note that δ_0 can be defined uniquely. For example, for $h = 0$ $a_0 = 1, \delta_0 = 0$, and for $h = \pi/2 - a_0 = 0, \delta_0 = 1$. Now we formulate several simple intermediate statements. Denote as γ a certain positive and independent of ε constant that exact value is not essential.

Lemma 1. Let $h > \frac{\pi}{2}$. Then for all sufficiently small ε Eq.(7) has root $\lambda(\varepsilon)$ which satisfies the inequality $Re\lambda(\varepsilon) \geq \gamma$.

Thereby, under the condition of Lemma 1 the problem concerning the dynamics of Eq.(4) (provided (3)) becomes non local: the attractor cannot exist in an arbitrarily fixed (but independent of ε) neighbourhood of the zero equilibrium state of Eq.(4). Therefore, below we assume the following inequality to be valid

$$0 < h < \frac{\pi}{2}. \tag{8}$$

Note that the condition $0 \leq h < \frac{\pi}{2}$ is necessary and sufficient for asymptotical stability of solutions of the more simple equation

$$\varepsilon\frac{dx}{dt} + x(t - \varepsilon h) = 0.$$

Lemma 2. Let $|a| > a_0$. Then for all sufficiently small ε there exists such root $\lambda(\varepsilon)$ of Eq.(7) for that $Re\lambda(\varepsilon) \geq \gamma$.

Lemma 3. Let $|a| < a_0$. Then for all sufficiently small ε the all roots of Eq.(7) satisfy the inequality $Re\lambda \leq -\gamma$.

Under the condition of Lemma 2 the problem concerning the dynamics of Eq.(4) is again non local, and under the condition of Lemma 3 it is trivial: the all solutions belonging to an arbitrarily fixed (but independent of ε) neighbourhood of $x = 0$ tend to zero for $t \to \infty$.

So, we need to study the situation when parameter a is "close" to a_0 by its modulus. Below we assume that

$$a = \pm a_0(1 + \varepsilon^2 a_1). \tag{9}$$

Thus, provided (3), (6), (8) and (9) we set the problem of studying the dynamics of Eq.(4) in an arbitrarily fixed (but independent of ε and μ) range of phase space $C_{[-1,0]}$.

Note that provided (6) and (9) (for $0 < h < \pi/2$) the trajectories of difference Eq.(5) tend to zero for $t \to \infty$.

The following statement describes the behaviour of the roots of the characteristic quasi-polynomial more completely.

Lemma 4. Let conditions (8) and (9) be valid. Then there exist infinitely many roots $\lambda_k(\varepsilon)$ and $\bar{\lambda}_k(\varepsilon)$ $(k = 0, \pm1, \pm2, \ldots)$ of characteristic Eq.(7) for that $Re\lambda_k(\varepsilon) \leq \varepsilon^2\gamma$ and $\lim\limits_{\varepsilon\to 0} Re\lambda_k(\varepsilon) = 0$, and all the other roots satisfy the inequality $Re\lambda \leq -\gamma$.

It follows from this Lemma that the problem concerning stability of the solutions of Eq.(4) has infinite dimension in the considered situation. To find asymptotical behaviour $\lambda_k(\varepsilon)$ we introduce several denotions. Denote as $\theta = \theta(\varepsilon)$ such value belonging to interval $[0, 2\pi)$ for that value $\delta_0\varepsilon^{-1} + \theta$ is integer multiple of 2π. Let $\Omega-$ be the value belonging to that interval $[0, 2\pi)$ of such root of the equation

$$\mathrm{tg}\Omega = [\sin(\delta_0 h) - \delta_0](\cos(\delta_0 h))^{-1},$$

for that value $a\cos(\Omega)$ is positive. Further, we denote $P(\delta) = \rho(\delta)\exp i\varphi(\delta)$. Then $\rho(\delta_0) = a_0, \rho'(\delta_0) = 0$ and

$$\rho''(\delta_0) > 0. \tag{10}$$

In conclusion of this item we give the asymptotical for $\varepsilon \to 0$ formulae for $\lambda_k(\varepsilon)$:

$$\lambda_k(\varepsilon) = i[\frac{\delta_0}{\varepsilon} + \theta + \Omega + 2\pi k] + \varepsilon\lambda_{k1} + \varepsilon^2\lambda_{k2} + O(\varepsilon^3) \tag{11}$$

where

$$\lambda_{k1} = i\varphi'(\delta_0)R_k, \quad R_k = \theta + \Omega + 2\pi k,$$

$$\begin{aligned}\lambda_{k2} = \; & -\tfrac{1}{2}R_k^2[\rho''(\delta_0)|a_0^{-1}| + i\varphi''(\delta_0)] - \\ & -iR_k(\varphi'(\delta_0))^2 + a_1.\end{aligned}$$

3. The normalized form construction. To investigate the nonlinear dynamics of Eq.(4)

we will use the standard approach according to that the established conditions can be formed "around" harmonics with "critical" frequencies. Denote in (6) $\mu = \sigma\varepsilon^2$ and consider the formal series

$$\begin{aligned} x &= \exp[i(\tfrac{\delta_0}{\varepsilon} + \theta + \Omega + \varepsilon\varphi'(\delta_0)(\theta + \\ &+\Omega))] \textstyle\sum_{k=-\infty}^{\infty} \xi_k(\tau) \exp(2k\pi i t_1) + \\ &+ \exp[-i(\tfrac{\delta_0}{\varepsilon} + \theta + \Omega + \varepsilon\varphi'(\delta_0)(\theta + \\ &+\Omega))] \textstyle\sum_{k=-\infty}^{\infty} \bar{\xi}_k(\tau) \exp(-2k\pi i t_1) + \\ &+\varepsilon^2 x_2 + \varepsilon^3 x_3 + \ldots \end{aligned} \quad (12)$$

where $t_1 = (1 + \varepsilon\varphi'(\delta_0))t$, $\xi_k(\tau)$ are slowly changing amplitudes: $\tau = \varepsilon^2 t$, and function $x_j = x_j(t\varepsilon^{-1}, \varepsilon t, \tau)$ is periodical with respect to the first three arguments. We substitute (12) in (4) and gather the coefficients at the same degrees of ε. Using the conditions of the obtained equations solvability with respect to x_2, we come to the normalized form that is the infinite equations system with respect to $\xi_k(\tau)$:

$$\frac{d\xi_k}{d\tau} = \lambda_{k2}\xi_k + \varphi_k \quad (13)$$

where φ_k is the coefficient at $\exp(i(\tau + 2k\pi t))$ in Fourier-series expansion of the function

$$\begin{aligned} &(P(\delta_0))^{-1} f(\exp(i\Omega)\exp(i\tau)\times \\ &\times \textstyle\sum_{m=-\infty}^{\infty} \xi_k \exp(2m\pi i t) + \\ &+ \exp(-i\Omega)\exp(-i\tau)\times \\ &\times \textstyle\sum_{m=-\infty}^{\infty} \bar{\xi}_k \exp(-2m\pi i t)). \end{aligned}$$

Then we introduce into consideration the parabolic boundary value problem

$$\begin{aligned} &\tfrac{\partial z}{\partial \tau} = d_1 \tfrac{\partial^2 z}{\partial s^2} + d_2 \tfrac{\partial z}{\partial s} + d_3 z + f_0(z), \\ &z(\tau, s+1) \equiv z(\tau, s), \end{aligned} \quad (14)$$

where

$$d_1 = \frac{1}{2}[\rho''(\delta_0)|a_0|^{-1} + i\varphi''(\delta_0)] \quad (Re d_1 > 0),$$

$$d_2 = -(\varphi'(\delta_0))^2 + 2id_1(\theta + \Omega),$$

$$d_3 = a_1 - i(\varphi'(\delta_0))^2(\theta + \Omega) - d_1(\theta + \Omega)^2,$$

$$\begin{aligned} f_0(z) &= P^{-1}(\delta_0)(2\pi)^{-1}\sigma \int_0^{2\pi} f(\exp(-i\Omega)z\times \\ &\times \exp(i\tau) + \exp(-i\Omega)\bar{z}\exp(-i\tau))\exp(-i\tau)d\tau. \end{aligned}$$

The formal Fourier- series expansion of function $z(\tau, s)$
$z = \sum_{k=-\infty}^{\infty} \xi_k(\tau)\exp(2k\pi i s)$ leads to system (13). Therefore, it is natural to call boundary value problem (14) as a normalized form. Thereby, the solutions of this boundary value problem determine the behaviour of slowly changing amplitudes $\xi_k(\tau)$ in (12) and, then, determine the dynamics of Eq.(4). So, a non-zero equilibrium state in (14) is corresponded by a periodical solution in Eq.(4) of the same stability. A cycle in (14) is corresponded by a torus in Eq.(4) etc. We know that boundary value problem (14) can have attractors with complicated (non-regular) structure. Correspondingly, the same conclusions hold for Eq.(4).

4. The results generalization. Above we assumed that function $F(x)$ was quasi-linear. However, often, in problems of the nonlinear oscillations theory we have to deal with nonlinear functions that are not quasi-linear. Then, instead of the quasi-linearity condition we put the locality condition: we consider Eq.(4) in a certain sufficiently small (but independent of ε) neighbourhood of the zero equilibrium state. In this connection we briefly dwell on the case when $F(x) = ax + bx^2 + cx^3 + O(x^4)$ in Eq.(4). The changes are not great in comparison with the analyzed case. We only should to write factor ε before the first two summands in the right-hand part in the formula analogous to (12). The concluding normalized equation can be represented in the form of Ginsburg-Landau equation

$$\begin{aligned} &\tfrac{\partial z}{\partial \tau} = d_1 \tfrac{\partial^2 z}{\partial s^2} + d_2 \tfrac{\partial z}{\partial s} + d_3 z + d|z|^2 z, \\ &z(\tau, s+1) \equiv z(\tau, s), \end{aligned} \quad (15)$$

where

$$\begin{aligned} d &= \exp(i\Omega)3c + 2b(1 - a_0)^{-1} + \\ &+2b[P(2\delta_0) - a_0\exp(-2i\Omega)]^{-1}. \end{aligned}$$

The dynamics of boundary value problem (15) is rather complicated. It is studied in detail in Ref.[5].

It is important to note that coefficients d_2 and d_3 in (14) and (15) depend on "interior" parameter θ. For $\varepsilon \to +0$ this parameter changes from 0 to 2π infinitely many times. For different θ the number of established conditions and their stability properties can change. Hence it follows that there are possible the situations when for $\varepsilon \to +0$ unbounded alternation of the processes of "birth" and "death" of certain established conditions occurs.

In conclusion we compare the obtained results with those that hold for Eq.(2), i.e. for $h = 0$ $(a_0 = 1, \delta_0 = 0)$. In Ref.[3] it was shown that the following boundary value problems are normalized equations for $a = 1 + \varepsilon a_1$ and $a = -(1 + \varepsilon a_1)$ (provided (6)) correspondingly

$$\frac{\partial z}{\partial \tau} = \frac{1}{2}\frac{\partial^2 z}{\partial s^2} + \frac{\partial z}{\partial s} + f(z), \quad z(\tau, s+1) \equiv z(\tau, s) \tag{16}$$

and

$$\frac{\partial z}{\partial \tau} = \frac{1}{2}\frac{\partial^2 z}{\partial s^2} + \frac{\partial z}{\partial s} + \frac{1}{2}(f(-z) - f(z)), \quad z(\tau, s+1) \equiv -z(\tau, s). \tag{17}$$

The dynamics of boundary value problem (16) is trivial. Only homogeneous equilibrium states can be its stable solutions. In (17) heterogeneous equilibrium states can be stable. Unlike (14), variable z and the coefficients in (16) and (17) are real and, then, the dynamics in the case $h = 0$ is essentially simpler.

5. The neutral equations normalization. The neutral equation given in Introduction after norming time $t \to Tt$ takes the form

$$\varepsilon[\frac{dx}{dt} - ax(t-1) - \mu f(x(t-1))] + x(t - \varepsilon h) = 0. \tag{18}$$

To study its dynamics for $0 < \varepsilon, \mu \ll 1$ we assume $\mu = 0$ and consider the characteristic quasi-polynomial of the obtained linear equation

$$\varepsilon\lambda[1 - a\exp(-\lambda)] + \exp(-\varepsilon\lambda h) = 0.$$

For the roots of this equation the statements of Lemmas 1-4 where $a_0 = \min_{0 \le \delta < \infty} |P(\delta)| = |P(\delta_0)|$, $P(\delta) = 1 - i\delta^{-1}\exp(-i\delta h)$ are valid. Appearing in (11) value $\Omega \in [0, 2\pi)$ can be defined from the inequality $\exp(-i\Omega) = a_0^{-1}P(\delta_0)$, and for λ_{k1} and λ_{k2} $(k = 0, \pm 1, \ldots)$ we obtain the relations

$$\lambda_{k1} = -i\varphi'(\delta_0)R_k, \quad R_k = \Omega + \theta + 2k\pi, \quad (P(\delta) = \rho(\delta)\exp(i\varphi(\delta)),$$

$$\lambda_{k2} = \frac{1}{2}R_k^2[\rho''(\delta_0)|a_0^{-1}| - i(2\delta_0^{-1}\varphi'(\delta_0) - \varphi''(\delta_0))] + R_k[(\delta_0)P(\delta_0)\exp(i\delta_0 h)]^{-1}\varphi'(\delta_0)[i\delta_0)^{-1} - h] + a_1.$$

In the capacity of the normalized equations we again come to parabolic (by virtue of inequality (10)) boundary value problem (14) where

$$d_1 = \tfrac{1}{2}[\rho''(\delta_0)|a_0|^{-1} + i(\varphi''(\delta_0) - 2\delta_0^{-1}\varphi'(\delta_0))] \quad (Re d_1 > 0),$$

$$d_2 = (\delta_0 P_0(\delta_0)\exp(i\delta_0 h)]^{-1}\varphi'(\delta_0)[\delta_0 + ih] + i2d_1(\theta + \pi),$$

$$d_3 = a_3 + d_1(\theta + \Omega)^2 + id_2(\theta + \Omega),$$

$$f_0(z) = \sigma[2\pi\delta_0 P(\delta_0)]^{-1}\int_0^{2\pi} f(\exp(-i[-i(\Omega + \tau)] + \bar{z}\exp(i(\Omega + \tau)])\exp(-i\tau)d\tau.$$

Draw attention to importance of the condition $h > 0$ for considering Eq.(18). The point is that for $h = 0$ we have $a_0 = 1, \delta_0 = \infty$ and $Re d_1 = 0$.

References

1. Sharkovsky A.N., Maystrenko Yu.L., Romanenko E.Yu. Difference equations and their applications. Kiev: Naukova dumka, 1981.

2. Dmitriev A.S., Kislov V.Ya. Stochastic oscillations in radio engineering. Moscow, Nauka. 1989.

3. Kaschenko S.A. The normalization method application to studying the dynamics of difference differential equations with a small coefficient at derivative// Differential equations, 1989, 25, N 8. 4 pgs.

4. Kaschenko S.A. Normalization in the systems with small diffusion// International Journal of Bifurcations and chaos. 6. No. 7, 1996. 17 p.

5. Akhromeeva T.S., Kurdyumov S.P., Malinetskiy G.G., Samarsky A.A. Unstationary structures and diffusional chaos. Moscow: Nauka, 1992.

Yaroslavl State University

CONDITION OF "FAST" AND "SLOW" SYNCHRONIZATION IN THE SYSTEM OF TWO AUTOGENERATORS WITH RELAY DELAYING FEEDBACK

Dmitry S. Kaschenko
Yaroslavl State University, Department of Mathematics,
14 Sovetskaya st. 150000 Yaroslavl, Russia.
e-mail: dimak@uniyar.ac.ru

Studying behavioral synchronization of interacting dynamical systems phenomena is of great interest at present. Up to now there is a considerable number of analytical and experimental investigations of the phenomenon of chaotic synchronization of dynamical systems with different nature [1-5]. In spite of that rather general results have been obtained in some works characterizing regularities of arising the chaotic synchronization, further studying this phenomenon is theoretically and practically interested.

In this paper we study the phenomenon of synchronization in system of two coupled simple autogenerators of the first order with different types of nonlinear delaying relay feedback. Two types of feedbacks - diffusional one and through nonlinearity $f(x)$ - are most interested for us. Such systems are widely used in a number of concrete applications, for example, in elecrotechnics [6,7].

Consider the system of the equations with diffusional feedback:

$$\begin{aligned} \dot{x}+x &= f(x(t-T))+d_1(y-x), \\ \dot{y}+y &= f(y(t-T))+d_2(x-y), \end{aligned} \qquad (1)$$

where coefficients of diffusional feedback d_1, d_2 are non-negative; and the system of equations connected through nonlinear function:

$$\begin{aligned} \dot{x}+x &= f[x(t-T))+d_1(y(t-T)-x(t-T))], \\ \dot{y}+y &= f[y(t-T))+d_2(x(t-T)-y(t-T))], \end{aligned} \qquad (2)$$

In (1) and (2) function $f(s)$ is of relay type:

$$f(s)=\begin{cases} 1, & s<\gamma \\ 0, & s\geq\gamma \end{cases}, \quad 0<\gamma<1. \qquad (3)$$

Note that the differential equation

$$\dot{x}+x=f(x(t-T)), \qquad (4)$$

has no equilibrium states. We have established by analytical methods that there exists a denumerable set of periodical solutions $x_m(t)$ with $2m$ $(m=0,1,\ldots)$ intersections by straight line $x=\gamma$. And all of them, except for $x_0(t)$ are unstable.

The results of numerical experiments show that all solutions become equal to $x_0(t)$ beginning from a certain time moment L. Standart numerical methods that were used for studying the question concerning dependence of L on parameter T show that this dependence is exponential.

Thus solution $x_0(t)$ is the only stable condition of Eq.(4), but the time required for an arbitrary solution to fall into its "small" neighbourhood essentially depends on parameter T and on "complexity" of the initial conditions: the "more complicated" initial condition the longer corresponding solution demonstrates "chaotic" properties.

Further, provided

$$T\gg 1. \qquad (5)$$

we will study the question concerning synchronization of solutions of systems (1) and (2).

It is comfortable to make change $t\to Tt$ in systems (1) and (2). Then formally assuming $T=\infty$ and analyzing the obtained difference equations we get following synchronization condition for feedback coefficients d_1 and d_2 in system (1):

$$\min\{\gamma,1-\gamma\}<(1+d_1+d_2)^{-1}\max\{d_1,d_2\}, \qquad (6)$$

and in system (2):

$$\min\{\gamma,1-\gamma\}<\max\{d_1,d_2\}. \qquad (7)$$

Numerical experiments show: if inequality (6) in system (1) or inequality (7) in system (2) is valid then there happens "fast" synchronization at a comparatively short time that does not increases if T increases.

However if condition (6) for system (1) or condition (7) for system (2) does not hold the synchronization also takes place, but the time which the synchronization takes up is essentially greater and unboundedly linearly increases for $T \to \infty$.

It is important to note that synchronization in systems (1) and (2) accomplishes essentially faster than establishment of cycle $x_0(t)$.

We already have noted that Eq.(4) has a countable number of unstable cycles $x_m(t)$ ($m = 1, 2, \ldots$). Now we fix arbitrarily $m \geq 1$. Systems of equations (1) and (2) have unstable periodical solution $x(t) = y(t) = x_m(t)$. Then we consider separately second equations in systems (1) and (2) for $x(t) = x_m(t)$:

$$\dot{y} + y = f(y(t-T)) + d(x_m(t) - y). \tag{9}$$

and

$$\dot{y} + y = f[y(t-T) + d(x_m(t-T) - y(t-T))] \tag{10}$$

It is obvious that solution

$$y(t) = x_m(t) \tag{11}$$

is unstable for small d. By analytical methods we can easily establish existence of such value d_0 that for $d > d_0$ periodical solution (11) is asymptotically stable. But using these methods we can obtain just rough estimation of d_0. Therefore we need to find value d_0 and to establish its dependence on parameters m and T numerically. Besides this there arises the problem of describing attraction basin of periodical solution (11) for $d > d_0$. Such problems are called as synchronization problems on unstable cycle.

Now we give main results of numerical investigation of the indicated questions.

1. For fixed T, γ, m value d_0 is greater if the initial conditions are more complicated (i.e. they have more intersections by straight line $y = \gamma$ in interval $(-T, 0)$). Threshold value d_{max} of d_0 such that for $d \geq d_{max}$ solution (11) is globally stable corresponds to initial condition $y(t)$ that does not intersect straight line $y = \gamma$ in interval $(-T, 0)$. Note that this initial condition corresponds to the simplest cycle $x_0(t)$.

2. The dependencies of d_{max} on delay time T and on parameter m for $\gamma = 0.3$ are given for Eq.(9) in table 1, and for Eq.(10)– in table 2:

Table 1

	$m = 1$	$m = 2$
$T = 10$	0.5301	0.8171
$T = 20$	0.4335	0.4902

Table 2

	$m = 1$	$m = 2$
$T = 10$	0.3290	0.4143
$T = 20$	0.3011	0.3166

Conclusion: d_{max} decreases when delaying time T increases and increases when m increases.

Remark. If solution (11) is unstable, then solutions of Eqs.(9) and (10) tend to more complicated solution that is close to periodical one.

Numerical results show that when coefficients d_1, d_2 decrease the structure of solutions of systems (1) and (2) becomes more complicated. Therefore studying the question concerning dynamics of two weakly coupled generators of form (4) is important for us. Analytical results concerning dynamics of such systems under additional condition when parameter γ is small are given below.

These results explain a number of complex effects discovered under numerical analysis of the systems with weak feedback.

Note that obtained results are valid also when γ is close to 1, because this case can be reduced to the first one if we replace $\bar{x} = 1 - x$, $\bar{y} = 1 - y$.

It is comfortable to make changes $x \to \gamma x$, $y \to \gamma y$ in system (1), as a result we obtain the system

$$\begin{cases} \dot{x} + x = \lambda \Phi(x(t-T)) + d_1(y - x), \\ \dot{y} + y = \lambda \Phi(y(t-T)) + d_2(x - y), \end{cases} \tag{12}$$

where $\lambda = \gamma^{-1}$,

$$\Phi(s) = \begin{cases} 1, & \text{for } s < 1, \\ 0, & \text{for } s \geq 1, \end{cases}$$

Note that system (12) has homogeneous cycle

$$y(t) = x(t) = \lambda x_0(t, T). \tag{13}$$

Applying the method of large parameter we obtain following results:
Statement. *for any fixed d_1, d_2 (i.e. independent of λ) and for sufficiently large λ functions $x(t)$ and $y(t)$ tend to cycle (13) for $t \to \infty$.*

Results become more interesting when coefficients d_1, d_2 are small. Depending on their smallness we can distinguish two cases, for which dynamics of the considered system differs principally.

In first case coefficients d_j have order $O(|\ln \lambda|^{-1})$, and in second case: $d_j = O(\lambda^{-1})$. We consider each of these cases separately.

Let $d_j = O(|\ln \lambda|^{-1})$, $j = 1, 2$. i.e.

$$d_1 = \frac{\tilde{d}_1}{\ln \lambda}, \quad d_2 = \frac{\tilde{d}_2}{\ln \lambda}. \tag{14}$$

We introduce into consideration auxiliary function $q(\delta)$:

$$q(\delta) = (1+\delta)(X-Y)\sigma[(1+\delta)X-(X-Y)(1+\delta\sigma)]^{-1},$$

where $\sigma = \exp(-(\tilde{d}_1 + \tilde{d}_2))$,

$$X = 1-\exp(-T),$$
$$Y = \begin{cases} 0, & 1+|z| \geq \exp(T), \\ 1-(1+|z|)\exp(-T), & 0 \leq 1+|z| < \exp(T). \end{cases} \quad (15)$$

The main statement implies that function $g(z)$ appearing in mapping $\bar{z} = g(z)$ (its dynamics determine the behaviour of solutions $x(t)$ and $y(t)$ for $t \to \infty$) has the following form with accuracy up to $o(1)$ for $\lambda \to \infty$

$$g(z) = \begin{cases} -q(\delta) & \text{for } z \geq 0, \\ q(\delta^{-1}) & \text{for } z < 0, \end{cases} \quad (16)$$

where $\delta = \frac{d_2}{d_1}$. Analyzing mapping $g(z)$ we obtain the result:

Theorem. *Let*

$$\frac{\exp(-(\tilde{d}_1+\tilde{d}_2))}{\exp(T)-1} < 1 \quad (>1).$$

Then zero equilibrium state of mapping $g(z)$ (that homogeneous cycle (13) corresponds to) is asymptotically stable (unstable).

Then, provided

$$\text{and} \quad |g(\exp(T))| = \frac{(\delta^{-1}+1)\exp(-(\tilde{d}_1+\tilde{d}_2))}{1-\exp(-(\tilde{d}_1+\tilde{d}_2))} \geq \exp(T)$$

and

$$|g(-\exp(T))| = \frac{(1+\delta)\exp(-(\tilde{d}_1+\tilde{d}_2))}{1-\exp(-(\tilde{d}_1+\tilde{d}_2))} \geq \exp(T)$$

mapping $g(z)$ has superstable cycle of period 2: $(g(\exp(T)),\ g(-\exp(T)))$ and there are no cycles of other periods. Note that cycle of system (12) corresponds to cycle of mapping $g(z)$ and these cycles have the same stability.

Now we consider important special case when one of the feedback coefficients is equal to zero. Let

$$d_1 = 0 \quad d_2 = \frac{\tilde{d}_2}{\ln \lambda}.$$

Then for $z \geq 0$ mapping (16) has the form

$$g(z) = -\frac{(X-Y)\exp-(\tilde{d}_2)}{X-(X-Y)\exp-(\tilde{d}_2)}, \quad (17)$$

where X and Y are defined by formulae (15).

If $z < 0$ then

$$g(z) = -\frac{|z|\exp(-\tilde{d}_2)}{1+|z|}. \quad (18)$$

Taking into account formulae (17) and (18) we conclude that the zero equilibrium state of mapping $g(z)$ (that homogeneous cycle (13) corresponds to) is globally stable for any d.

As it turns out, essential reconstruction of the phase picture of the original system can occur when feedback coefficients d_1, d_2 become of order $O(\lambda)$. We can show that in this situation the dynamics of the original system can be also described by dynamics of analytically constructed one-dimensional mapping.

Thus complexity of the dynamics of the considered system increases if the feedback coefficients decrease.

References

[1] Dmitriev A.S. "Chaos and information processing in nonlinear dynamical systems." // Radio technics and electronics. 1993. V.38. N1. pg1.

[2] Parlitz U., Chua L.O., Kocarev L., Halle K., Shang A. "Transmission of Digital Signals by Chaotic Synchronization." //International Journal of Bifurcation and Chaos. 1992. V.2. N 4. pg 973.

[3] Belsky Yu.L., Dmitriev A.S. "Information transmission with the help of determined chaos." // Radio technics and electronics. 1993. V.38. N7. pg 1310.

[4] Fujisaka H, Yamada T. "Stability Theory of Synchronized Motion in Coupled-Oscillator Systems." // Prog. Theor. Phys. 1983. V. 69. N 1. pg 32.

[5] Aranson I.S., Gaponov-Grekhov A.V., Rabinovich M.I., Starobinets I.M. "Dynamical model of space development of turbulence." // Letters in GETPh. 1984. V. 39. N 12. pg 561.

[6] T. Kilias, K. Kutzer, A. Moegel, W. Schwarz "Electronic chaos generators - design and applications." International Journal of Electronics, vol. 79, No. 6, pages 737-753, Nov. 1995.

[7] A. Moegel, W. Schwarz, S. Kaschenko: "Analysis and simulation principles for chaotic systems containing delay elements." (NDES '96) Seville, Spain, 1996.

THE SOLUTION OF THE SINGULARLY PERTURBED DIFFERENTIAL EQUATION OF THE SECOND ORDER HAVING IN THE LIMIT THE CORNER POINTS *

Vasil'eva A.B., Nikitin A.G.

Moscow State University
Faculty of Physics, Departments of Mathematics
119899 Moscow, Russia
E-mail: abvas@mathabv.phys.msu.su

Abstract: The singularly perturbed second order differential equation is considered .The degenerate equation (of the first order) has a solution L that can change its stability on the curve C defined from some algebraic equation. Then may be realized the solution consisting(in the limit) from the solution L (before the cross point M with the curve C) and the curve C (after the cross point M).

Keywords: Nonlinear Systems, Singular Perturbations, Internal and Boundary Layers.

The quasilinear singularly perturbed problem

$$\varepsilon\ y'' = A(y,x)y' + B(y,x), \qquad (1)$$
$$y(0,\varepsilon) = y^0,\ y(1,\varepsilon) = y^1$$

is considered.

Let be realized following conditions

1^0. There are the solutions $\bar{y}^l(0) = y^0, \bar{y}^r(1) = y^1$ of differential equation

$$A(y,x)y' + B(y,x) = 0,\ \ 0 < x < 1. \qquad (2)$$

2^0. $A(\bar{y}^l,x) > 0,\ \ A(\bar{y}^r,x) < 0.$ The equation $A(y_0(x),x) = 0$ has the solution $y = y_0(x)$ such that

$$\bar{y}^l(x) > y_0(x) > \bar{y}^r(x),\ \ 0 < x < 1.$$

Then there is form the solution of the problem (1) of the of the step (fig.1)(the step type contrast structure) and following limit relation is valid (Vasil'eva,1995a)

$$\lim_{\varepsilon\to 0} y(x,\varepsilon) = \begin{cases} \bar{y}^l(x) & ,0 \le x \le x_0, \\ \bar{y}^r(x) & ,x_0 \le x \le 1, \end{cases}$$

where x_0 can be found from the equation

$$\int_{\bar{y}^l(x)}^{\bar{y}^r(x)} A(y,x)dy = 0.$$

There is another case when $y = y_0(x)$ intersects the graphs of $\bar{y}^l$ and $\bar{y}^r$. Consider the problem

* This work is supported by Competitive Centre of Fundamental Science under the St. Petersburg State University , pr. 95-0-1.8-112.

$$\varepsilon\, y'' = (y'+1)(y-\gamma(x)), \qquad (2)$$
$$y(0,\varepsilon) = y^0,\, y(1,\varepsilon) = y^1.$$

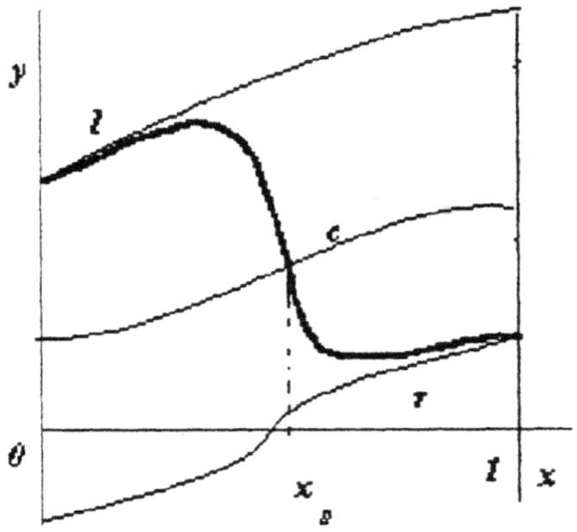

Fig. 1

In this case the solution (fig.2) for that the following limit relation is valid may exist

$$\lim_{\varepsilon\to 0} y(x,\varepsilon) = \begin{cases} \bar{y}^l = x_1 - x + \gamma(x_1), 0 \le x \le x_1 \\ \gamma(x), x_1 \le x \le x_2, \\ \bar{y}^r = x_2 - x + \gamma(x_2), x_2 \le x \le 1. \end{cases} \qquad (3)$$

For this problem using the boundary function method (Vasil'eva, *at al.*,1995b) the asymptotic containing the terms of the order $O(\varepsilon^0), O(\sqrt{\varepsilon})$ and the remainder term of $O(\varepsilon)$ is constructed . This is

$$y(x,\varepsilon) = \begin{cases} \bar{y}^l(x) + \sqrt{\varepsilon}(\bar{y}_1^l(x) + \Pi_1 y^l) + \\ O(\varepsilon), 0 \le x \le x_1, \\ \gamma(x) + \sqrt{\varepsilon}(Q_1 y + R_1 y) + \\ O(\varepsilon), x_1 \le x \le x_2, \\ \bar{y}^r(x) + \sqrt{\varepsilon}(\bar{y}^r(x) + \Pi_1 y^r) + \\ O(\varepsilon), x_2 \le x \le 1, \end{cases} \qquad (4)$$

where

$$\Pi_1 y^r = \int_{-\infty}^{\tau_1} \Pi_0 z\, d\tau, \tau_1 = (x - x_1)/\sqrt{\varepsilon}.$$

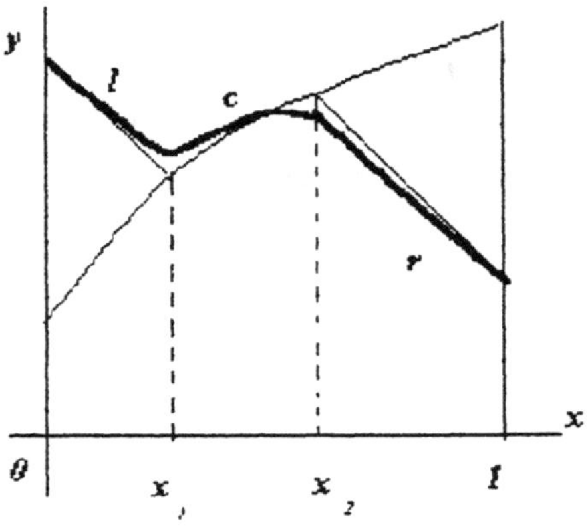

Fig. 2.

Functions $\Pi_0 z$ and $\Pi_1 y^l$ satisfy the system

$$\frac{d}{d\tau_1}\Pi_0 z = \Pi_0 z(\bar{y}_1^l(x_1) + \Pi_1 y^l - (1 + \gamma'(x_1))\tau_1), \qquad (5)$$

$$\frac{d}{d\tau_1}\Pi_1 y^l = \Pi_0 z,$$

and the boundary conditions

$$\Pi_0 z(0) = 1 + z_{10}, \Pi_1 y^l(0) = y_{11},$$
$$\Pi_0 z(-\infty) = \Pi_1 y^l(-\infty) = 0$$
$$(\bar{y}_1^l(x_1) = y_{11} - \int_{-\infty}^{0} \Pi_0 z\, d\tau).$$

The $Q_0 z$ and

$$Q_1 y = \int_{\infty}^{\tau_1} Q_0 z\, d\tau$$

satisfy the system

$$\frac{d}{d\tau_1} Q_0 z = (Q_0 z + 1) Q_1 y,$$
$$\frac{d}{d\tau_1} Q_1 y = Q_0 z,$$
$$Q_0 z(0) = z_{10} - \gamma'(x_1), Q_1 y(0) = y_{11}, \qquad (6)$$
$$Q_0 z(-\infty) = Q_1 y(-\infty) = 0.$$

The $R_1 z, R_1 y$ may be obtained analogously. The representation in the interval $[x_2, 1]$ is analogously to $[0, x_1]$.The system (5) may be integrated in the quadratures but it contains the unknown values y_{11}, z_{10} - the main terms of the asymptotic expansion for $y(x_1) - \gamma(x_1)$ and $z_1(x_1) = \frac{dy}{dx}(x_1)$:

$$y(x_1) - \gamma(x_1) = \sqrt{\varepsilon} y_{11} + \ldots, \qquad (7)$$
$$z(x_1) = z_{10} + \ldots$$

The conditions $\Pi_0 z \to 0, \Pi_1 y \to 0$ by $\tau_1 \to -\infty$ give the connection between z_{10} and y_{11}.The system (6) is also integrated in the quadratures and the condition $Q_0 z \to 0$, $Q_1 y \to 0$ by $\tau_1 \to \infty$ gives the second connection between z_{10} and y_{11}. Analogously y_{21}, z_{20} - the same values as in (7) but corresponding to the point x_2 may be found.. So the parameters $y_{11}, y_{21}, z_{10}, z_{20}$ is presented and therefore all terms in the representation (4) are obtained. The proof of the problem (2) is led.

The quadratic case

$$\varepsilon\, y'' = (1 - y'^2)(y - \gamma(x)),$$
$$y(0,\varepsilon) = y^0, y(1,\varepsilon) = y^1$$

may be considered in the same way (Fig. 3).

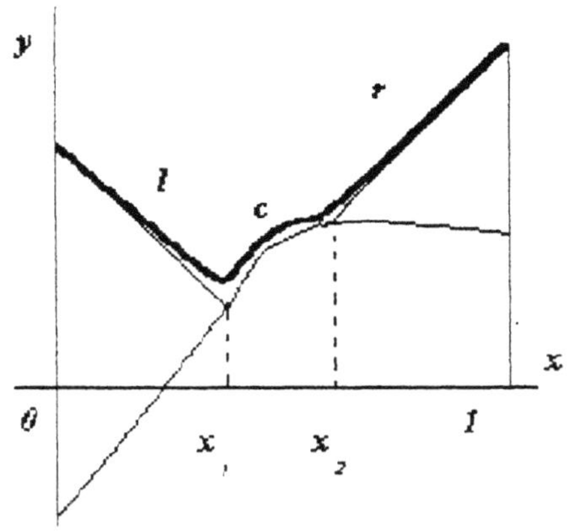

Fig. 3.

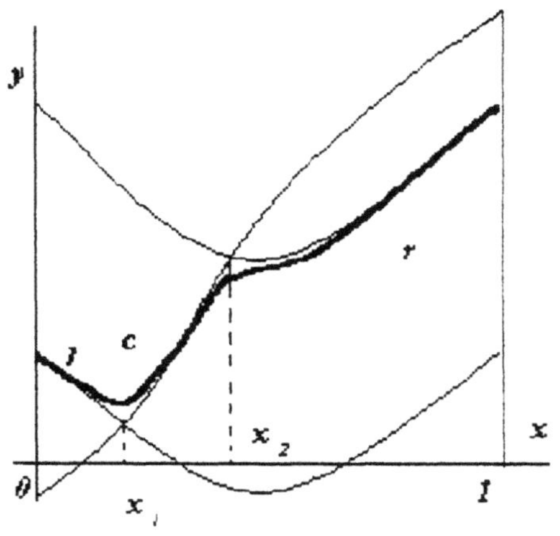

Fig. 4.

The nonlinear problem

$$\varepsilon\, y'' = f(y', x)\, g(y, x), \qquad (8)$$
$$y(0,\varepsilon) = y^0, \quad y(1,\varepsilon) = y^1.$$

is considered also. Let the equations $f(y', x) = 0$ and $g(y, x) = 0$ have isolated solutions $y' = \alpha(x)$, $y = \gamma(x)$ within the interval [0, 1] (Fig.4). The asymptotic expansion of y is similar to (4).

The regular part of the asymptotic

$$\bar{y} = \begin{cases} \int_{x_1}^{x} \alpha(t)dt + \gamma(x_1), 0 \le x \le x_1, \\ \gamma(x), x_1 < x < x_2, \\ \int_{x_2}^{x} \alpha(t)dt + \gamma(x_2), x_2 \le x \le 1. \end{cases}$$

The boundary part of the asymptotic on the interval $[0, x_1]$ satisfies the system

$$\frac{d}{d\tau_1}\Pi_0 z = f(\alpha(x_1) + \Pi_0 z, x_1)\{g'_y(\gamma(x_1), x_1)$$
$$(\Pi_1 y^l + y_1^l(x_1) + \alpha(x_1)\tau_1) + \qquad (9)$$
$$g'_x(\gamma(x_{1^-}), x_1)\tau_1\},$$
$$\frac{d}{d\tau_1}\Pi_1 y^l = \Pi_0 z.$$

The left boundary layer on the interval $[x_1, x_2]$ is found from the system

$$\frac{d}{d\tau_1}Q_0 z = f(Q_0 z, x_1)\, g'_y(\gamma(x_1), x_1)\, Q_1 y, \qquad (10)$$
$$\frac{d}{d\tau_1}Q_1 y = Q_0 z.$$

The systems (9) and (10) may be integrated in quadratures as the systems (5) and (6) but the forms of the solutions are not allowed to make use of the method of the proof of the problem (2). Now for the problem (8) the formal construction is presented only.

REFERENCES

Vasil'eva, A.B. (1995a). Step type contrast structures for the second order differential equation that is linear concerning derivatives. *Journal of Computational Mathematics and Mathematical Physics* **35 N 4**, 520-531,(in Russian).

Vasil'eva, A.B, V.F. Butuzov and L.V. Kalachev (1995b). *The Boundary Function Method for Singular Perturbation Problems*, SIAM, Philadelphia PA.

CONTRAST STRUCTURES OF ALTERNATING TYPE*

A.B.Vasilieva, A.P.Petrov, A.A.Plotnikov

MSU, Faculty of Physics, Department of Mathematics,
119899 Moscow, Russia

Abstract: Singularly perturbed parabolic equation with nonlinear right-hand part depending periodically on t is considered. The solution changing periodically his form from pure boundary layer type to the step-type is studied. A comparison is made between theoretical values and numerical data values.

Keywords: nonlinear system, singular perturbations, boundary layer, interior layer, contrast structure.

1. Consider singularly perturbed boundary value problem ($\varepsilon > 0$ is a small parameter)

$$\begin{aligned} \varepsilon^2 u'' = F(u,x), \quad 0 < x < 1; \\ u(0,\varepsilon) = u^0, \quad u(1,\varepsilon) = u^1. \end{aligned} \tag{1}$$

1^0 If the degenerated equation $F(u,x) = 0$ has a solution

$$u = \varphi(x) \text{ and } F_u(\varphi(x),x) > 0 \, (0 \le x \le 1),$$

then the solution $u(x,\varepsilon)$ of the problem (1) with two boundary layers: near $x = 0$ and near $x = 1$ may exist (Vasil'eva, *et al.*, 1995) , (Butuzov and Vasilieva, 1997) .

2^0 If the degenerated equation $F(u,x) = 0$ has three solutions:

$$\begin{aligned} &\varphi_1(x) < \varphi_2(x) < \varphi_3(x); \\ &F_u(\varphi_i,x) > 0 \, (i = 1,3), \\ &F_u(\varphi_2(x),x) < 0, \end{aligned}$$

then the solution $u(x,\varepsilon)$, that has not only boundary layers, but one or more interior layers, may exist (Vasil'eva, *et al.*, 1995) , (Butuzov and Vasilieva, 1997) . Solutions with interior layers are called *contrast structures.* Consider the contrast structure of *step-type*.

Consider now the partial equation

$$\begin{aligned} &\varepsilon^2(u_{xx} - u_t) = F(u,x,t), \\ &0 < x < 1, \ -\infty < t < \infty; \\ &u(0,t,\varepsilon) = u^0(t), \quad u(1,t,\varepsilon) = u^1(t). \end{aligned} \tag{2}$$

*This work is supported by Grant $N^{\underline{o}}$ 96-01-00694 from the Russian Foundation for Fundamental Researches.

If $F(u,x,t)$ is periodic with respect to t, then the periodic solution may exist. There are two possibility: a) the graph of of step-type remains as the step; b) the graph of step-type turns into the graph without step (with boundary layers only) and backwards. Authors will call the contrast structure in the case b) as the *contrast structure of alternating type*.

The asymptotic investigation of such contrast structures is very difficult problem. It may be considered for the present only by stages with respect to t. Nevertheless authors can obtain certain results by combination of asymptotic and numerical methods.

2. First consider the contrast structure in the case a). Let the following conditions be fulfilled:

I. The function $F(u,x,t)$ is periodic with respect to t and continuous with its first derivatives.

II. The degenerated equation $F(u,x,t) = 0$ has three roots:

$$u = \varphi_i(x,t), \quad (\varphi_1(x,t) < \varphi_2(x,t) < \varphi_3(x,t)),$$

so that $F_u(\varphi_i,x,t) > 0 \, (i = 1,3), F_u(\varphi_2,x,t) < 0$ for all $0 \le x \le 1$ and $-\infty < t < +\infty$.

III. The equation

$$I(x,t) = \int_{\varphi_1(x,t)}^{\varphi_3(x,t)} F(u,x,t)\,du = 0$$

has the solution $x = x_0(t)$, where $\delta < x_0(t) < 1-\delta$ (δ is arbitrary small but doesn't depend on ε).

IV. $\quad I'_x(x_0(t), t) < 0$.

V. In the phase plane of the equation

$$\frac{d^2u}{d\tau^2} = F(u, 0, t)$$

the straight line $u = 0$ for any t intersects at the point $M(\tau = 0)$ the separatrix entering the saddle point $\left(\varphi_1(0,t), 0\right)$ as $\tau \to \infty$, so that $\frac{du}{d\tau} \neq 0$ for $0 < \tau < \infty$. In the phase plane of the equation

$$\frac{d^2u}{d\tau^2} = F(u, 1, t)$$

the straight line $u = 0$ for any t intersects at the point $N(\tau = 0)$ the separatrix entering the saddle point $\left(\varphi_3(1,t), 0\right)$ as $\tau \to -\infty$, so that $\frac{du}{d\tau} \neq 0$ for $0 > \tau > -\infty$.

Then authors can prove the existence of periodic solution $u(x, t, \varepsilon)$ of the problem (2) of step–type, that remains as a step by changing of t, and the limit relation

$$\lim_{\varepsilon \to 0} u(x, t, \varepsilon) = \begin{cases} \varphi_1(x,t), \ 0 < x < x_0(t) \\ \varphi_3(x,t), \ x_0(t) < x < 1 \end{cases} \tag{3}$$

is valid.

The value $x_0(t)$ (obtained from III) is the main term of the transition point (of its x–coordinate). The transition point is defined as the cross point of the solution $u(x, t, \varepsilon)$ with the curve $u = \varphi_2(x, t)$.

The proof can be realised by differential inequalities method (as in (Nefedov, 1995) ,(Vasil'eva and Nikitin, 1996)).

Authors can also prove the existence of the solution with transition from φ_3 to φ_1.

3. Consider the problem

$$\begin{aligned} &\varepsilon^2 u_{xx} = F(u, x, t), \quad 0 < x < 1, \\ &-\infty < t < +\infty; \\ &u(0, t, \varepsilon) = 0, \ u(1, t, \varepsilon) = 0. \end{aligned} \tag{4}$$

(in comparison with (2) the u_t in the left–hand side is absent), that authors will call the *quasistationary* problem.

Under conditions I – V authors can prove for the problem (4) the same existence theorem and the same limit relation (3). So the main relation (3) for the problems (2) and (4) is the same.

Formulate also two more statements about the problem (4). Introduce the conditions:

V′. It is the condition V containing $\varphi_1(1, t)$ instead of $\varphi_3(1, t)$.

V″. It is the condition V containing $\varphi_3(0, t)$ instead of $\varphi_1(0, t)$.

By the condition V′ or V″ there exists the solution of the problem (4) with two boundary layers (pure boundary layer solution, without interior layers). This solution is stable in the sence of Lyapunov (Fife, 1973).

The step–type solution of the problem (4) with the transition from φ_1 to φ_3 is stable in the sence of Lyapunov. The solution with the transition from φ_3 to φ_1 is stable if in the condition IV the sign $<$ is changed to the sign $>$ (Vasil'eva, 1991).

4. Now will be given the results of some numerical experiments. Consider one example of the problem (2):

$$\begin{aligned} &\varepsilon^2(u_{xx} - u_t) = \\ &= (u^2 - 1)(u - 0.5x + 0.73 \sin t + 0.25), \\ &0 < x < 1, \quad 0 < t; \\ &u(0, t, \varepsilon) = u(1, t, \varepsilon) = 0, \\ &u(x, 0, \varepsilon) = \sin 2\pi x; \quad \varepsilon^2 = 10^{-3}. \end{aligned} \tag{5}$$

The main positions are:

1): $t = 0$, the initial function $u(x, 0, \varepsilon) = \sin 2\pi x$;

2), 3): the step moves to the right boundary point $x = 1$;

3): The lower half–wave vanishes and the pure upper boundary layer solution appears;

4)–8): the form of the solution remains;

9): the lower half–wave appears;

9), 10), 11): the step moves to the left boundary point $x = 0$;

11): the upper half–wave vanishes and the pure lower boundary layer solution appears;

12)–14): the form of the solution remains;

15): the upper half–wave appears;

16) : the step moves to the right boundary point $x = 1$;

and so on.

The stages 9)—11) we call the "run". The stages 4)—8) authors call the "upper halt", the stages 12)—14) - the"lower halt".

As the results of computation is obtained:

$$3)\ t_3^c = 0.3\,;\quad 9)\ t_9^c = 3.4\,; \qquad\qquad (6)$$
$$11)\ t_{11}^c = 3.6\,;\quad 15)\ t_{15}^c = 6.7\,;$$

(upper index means: obtained by means of computer).

Let us try to obtain these data analitically considering this process as quasistationary process. The corresponding t_i authors denote by t_i^a (upper index means: obtained analitically). The stages 1)—3) authors can consider as in the section 2. Here $x_0(t) = 2 \cdot (0.73 \sin t + 0.25)$. The theory of the section 2 is not valid when $x_0(t) = 1$. From this equation authors obtain $t_3^a = 0.34$. Now begins the upper halt. This situation exists as long as the pure boundary layer solution near $\varphi_3 = 1$ exists. According to the section 3 such solution exists while the separatrices connecting $u = 0$ with $\varphi_3(0, t)$ and $u = 0$ with $\varphi_3(1, t)$ exist.

In the phase plane by $x = 1$ there is a series of pictures:

$a)\ \varphi_2 > 3/8\,,\quad b)\ \varphi_2 = 3/8\,,\quad c)\ 3/8 > \varphi_2 > 0\,,$
$d)\ \varphi_2 = 0\,,\quad e)\ \varphi_2 < 0\,.$

In the cases $e)$, $d)$, $c)$, $b)$ there exists the separatrix connecting $u = 0$ with $u = 1$. When $\varphi_2 > 3/8$ the connecting separatrix does not exist.

By $t = t_3^a = 0.34$ authors have the case d). Then by increasing of t we have c) and at last the case b). The corresponding value of t can be find from the equation $0.25 - 0.73 \sin t = 3/8$, and authors have $t = t_9^a = 3.31$. In the interval $(t_3^a, t_9^a) = (0.34, 3.31)$ the connecying separatrix at $x = 0$ also exists. So the upper halt exists in the interval $(t_3^a, t_9^a) = (0.34, 3.31)$. Then, by $t > t_9^a$ the pure boundary solution near $u = 1$ vanishes and begins the run as long as $x_0(t) > 0$. By $x_0(t) = 0$ begins the lower halt. The corresponding $t = t_{11}^a = 3.48 = t_3^a + \pi$ is calculated from the equation $\varphi_3(0, t) = -0.73 \sin t - 0.25 = 0$. Then authors calculate t_{15}^a from the equation $\varphi_2(0, t) = -0.73 \sin t - 0.25 = -3/8$ and authors obtain $t_{15}^a = 6.45 = t_9^a + \pi$. And so on. The duration of the run is $t_{11}^a - t_9^a = 0.17$. The duration of the halt is $t_{15}^a - t_{11}^a = 2.97$ $(0.17 + 2.97 = \pi)$.

The theoretical values

$$t_3^a = 0.34\,,\qquad t_9^a = 3.31\,, \qquad\qquad (7)$$
$$t_{11}^a = 3.48\,,\qquad t_{15}^a = 6.45\,,$$

agree well with values obtained by means of computer (6). The theoretical duration of the run $(t_{11}^a - t_9^a = 0.17)$ is less than the numerical $(t_{11}^c - t_9^c = 0.2)$ because authors obtain this duration according to the theory of section 2 from t_9^a (where $x_0 = 0.25$) to t_{11}^a (where $x_0 = 0$). Authors haven't taken into account the time interval when the solution is attracted from the step moving to left. Authors tell below (in the section 5) how we can estimate the duration of the attraction.

5. In this section authors inwestigate more carefully the stage of run.

Integrate the equation

$$\varepsilon^2 (u_{xx} - u_t) = (u^2 - 1)(u - \varphi(x, t)) \qquad (8)$$

over x on $(0, 1)$:

$$\varepsilon^2 u_x|_{x=1} - \varepsilon^2 u_x|_{x=0} - \varepsilon^2 \int_0^1 \frac{\partial u}{\partial t}\, dx = \qquad (9)$$
$$= \int_0^1 (u^2 - 1)(u - \varphi(x, t))\, dx\,.$$

Let $r \in (0, 1)$ be a value of x at which $u(x, t, \varepsilon) = 0$ (i.e. $r(t, \varepsilon)$ can be considered as a coordinate of the wave). Suppose that the solution is close to automodelling and

$$\int_0^1 \frac{\partial u}{\partial t}\, dx = \frac{\partial}{\partial t} \int_0^1 u\, dx = 2 \frac{dr}{dt} \qquad (10)$$

(the variation of the area under the graph of the solution = variation of the coordinate of the wave multiplied by height of the wave).

Consider the integral in the right-hand side of (9):

$$\int_0^1 (u^2 - 1)(u - \varphi(x, t))\, dx = \int_0^{r/2} + \int_{r/2}^{(r+1)/2} + \int_{(r+1)/2}^1 .$$

For $x \in (0, r/2)$ during the run $\varepsilon^2 u_{xx} = (u^2 - 1)(u - \varphi(x, t))$. Therefore

$$\int_0^{r/2} (u^2 - 1)(u - \varphi(x, t))\, dx = \qquad (11)$$
$$= \varepsilon^2 \int_0^{r/2} u_{xx}\, dx = -\varepsilon^2 u_x|_{x=0} + Exp(\varepsilon).$$

(here and below $Exp(\varepsilon)$ denotes an exponentially small value when $\varepsilon \to 0$).

Analogously

$$\int_{(r+1)/2}^{1} (u^2-1)(u-\varphi(x,t))\,dx = \tag{12}$$
$$= \varepsilon^2 \int_{0}^{r/2} u_{xx}\,dx = \varepsilon^2 u_x|_{x=1} + Exp(\varepsilon)\,.$$

So authors have from (9)—(12)

$$-2\varepsilon^2 \frac{dr}{dt} = \tag{13}$$
$$= \int_{r/2}^{(r+1)/2} (u^2-1)(u-\varphi(x,t))\,dx + Exp(\varepsilon)\,.$$

Suppose that the wave looks like a typical solution of the singularly perturbed problem with the cubic nonlinearity :

$$u(x,t,\varepsilon) = \frac{1-\exp(\sqrt{2}(x-r)/\varepsilon)}{1+\exp(\sqrt{2}(x-r)/\varepsilon)}\,,$$

Then $\int_{r/2}^{(r+1)/2} (u^2-1)u\,dx = Exp(\varepsilon)$ and

$$\int_{r/2}^{(r+1)/2} (u^2-1)\varphi(x,t)\,dx = -2\sqrt{2}\varphi(r,t) + O(\varepsilon)\,.$$

So authors have from (13):

$$\varepsilon\frac{dr}{dt} = -\sqrt{2}\varphi(r,t) + O(\varepsilon)\,.$$

For equation (5) $\varphi(r,t) = 0.5r - 0.73\sin t - 0.25$. The initial condition: $r(3.31,\varepsilon) = 1$ (the start of the run). Then

$$r(t,\varepsilon) = 1.46\sin t + \frac{1}{2} + \frac{3}{4}\exp(\frac{3.31-t}{\varepsilon\sqrt{2}}) + O(\varepsilon)\,,$$

where $t \geq 3.31$. The expression $1.46\sin t + 1/2$ describes the movement of the point $x_0(t)$; the term $\frac{3}{4}\exp(\frac{3.31-t}{\varepsilon\sqrt{2}})$ describes the "falling" of the wave on this point. Duration of the run 0.19 calculated by this formula agrees closely with computational data (see in the section 4).

It is easy to verify that the function

$$\tilde{u}(x,t,\varepsilon) = \frac{1-\exp(\sqrt{2}(x-\tilde{r})/\varepsilon)}{1+\exp(\sqrt{2}(x-\tilde{r})/\varepsilon)}\,,$$

where $\tilde{r}$ is obtained from the equation

$$\varepsilon\frac{d\tilde{r}}{dt} = -\sqrt{2}\varphi(\tilde{r},t)\,,$$

satisfies the equation (8) with discrepancy $O(\varepsilon)$.

REFERENCES

Butuzov V.F. and Vasilieva A.B. (1997) Singularly Perturbed Problems with Boundary and Interior Layers: Theory and Applications. *Advances in Chemical Physics Series*, V.47, p.47–179. John Wiley and Sons, New York.

Fife P.C. (1973) Semilinear elliptic boundary value problem with small parameters. *Arch. Rat. Mech. and Anal.*, V.52, № 3, p.205–232.

Nefedov N.N. (1995) The Differential Inequality Method for Some Classes of Nonlinear Singularly Perturbed Problems with Interior Layers. *Differential equations* V.31, № 7, p.1132–1139 (in Russian).

Vasil'eva A.B. (1991) On the Stability of Contrast Structures. *Mathematical modelling*, V.3, № 4, p.114–123 (in Russian).

Vasil'eva A.B., Butuzov V.F. and Kalachev L.V. (1995), *The Boundary Function Method for Singular Perturbation Problems*, p.1—221, SIAM, Philadelphia.

Vasil'eva A.B. and Nikitin A.G. (1996) On the Stability of Periodic Contrast Structures in Two-dimensional Case with Respect to Spatial Variables.*Differential equations*, V.32, № 10, p.1355–1361 (in Russian).

PROBLEMS OF CONVERGENCE IN THE SECOND-ORDER METHODS OF CONTROL IMPROVEMENT

Vladimir A. Baturin, Dmitry E.Urbanovich

Irkutsk Computing Centre of Siberian
Branch of the Russian Academy of Sciences
134, Lermontov Str., Irkutsk, 664033, Russia
e-mail : rozen@icc.ru, tel:(395-2) 31-15-07

Abstract. The new second-order improving methods for optimal control problem, which based on Krotov's sufficiently conditions of optimality are under consideration. The algorithm of methods is presented. The relaxation and convergence problems are investigated.

Keywords. Optimal control, algorithms, matrix Riccati equation, convergence of numerical methods, convergence proofs, tests.

1. INTRODUCTION

This paper is aimed at investigating the properties of some of methods for improvement of optimal control problems. These methods are based on V.Krotov,s sufficient conditions of optimality (Krotov, 1996). Now, computational algorithms, which are based on necessary conditions of optimality are widely known (see, e. g. (Srochko, 1995; Vasiliev, 1996)). In the present case constructing these methods is based on the 1-st or 2-nd order approximation of the constructions of sufficient conditions of optimality, where the Krotov function is chosen from the class of linear or linear-quadratic functions (Gurman, *et al* 1983).

2. STATEMENT OF A PROBLEM

We consider the optimal control problem in following form. Consider the controlled system

$$\frac{dx}{dt} = f(t,x,u) \tag{1}$$

on the time interval $T = [t_0, t_1]$ with the initial condition $x(t_0) = x_0$ and control constraints $u(t) \in U$, where $x(t) \in R^n$ is continuous and piecewise differentiable.

The set of pairs $(x(t), u(t))$ satisfying the conditions above is called admissible and is denoted by D. Suppose that $D \neq \emptyset$. Let the functional

$$I(x,u) = \int_{t_0}^{t_1} f^0(t,x,u)dt + F(x(t_1))$$

be given. The problem is to find a sequence $\{x_s(t),\ u_s(t)\} \in D$ such that

$$I(x_s, u_s) \to \inf_D I.$$

Traditional algorithms of successive improvements limit either the state increment Δx or the control increment Δu. Corresponding algorithms are usually called the methods of strong and weak improvement. For example, gradient-type methods are the methods of weak improvement.

Let $\varphi(t,x)$ be the function defined for each t and continuously differentiable on both the arguments. Introduce the following constructions:

$$R(t,x,u) = \varphi_x'(t,x)f(t,x,u) - f^0(t,x,u) + \varphi_t(t,x),$$

$$G(x(t_1)) = \varphi(t_1, x(t_1)) - \varphi(t_0, x(t_0)) + F(x(t_1)),$$

$$L(x,u) = G(x(t_1)) - \int_{t_0}^{t_1} R(t,x,u)dt.$$

Let p be an n-dimensional vector, then

$$H(t,x,u,p) = \varphi_x'(t,x)f(t,x,u) - f^0(t,x,u),$$

$$\mathcal{H}(t,x,p) = \sup_{u\in U} H(t,x,u,p).$$

It can readily be seen, if (x(t),u(t)) is an admissible process then $L(x,u) = I(x,u)$.

3. METHODS OF IMPROVEMENT

Let $(x^I(t), u^I(t)) \in D$ be an initial approximation. It is necessary to find new process $(x^I I(t), u^I I(t)) \in D$ such that the value of the functional will be diminish.

Introduce a new functional

$$I_\alpha(x,u) = \alpha I(x,u) + (1-\alpha)J(x^I, u^I, x, u),$$

where $\alpha \in [0,1], J(x^I, u^I, x, u)$ is a positive definite functional. The idea of constructing of these algorithms is that investigation of the improvment problem for the initial functional I is reduced to investigation of the improvment problem for the new functional I_α. For this problem suppose that functional J is defined as:

$$J(x^I, u^I, x, u) = \frac{1}{2}\left(\int_{t_0}^{t_1} |\Delta x|^2 dt + |\Delta x(t_1)|^2\right),$$

$\Delta x = x(t) - x^I(t)$.

Note, if

$$\Delta I_\alpha(x,u) = I_\alpha(x,u) - I_\alpha(x^I, u^I) =$$

$$= \int_{t_0}^{t_1} o(|\Delta x|^2)dt + o(|\Delta x_1|^2),$$

then for sufficiently small $\Delta x \neq 0$, the value of $\Delta I_\alpha(x,u) < 0$ and improvment problem is solved.

Introduce the functional $L^\alpha(x,u)$ for the functional I_α likewise for the functional I. In this case, some of the functions assume the form of

$$R^\alpha(t,x,u) = \varphi_x'(t,x)f(t,x,u) - \alpha f^0(t,x,u) + \\ +\varphi_t(t,x),$$

$$G^\alpha(x(t_1)) = \varphi(t_1, x(t_1)) - \varphi(t_0, x(t_0)) + \alpha F(x(t_1)),$$

$$L^\alpha(x,u) = G^\alpha(x(t_1)) - \int_{t_0}^{t_1} R^\alpha(t,x,u)dt,$$

$$H^\alpha(t,x,u,p) = \varphi_x'(t,x)f(t,x,u) - \alpha f^0(t,x,u),$$

$$\mathcal{H}^\alpha(t,x,p) = \sup_{u\in U} H^\alpha(t,x,u,p).$$

Consider the increment of the functional L_α and its linear-quadratic approximation. The function $\varphi(t,x)$ is chosen from the class of linear or linear-quadratic functions and defined in the way to fulfill the following estimation

$$\Delta L^\alpha(x,u) = I_\alpha(x,u) - I_\alpha(x^I, u^I) =$$

$$= \int_{t_0}^{t_1} o(|\Delta x|^2)dt + o(|\Delta x_1|^2). \quad (2)$$

The conditions, which define this estimation constitutes the main constructions of the algorithm.

Let

$$\varphi(t,x) = \psi(t)'\Delta x + \frac{1}{2}\Delta x'\sigma(t)\Delta x,$$

where $\psi(t)$ is an n-dimensional vector-function, $\sigma(t)$ is an $n \times n$ symmetric matrix-function.

In this manner the algorithm of the strong improvement method assumes the following form.

Algorithm.

1. Take a control $u = u^I(t)$, integrate "from the left to the right" the system $\dot{x} = f(t,x,u^I(t))$, $x(t_0) = x_0$, and find $x^I(t)$.

2. Take a parameter $\alpha \in (0,1]$ of the algorithm.

3. Integrate "from the right to the left" the system

$$\dot{\psi} = -\mathcal{H}_x^\alpha - \sigma(\mathcal{H}_p^\alpha - H_p^\alpha), \quad (3)$$

$$\dot{\sigma} = -\mathcal{H}_{xx}^\alpha - \sigma\mathcal{H}_{px}^\alpha - \mathcal{H}_{xp}^\alpha\sigma - \\ -\sigma\mathcal{H}_{pp}^\alpha\sigma + (1-\alpha)E, \quad (4)$$

$$\psi(t_1) = -\alpha F_x(x^I(t_1)), \quad (5)$$

$$\sigma(t_1) = -\alpha F_{xx}(x^I(t_1)) - (1-\alpha)E. \quad (6)$$

4. Integrate the system

$$\dot{x} = f(t,x,\tilde{u}(t,x,p)), x(t_0) = x_0,$$

$\tilde{u}(t,x,p) = \text{argmax} H^\alpha(t,x,p,u), u \in U, p = \psi + \sigma\Delta x$, and find $x^{II}(t)$ and

$$u^{II}(t) = \tilde{u}(t, x^{II}(t), \psi(t) + \sigma(t)(x^{II}(t) - x^I(t))).$$

5. If $I(x^{II}, u^{II}) \geq I(x^I, u^I)$, then the parameter α is diminished and the process is repeated starting from step 3.

Similarly it is possible to obtain the weak improvement method using the auxiliary functional J of the form:

$$J(x^I, u^I, x, u) = \frac{1}{2}\left(\int_{t_0}^{t_1} |\Delta u|^2 dt + |\Delta x(t_1)|^2\right),$$

$\Delta u = u(t) - u^I(t)$.

Suppose that in the optimal control problem $U = R^m$.

Algorithm.

1. Take a control $u = u^I(t)$, integrate "from the left to the right" the system $\dot{x} = f(t,x,u^I(t))$, $x(t_0) = x_0$, and find $x^I(t)$.

2. Take a parameter $\alpha \in (0,1]$ of the algorithm.

3. Integrate "from the right to the left" the system

$$\dot{\psi} = -H_x^\alpha - (H_{xu}^\alpha + \sigma H_{\psi u}^\alpha)(H_{uu}^\alpha - (1-\alpha)E)^{-1} H_u^\alpha,$$

$$\dot{\sigma} = -H_{xx}^\alpha - \sigma H_{\psi x}^\alpha - H_{x\psi}^\alpha \sigma + (H_{xu}^\alpha + \sigma H_{\psi u}^\alpha)(H_{uu}^\alpha - -(1-\alpha)E)^{-1}(H_{xu}^\alpha + \sigma H_{\psi u}^\alpha)',$$

$$\psi(t_1) = -\alpha F_x(x^I(t_1)),$$

$$\sigma(t_1) = -\alpha F_{xx}(x^I(t_1)) - (1-\alpha)E.$$

4. Integrate the system $\dot{x} = f(t,x,\bar{u}), x(t_0) = x_0$, for $\bar{u} = u^I(t) + \Delta u(t, \Delta x)$, where $\Delta u(t, \Delta x)$ is defined by

$$\Delta u(t, \Delta x) = -(H_{uu}^\alpha - -(1-\alpha)E)^{-1}[H_u^\alpha + (\sigma H_{\psi u}^\alpha + H_{xu}^\alpha)\Delta x].$$

Define $x^{II}(t)$ and $u^{II}(t) = u^I(t) + \Delta u(t, x^{II}(t) - x^I(t))$.

5. If $I(x^{II}, u^{II}) \geq I(x^I, u^I)$ then the parameter α is diminished and the process is repeated starting from step 3.

In the 2-nd order improvement procedure , the most complicated process is integrating of an auxiliary vector-matrix system for ψ, σ. Therefore, consider the constructions of algorithms in the case when the function $\varphi(t,x)$ of linear form is sought, i. e. $\sigma(t) = 0$.

For the first order method, the auxiliary system is transformed to the following one:

$$\dot{\psi} = -H_x^\alpha - H_{xu}^\alpha (H_{uu}^\alpha - (1-\alpha)E)^{-1} H_u^\alpha,$$

$$\psi(t_1) = -\alpha F_x(x^I(t_1)),$$

and the formula for calculating of the control increment is:

$$\Delta u(t, \Delta x) = -(H_{uu}^\alpha - (1-\alpha)E)^{-1}[H_u^\alpha + H_{xu}^\alpha \Delta x].$$

The new control and trajectory are found by integrating the original system:

$$\dot{x} = f(t, x, u^I + \Delta u(t, x - x^I(t))), x(t_0) = x_0.$$

4. RELAXATION AND CONVERGENCE OF ALGORITHMS

Suppose that following conditions are satisfied:

(1^0) the function $\mathcal{H}(t,x,p)$ is continuous and twice continuously-differentiable with respect to x and p;

(2^0) There exists a function $\tilde{u}(t,x,p)$ continuous and twice differentiable with respect to p such that $H(t,x,p,\tilde{u}(t,x,p)) = \mathcal{H}(t,x,p)$.

To obtain the estimation of increment of the functional I execuye the following transformations:

$$\alpha I(x^{II}, u^{II}) = G^\alpha(x^{II}(t_1)) - \int_{t_0}^{t_1} \max_{u \in U} R^\alpha(t, x^{II}(t), u) dt,$$

$$\alpha I(x^I, u^I) = G^\alpha(x^I(t_1)) - \int_{t_0}^{t_1} R^\alpha(t, x^I(t), u^I(t)) dt.$$

Taking into account, that

$$R^\alpha(t, x^I(t), u^I(t)) \leq \max_{u \in U} R^\alpha(t, x^I(t), u) dt. \quad (7)$$

$$\alpha I(x^I, u^I) \geq G^\alpha(x^I(t_1)) - \int_{t_0}^{t_1} \max_{u \in U} R^\alpha(t, x^I(t), u) dt. \quad (8)$$

Substitution of the inequality (7-8) into equation (2) gives:

$$\alpha I(x^{II}, u^{II}) - \alpha I(x^I, u^I) \leq \Delta G^\alpha(x(t_1)) - \int_{t_0}^{t_1} \Delta \max_{u \in U} R^\alpha(t, x^{II}(t), u) dt.$$

As a result decomposing the increment of the functions G^α and R^α in Taylor series and taking into account the form of the auxiliary system for strong improvement method (3-6), the latter inequality transformes to the form:

$$\alpha I(x^{II}, u^{II}) - \alpha I(x^I, u^I) \leq -\frac{(1-\alpha)}{2} \left(\int_{t_0}^{t_1} |\Delta x|^2 dt + |\Delta x(t_1)|^2\right) + \left(\int_{t_0}^{t_1} o(|\Delta x|^2) dt + o(|\Delta x(t_1)|^2)\right). \quad (9)$$

Hence, if a function $F(x)$ is continuous and twice differentiable; the functions $\mathcal{H}, H$ and $\tilde{u}(t,x,p)$ satisfy the conditions (1^0)-(2^0); if $(x^I(t), u^I(t))$ is not Pontryagin's extremum, then the algorithm determines new $u^{II}(t)$ and corresponding trajectory $x^{II}(t)$ such that $I(x^{II}, u^{II}) < I(x^I, u^I)$.

The proof of this result contains in (Gurman, *et al* 1983).

Consider the convergence problem for this method. Let $\{x^k, u^k\}$ be a sequence of controls and trajectory generated by the algorithm.

Take a second look at the estimate (9). Let the functional I be lower bounded by the value I^*. Since $I(x^k, u^k)$ is monotonically decreasing and bounded then

$$\lim_{k \to \infty} (I(x^k, u^k) - I(x^{k+1}, u^{k+1})) = 0.$$

Then the estimate (9) implies:

$$\int_{t_0}^{t_1} | x^{k+1} - x^k |^2 dt \to 0,$$

$$| x^{k+1}(t_1) - x^k(t_1) |^2 \to 0,$$

when $k \to \infty$. Consequently

$$\lim_{k\to\infty} \int_{t_0}^{t_1} | x^{k+1} - x^k |^2 dt = 0,$$

$$\lim_{k\to\infty} | x^{k+1}(t_1) - x^k(t_1) |^2 = 0.$$

Theorem 1. Let a vector-function $f(t, x, u)$ be continuous with recpect to the family of arguments; the function $\mathcal{H}$ be twice continuously differentiable with recpect to x and p; the functions F and H be twice continuously differentiable with recpect to x; the functional be lower bounded. Then the sequence $\{x^k, u^k\}$ generated by the algorithm converges to the realization of maxumum principle in the sense of

$$\lim_{k\to\infty} \int_{t_0}^{t_1} (\max_{u\in U} H(t, x^k, u, \psi) -$$

$$-H(t, x^k, u^k, \psi))dt = 0.$$

Proof. Consider an increment of functional I

$$I(x^k, u^k) - I(x^{k+1}, u^{k+1}) = G(x^k(t_1)) -$$

$$-G(x^{k+1}(t_1)) - \int_{t_0}^{t_1} (R(t, x^k(t), u^k(t)) -$$

$$-R(t, x^{k+1}(t), u^{k+1}(t)))dt.$$

Since

$$\lim_{k\to\infty} (I(x^k, u^k) - I(x^{k+1}, u^{k+1})) = 0.$$

it follows that $G(x^k(t_1)) - G(x^{k+1}(t_1)) \to 0$, and

$$\int_{t_0}^{t_1} (R(t, x^k(t), u^k(t)) -$$

$$-R(t, x^{k+1}(t), u^{k+1}(t)))dt \to 0.$$

The function R may be written using the function H, then decompose the increment of function R in Taylor series to obtain the following formula.

$$\int_{t_0}^{t_1} (R(t, x^k(t), u^k(t)) - R(t, x^{k+1}(t), u^{k+1}(t)))dt =$$

$$= \int_{t_0}^{t_1} (H(t, x^k(t), u^k(t), \psi(t)) -$$

$$-\mathcal{H}(t, x^k(t), \psi(t)))dt + \int_{t_0}^{t_1} (\mathcal{H}_x(t, x^k(t), \psi(t)) +$$

$$+\sigma(t)\mathcal{H}_\psi(t, x^k(t), \psi(t)))'(x^{k+1}(t_1) - x^k(t_1))dt +$$

$$+ \int_{t_0}^{t_1} (\varphi_t(t, x^k(t)) - \varphi_t(t, x^{k+1}(t)))dt.$$

The passage to the limit in latter formula yields:

$$\lim_{k\to\infty} \int_{t_0}^{t_1} (\max_{u\in U} H(t, x^k, u, \psi) -$$

$$-H(t, x^k, u^k, \psi))dt = 0. \text{ Q.E.D.}$$

Suppose that there exists a pair of functions $(x^*(t), u^*(t))$, satisfying initial control system (1) and

$$\lim_{k\to\infty} x^k(t) = x^*(t), \lim_{k\to\infty} u^k(t) = u^*(t).$$

Theorem 2. Let the conditions of Theorem 1 be satisfied and λ, λ_F be either negative or $\lambda \to C_1 < 0$, $\lambda_F \to C_2 < 0$ as $k \to \infty$, where λ, λ_F are maximal eigenvalues of matrixes $\mathcal{H}^k_{xx}$ and $-F^k_{xx}$, correspondingly.

Then the sequence $\{x^k, u^k\}$ generated by the algorithm converges to realization of strong local minimum conditions in the sense of:

1. $\lim_{k\to\infty} \int_{t_0}^{t_1} (\max_{u\in U} H(t, x^k, u, \psi) -$

$$-H(t, x^k, u^k, \psi))dt = 0.$$

2. Limit Riccati equation has a solution on $[t_0, t_1]$.

Proof. The first condition has been proved in Theorem 1. Non-negative definiteness of the matrices $-\mathcal{H}^\alpha_{xx} + (1-\alpha)E$ and $\alpha F_{xx} + (1-\alpha)E$ is sufficient for existance of the solution of Riccati equation in each iteration of the method. Since these matrices are real, symmetric then for each t there exist matrices $W(t)$ and Z such that

$$W(t)\mathcal{H}^\alpha_{xx}W(t)' = \Lambda^\alpha_1(t), \alpha Z F_{xx} Z' = \alpha\Lambda_2,$$

where $\Lambda^\alpha_1(t), \Lambda_2$ are diagonal Jordan matrices. Let $\lambda_1(t)$ is maximal eigenvalue of matrix $\mathcal{H}^\alpha_{xx}$. Then it is possible to choose the parameter α such that the matrices $-\Lambda^\alpha_1 + (1-\alpha)E$ and $-\mathcal{H}^\alpha_{xx} + (1-\alpha)E$ be positive definite. It guarantees that solution of Riccati equation exists in each k iteration. Since maximal eigenvalues tend to negative value (according to the condition of the Theorem) for $k \to \infty$ then there exists limit non-negative definite matrices $\overline{\Lambda}_1{}^\alpha, \overline{\Lambda}_2$ and $\alpha = 1$. Consequently, there exists the solution of the limit Riccati equation. This fact in combination with conditions 1 and 2 of Theorem 2 corresponds to the conditions of strong local minimum. Q.E.D.

Consider the following case. Let for $\alpha = 1, u^I(t)$ satisfy the Maximum Principle. Then the system (3) - (6) is decomposed into two. The relations (4) and (6) are the matrix Riccati equation with the initial condition. If it has a solution on whole segment $[t_0, t_1]$ then $u^I(t)$ is not improved by our algorithm. Consider the cases when the solution exists on $(t^*, t_1]$, $t^* > t_0$. The latter fact means that $\lim_{t\to t^*} \sigma(t) = \infty$. Such point t^* is called singular.

In this case, the improvement algorithm is:

1. A parameter α is chosen that the singular point coincides with t_0.

2. Find a symmetric submatrix σ^I of the matrix σ with the property $\lim_{t\to t_0} \det(\sigma^I(t)) = \infty$.

3. Put $\chi = \lim_{t\to t_0} (\sigma^I)^{-1}$.

4. Solve the system $\chi b = 0, | b |= \epsilon > 0$.

5. Integrate the system:

$$\dot{x} = f(t, x, u(t, x, \psi(t) + \sigma(x - x^I(t))), t \in (t_0, t_1],$$

$$\dot{x}(t_0) = f(t_0, x_0, u(t, x, \psi + (b', 0)')), x(t_0) = x_0,$$

and find $x^{II}(t)$ and

$$u^{II}(t) = u(t, x, \psi + \sigma(x^{II}(t) - x^I(t))).$$

6. If $I(x^{II}, u^{II}) \geq I(x^I, u^I)$ then ϵ deminishes and the process starting from step 4.

In this case algorithm can improve the initial contol too.

5. A NUMERICAL EXAMPLE.

Example 1.
Consider the following optimal control problem:

$$\dot{x} = u, x(0) = 0,$$

$$I = \int_0^\pi (u^2 - x^2)dt.$$

As the initial approximation, the control $u^I(t) = 0$ is chosen. The corresponding trajectory is $x^I(t) = 0$. This pair satisfies the Maximum Principle, the stationary condition ($H_u = 0$) and the strong Legendre-Klebsh condition ($H_{uu} < 0$). Consider the algorithm of strong improvement. In the case of weak improvement the result is similar. We have

$$\mathcal{H}^\alpha(t, x, p) = \frac{p^2}{4\alpha} + \alpha x^2; \tilde{u}(t, x, p) = \frac{p}{2\alpha}.$$

The equation for σ has the form:

$$\dot{\sigma} = 1 - 3\alpha - \frac{\sigma^2}{2\alpha}, \sigma(\pi) = 0.$$

Its solution is expressed by the formula:

$$\sigma(t) = \text{tg}(\frac{a(\pi - t)}{2\alpha}a), a = \sqrt{(3\alpha - 1)2\alpha}.$$

For $\alpha = 1$ at the point $t^* = \frac{\pi}{2}, \sigma \to \infty$. Let $\alpha = \frac{2}{5}$, then $\sigma(t) = \frac{2}{5}\text{ctg}\frac{t}{2}$ and $\sigma(t) \to \infty$ as $t \to 0$, that is by choosing α the singular point is moved to the initial point of the sequent. Furthermore, the improvement algorithm yields:

$$u(t, x) = \frac{x}{2}\text{ctg}\frac{t}{2}, x(t) = \epsilon \sin\frac{t}{2},$$

and the value of the functional is

$$I = -\frac{3\pi}{8}\epsilon^2.$$

Consequently, for any $\epsilon \neq 0, I < 0$.

Example 2.

$$\dot{x}_1 = (1 - x_2^2)x_1 - x_2 + u, \quad x_1(0) = 1.5,$$
$$\dot{x}_2 = x_1, \qquad x_2(0) = 1.5,$$
$$t \in [0, 5.], | u(t) |\leq 1.5,$$
$$I = \frac{1}{2}\int_0^5 (x_1^2 + x_2^2 + u^2)dt \to \min.$$

It is developed software for optimal control problem There are methods of strong and weak improvement, their modifications and some gradient-type methods.

The solution for this example, using software, is I=4.3734. This is better then in (Fukushima, 1986).

ACKNOWLEDGMENTS

This research supported by the Russian Fund of Fundamental Researches, Grant No 97-01-00047.

REFERENCES

Fukushima, M., Y.Yamamoto. (1986) A second-order algorithm for continuous-time nonlinear optimal control problems. *IEEE Transactions on Automatic Control*, **Volume 31, No. 7**, p.p. 673-676.

Gurman, V.I., V.A. Baturin, I.V. Rasina. (1983) *Improving methods of optimal control.* Irkutsk State University, Irkutsk.

Krotov, V.F. (1996) *Global methods in optimal control theory.* Marcel Dekker Inc., N.Y.

Srochko V.A. (1995) Methods of linear-quadratic approximations in optimal control problems *Optimization, Control, Intellect.* **Volume No 1**, p.p. 110-135.

Vasiliev O.V. (1996) *Optimization methods.* World Federation Publisher Company, USA, Florida.

SECOND-ORDER IMPROVEMENT METHODS FOR LINEAR UNLIMITED CONTROL PROBLEMS

Vladimir A. Baturin, Irina O. Verkhozina

*Irkutsk Computing Centre of
Siberian Branch of the Russian Academy of Sciences
134, Lermontov Str., Irkutsk, 664033, Russia
e-mail: rozen@icc.ru*

Abstract. Optimal control problem in which differential system is linear on control is considered. Such problem has some singularities. The method of improvement based on special transformations is suggested. Algorithms of the successive improvement using the second order method of improvement for regular problem are considered.

Keywords. Differential equations, discontinuous trajectories, optimization problems, optimality, singular control, transformations.

1. INTRODUCTION

Let the controlled system be given

$$\frac{dx}{dt} = g(t,x) + h(t,x)u, \tag{1}$$

$$t \in [t_0, t_1], \quad x(t) \in R^n, \quad u(t) \in R^k, \quad x(t_0) = x_0, \tag{2}$$

where $x(t)$ is a piecewise differential function, $u(t)$ is a piecewise continuous.

Minimize the functional

$$I = F(x(t_1)), \tag{3}$$

on a set D of pairs of functions $v = (x(t), u(t))$ satisfying the conditions above.

Let the control $u^I(t)$ and the corresponding trajectory $x^I(t)$ be given. By the improvement problem it means the problem of finding $(x^{II}(t), u^{II}(t))$ such as $I(x^{II}(t), u^{II}(t)) < I(x^I(t), u^I(t))$.

The problem $(1)-(3)$ has some singularities. Procedures based on maximum principle, on Bellman's dynamical programming method and the linearization ones cannot be used directly without preliminary regularization of the problem. As a rool, the optimal control in such systems has impulse character. Regularization of the problem gives a significant increase of iteration number; round-off errors are accumulated which do not make possible find the optimum point.

The problem $(1)-(3)$ has been considered in numberous works directed to qualitative research (Krotov, Gurman, 1969; Gurman, 1977), or to formulating of the local minimum necessary and sufficient conditions. There are not many papers in which the calculating algorithms are constructed (Gurman, 1983; Gurman, 1987). These papers use either the derivative problem method (Gurman, 1977) or the methods of integral transformation of control which is a nonlinear analogs of Goh transformations.

Suppose that

1. $F(x(t_1))$ is twice continuously differentiable;

2. $g(t,x)$, $h(t,x)$ are twice continuously differentiable in (t,x);

3. matrix h satisfyes the Frobenius condition, that is if $h_1, ..., h_k$ are its columns then for any $i,j = \overline{1,k}$ the equalities

$$h_{ix}(t,x)h_j(t,x) = h_{jx}(t,x)h_i(t,x) \tag{4}$$

hold.

2. EQUIVALENT REGULAR PROBLEM

Introduce the equation

$$\frac{dz}{dt} = u, \quad z(t_0) = 0 \tag{5}$$

in addition to the system (1) and consider a system of equations with partial derivatives

$$\frac{\partial \chi}{\partial z} = h(t, \chi), \quad \chi|_{z=z^I(t)} = y. \tag{6}$$

Existence of its local solution is guaranteed by suppositions 2-3; $z^I(t)$ corresponds to approximation $u^I(t)$ as a solution of the system (5).

Let $y = \eta(t, x, z)$ be a set of the independent first integrals of the system (6). Pass from variables (x, u) to (y, z) at the problem (1) - (3). Consider a total derivative of $\eta(t, x, z)$ in t at the system (1), (5).

Hence

$$\frac{d}{dt}y = \eta_x \frac{dx}{dt} + \eta_z \frac{dz}{dt} + \eta_t = \eta_x g + \eta_t + (\eta_x h + \eta_z)u.$$

Since $\eta(t, x, z)$ is the integral of (6), identity

$$\eta_x h + \eta_z = 0 \tag{7}$$

holds.

Thus, problem is reduced to

$$\frac{dy}{dt} = f(t, y, z), \; y(t_0) = x_0, \tag{8}$$

$$I = \mathcal{F}(y(t_1), s) \to inf, \tag{9}$$

where

$$f(t, y, z) = \eta_x(t, \chi(t, y, z))g(t, \chi(t, y, z)) + \eta_t(t, \chi(t, y, z)),$$

$$\mathcal{F}(y(t_1), s) = F(\chi(t_1, y(t_1), s)),$$

s is k-dimensional parameter, $y(t)$ is a piecewise differentiable function (new phase variable), $z(t)$ is piecewise continuous function (new control).

Equivalence between problems (1) - (3) and (8) - (9) in sense of coincidence of greatest lower bounds of functionals is showed in the works (Krotov, Gurman, 1969; Gurman, 1977).

Reduced problem (8), (9) is nonlinear in control, therefore the Maximum Principle and other conditions are already effective for this problem.

As a basic algorithm at the problem (8) - (9) the second order weak improvement method [3] be used. It consists of the following main steps.

Denote by $\mathcal{H}(t, y, \psi, z) = \psi' f(t, y, z)$, where ψ is n-dimensional vector. Let the control $z^I(t)$ and trajectory $y^I(t)$ be given.

1^0. Integrate auxiliary vector-matrix system of differential equations:

$$\dot{\psi} = -\mathcal{H}_y + (\mathcal{H}_{yz} + \sigma\mathcal{H}_{\psi z})(\mathcal{H}_{zz} - (1-\alpha)E)^{-1}\mathcal{H}_z, \tag{10}$$

$$\dot{\sigma} = -\mathcal{H}_{yy} - \mathcal{H}_{y\psi}\sigma - \sigma\mathcal{H}_{\psi y} + (\mathcal{H}_{yz} + \sigma\mathcal{H}_{\psi z}) \times (\mathcal{H}_{zz} - (1-\alpha)E)^{-1}(\mathcal{H}_{yz} + \sigma\mathcal{H}_{\psi z})', \tag{11}$$

$$\psi(t_1) = -\alpha\mathcal{F}_y + \alpha\mathcal{F}_{yz}(\mathcal{F}_{zz} + (1-\alpha)E)^{-1}\mathcal{F}_z, \tag{12}$$

$$\sigma(t_1) = -\alpha\mathcal{F}_{yy} + \alpha\mathcal{F}_{yz}(\mathcal{F}_{zz} + \frac{1-\alpha}{\alpha}E)^{-1}\mathcal{F}_{zy}, \tag{13}$$

where $\sigma(t)$ is $n \times n$ -symmetric matrix, derivatives of the function $\mathcal{H}$ are calculated at the point $(t, y^I(t), \psi, z^I(t))$, but derivatives of F - at the point $(y^I(t_1), z^I(t_1)), \alpha \in [0,1]$ is a parameter of the algorithm.

2^0. Solve the system (8) closed by control synthesis (or feedback control)

$$\tilde{z}(t, \Delta y) = z^I(t) + (\mathcal{H}_{zz} - (1-\alpha)E)^{-1}(\mathcal{H}_z + (\mathcal{H}_{yz} + \sigma\mathcal{H}_{\psi z})'\Delta y), \tag{14}$$

where $\Delta y = y - y^I(t)$. Hence find $y^{II}(t)$ and $z^{II}(t) = \tilde{z}(t, y^{II} - y^I(t)) + z^I(t)$.

The algorithm assume two cases.

1. The stationary condition is not hold for the initial control $z^I(t)$ with $\alpha = 1$. $\mathcal{H}_z(t, y^I(t), \psi(t), z^I(t)) \neq 0$. (the first case is considered).

2. The stationarity condition $\mathcal{H}_z(t, y^I(t), \psi(t), z^I(t)) = 0$ and the intensive Legendre-Klebtsh condition $\mathcal{H}_{zz}(t, y^I(t), \psi(t), z^I(t)) < 0$ are hold for the control $z^I(t)$ and for corresponding trajectory $y^I(t)$ with $\alpha = 1$. The matrix Riccati equation (11) contains a singular point $t^* \in (t_0, t_1]$, i.e. $\lim_{t \to t^*+0} \|\sigma(t)\| = \infty$.

Consider the second variant of algorithm:

1^0. Choose the parameter α such that the singular point is coincided with t_0.

2^0. Find symmetric submatrix $\sigma^I(t)$ of the matrix $\sigma(t)$, which has the property

$$\lim_{t \to t_0} |det(\sigma^I(t))| = \infty.$$

3^0. Find $\chi = \lim_{t \to t_0} (\sigma^I(t))^{-1}$.

4^0. Solve the system $\chi b = 0, \;\; |b| = \epsilon > 0$.

5^0. Integrate the system

$$\dot{y} = f(t, x, z^I(t) + \Delta\tilde{z}(t, \Delta x)),$$

$$y(t_0) = x_0, \Delta x = x - x^I(t),$$

where $\Delta\tilde{z}$ is calculated by formula (14) with $t \in (t_0, t_1]$, and with $t = t_0$:

$$\Delta\tilde{z} = -(\mathcal{H}^{\alpha}_{uu} - (1-\alpha)E)^{-1}\mathcal{H}_{u\psi}\begin{pmatrix} b \\ 0 \end{pmatrix}.$$

6^0. If $I(y^{II}, z^{II}) \geq I(y^I, z^I)$, then parameter ϵ should be decreased and the process is repeated starting from the 4^0.

One has succeeded not always in obtaining of explicite analytic expresion for the functions $\mathcal{F}$ and $\mathcal{H}$, because it requires to know explicite expression of first integral for $\eta(t,x,z)$. Notice that a complete information about the integrals $\eta(t,x,z)$ is not necessary for defining of coefficients of system (10) - (13), but it is sufficient to know their elements (derivatives along of the original trajectory).

3. REPRESENTATION OF BASIC CONSTRUCTIONS OF THE ALGORITHM IN TERMS OF THE ORIGINAL PROBLEM

Write some additional identities which are necessary for deciphering of the derivatives of functions $\mathcal{H}$ and $\mathcal{F}$.

By virtue of the identities $\eta(t,x,z) = y$, $\eta_x h + \eta_z = 0$ relations

$$\eta_{xx}h + \eta_x h_x + \eta_{zx} = 0, \eta_{xt}h + \eta_x h_t + \eta_{zt} = 0,$$

$$\eta_x \chi_y = E, \eta_t + \eta_x \chi_t = 0$$

hold.

Along with the original approximation put $\eta_x = E$.

The function $H(t,x,\psi,u) = \psi' g(t,x) + \psi' h(t,x)u$ is considered for the system (1). Write the derivatived $\mathcal{H}$:

$$\mathcal{H}_{z_i} = (\psi' f(t,y,z))_{z_i} = \psi'(\eta_x g + \eta_t)_{z_i} =$$
$$= \psi'(\eta'_{xx_i}\frac{\partial x}{\partial z_i}g + \eta'_{xz_i}g + [\eta_x g_x \frac{\partial x}{\partial z_i}]' + \eta_{tz_i} + \eta_{tx}\frac{\partial x}{\partial z_i}).$$

Then, with regard to $\frac{\partial x}{\partial z} = h(t,x)$ and the auxiliary identities

$$\mathcal{H}_{z_i} = \psi'((-\eta'_x)gh_x - (-\eta'_x)g_x h_i + (-\eta_x)h_t) =$$
$$= -((\psi' g)_\psi H_{u_i x} - (\psi' g)_x H_{u_i\psi} + H_{u_i t}) =$$
$$= -(-H_{u_i\psi}H_x + H_{u_i x}H_\psi + H_{u_i t}) - (H_{u_i x}H_{\psi u_i} -$$
$$-H_{u_i\psi}H_{xu_i})u = -\frac{d}{dt}H_{u_i} = -\dot{H}_{u_i},$$

where $\frac{d}{dt}H_u = \dot{H}_u$ is the total derivative in virtue of usual derivatives for x and ψ, that is when $H_x = -\dot{\psi}, H_\psi = \dot{x}$. The relation $(H_{u_i x}H_{\psi u_i} - H_{u_i\psi}H_{xu_i}) = 0$ holds by the Frobenius condition.

Let r-vector-functions $R(t,x,p)$ and $L(t,x,p)$ be given. Denote by $\{R, L\}$ a Poisson bracket for functions R and L, which is calculated by formula $\{R, L\} = R_x L_p - R_p L_x$.

Find the second derivative of $\mathcal{H}$ in z.

$$\mathcal{H}_{z_i z_j} = \psi'(-\eta_{xx}h_j + \eta_{xz_j})(gh_x - g_x h_{ix} + h_{tx_j}) -$$
$$-\psi'\eta_x(gh_x - g_x h_i + h_i)_x = \psi'(\eta_x h_j) -$$
$$-(gh_x - g_x h_i + h_t) - -\psi\eta'_x(gh_x - g_x h_i + h_i)h_j =$$
$$= h_{x_j}\frac{d}{dt}H_{u_i} - [\frac{d}{dt}H_{u_i}]_x =$$
$$= [\frac{\partial}{\partial\psi}[\frac{d}{dt}H_{u_i}]H_x - \frac{d}{dt}H_{u_i x}H_{u_i\psi}] =$$
$$= -\{\frac{d}{dt}H_{u_i}, H_{u_j}\} = \{H_{u_j}, \dot{H}_{u_i}\}.$$

Determine expressions for other derivatives by the similar way:

$$\mathcal{H}_y = H_x; \mathcal{H}_{yy} = H_{xx};$$
$$\mathcal{H}_{yz} = -\frac{d}{dt}H_{ux} = -\frac{\partial}{\partial t}H_{ux}; \mathcal{H}_{y\psi} = H_{x\psi};$$
$$\mathcal{H}_{z\psi} = -\frac{\partial}{\partial\psi}\frac{d}{dt}H_u = -\frac{\partial}{\partial t}H_{u\psi};$$
$$\mathcal{F}_y = F_x; \mathcal{F}_{ys} = (h' F_x)_x; \mathcal{F}_s = h' F_x; \mathcal{F}_{yy} = F_{xx};$$
$$\mathcal{F}_{zz} = h' F_{xx}h.$$

Thus, the system (10) - (13) and formula (14) have the form:

$$\dot{\psi} = -H_x + \left[\left(\frac{dH_u}{dt}\right)'_x + \sigma\left(\frac{dH_u}{dt}\right)'_\psi\right] \times$$
$$\times(\{H_u, \dot{H}_u\} - (1-\alpha)E)^{-1}\dot{H}_u, \tag{15}$$

$$\dot{\sigma} = -H_{xx} - H_{x\psi}\sigma - \sigma H_{\psi x} +$$
$$\left[\left(\frac{dH_u}{dt}\right)'_x + \sigma\left(\frac{dH_u}{dt}\right)'_\psi\right] \times$$
$$\times(\{H_u, \dot{H}_u\}(1-\alpha)E)^{-1} \times \tag{16}$$
$$\times\left[\left(\frac{dH_u}{dt}\right)_x + \left(\frac{dH_u}{dt}\right)_\psi \sigma\right],$$

$$\psi(t_1) = -(F_x + (h' F_x)'_x(h' F_{xx}h +$$
$$+\frac{(1-\alpha)}{\alpha}E)^{-1}h' F_x))\alpha, \tag{17}$$

$$\sigma(t_1) = -(F_{xx} + (h' F_x)'_x(hF'_{xx}h +$$
$$+\frac{(1-\alpha)}{\alpha}E)^{-1}(h' F_x)_x)\alpha, \tag{18}$$

$$\tilde{z}(t, \Delta y) = z^I(t) + (\{H_u, \dot{H}_u\} -$$
$$-(1-\alpha)E)^{-1}\left(\dot{H}_u + \left[\left(\frac{dH_u}{dt}\right)_x + \right.\right. \tag{19}$$
$$\left.\left.\left(\frac{dH_u}{dt}\right)_\psi \sigma\right]\Delta y\right).$$

The second step of algorithm, consisting of integration of system (8) closed by the synthesis (14), can be substituted by integration of the linearized system (8) in form:

$$\delta\dot{y} = H_{\psi x}\delta y + \left(\frac{dH_u}{dt}\right)'_{\psi} (\{H_u, \dot{H}_u\} - \\ -(1-\alpha)E)^{-1}\left(\dot{H}_u + \left[\left(\frac{dH_u}{dt}\right)_x + \\ + \left(\frac{dH_u}{dt}\right)_{\psi}\sigma\right]\delta y\right), \ \delta y(t_0) = 0. \quad (20)$$

Hence, Discontinuous function $\xi(t)$ may be calculated by the formula:

$$\xi(t) = \begin{cases} z^I(t), & t = t_0; \\ \tilde{z}(t, \delta y(t)), & t \in (t_0, t_1); \\ -(h'F_{xx}h + \frac{1-\alpha}{\alpha}E)^{-1}\times & \\ \times((h'F)_x + (h'F_x)_x\delta y(t_1)), & t = t_1. \end{cases}$$

Construct a sequence of continuous functions $\xi_q(t)$ using function $\xi(t)$ by the following way. Let $\{\tau_i\}$, $i = \overline{1,p}$, be a set of discontinuity points for the function $\xi(t)$. Denote $\tau_i^+ = \tau_i + 1/q$, $\tau_i^- = \tau_i - 1/q$, $\tau_1 = t_0$, $\tau_p = t_1$. Then

$$\xi_q(t) = \begin{cases} \xi(t), & t \notin [\tau_i^-, \tau_i^+]; \\ a_i + b_i t, & t \in [\tau_i^-, \tau_i^+], \end{cases} \quad (21)$$

where $b_i = (\xi(\tau_i^+) - \xi(\tau_i^-))/(\tau_i^+ - \tau_i^-)$, $a_i = \xi(\tau_i^+) - b_i\tau_i^+$. Further, construct a sequence of controls $u_q^{II}(t) = u^I(t) + \frac{d}{dt}\xi(t)$ and corresponding sequence of trajectories $x_q(t)$ defined of the system (1).

4. ALGORITHM OF METHOD

Formulate one iteration of the second order succesive improvement method.

1^0. Let $u^I(t)$ be given. Integrate the system (1) and obtain the trajectory $x^I(t)$.

2^0. Fix the parameter $\alpha \in (0, 1]$.

3^0. Solve the vector-matrix system (15)-(18).

4^0. $\xi(t)$ is obtained by integrating the system (19) - (20).

5^0. Piecewise continuous function $\xi(t)$ is approximated by sequence of continuous functions $\xi_q(t)$ by the formula (21).

6^0. For sufficiently large q calculate new control $u^{II}(t) = u^I(t) + \frac{d}{dt}\xi_q(t)$ and corresponding trajectory $x^{II}(t)$.

7^0. Compare the values of functional. If $I^{II} \geq I^I$ then parameter α should be dicreased and the process is repeated starting from step 2^0.

In the case when initial approximation satisfies of the stationarity conditions in the derivative problem, the algorithm has analogycal structure (see step 2).

Example.

$$\dot{x}_1 = u; \ \dot{x}_2 = x_1; \ \dot{x}_3 = x_1^2;$$

$$x_1(0) = x_2(0) = x_3(0) = 0;$$

$$I = -x_2^2(T) + x_3(T).$$

Take $u^I = 0$ as an initial approximation. This control satisfyes of the stationarity conditions. Because of it we can use second variant of our algorithm.

The system (15) – (17) has the solution:

$$\psi_1(t) = 0; \ \psi_2(t) = 0; \ \psi_3(t) = -\alpha;$$

$$\sigma_{23}(t) = 0; \ \sigma_{13}(t) = 0;$$

σ_{11}, σ_{12}, σ_{22} satisfy of the equations:

$$\dot{\sigma}_{11} = 2\alpha - \sigma_{21}(\sigma_{12} - 2\alpha)^2/(\alpha+1), \quad \sigma_{11}(T) = 0;$$

$$\dot{\sigma}_{12} = -\sigma_{22} - \sigma_{22}(\sigma_{12} - 2\alpha)/(\alpha+1), \sigma_{12}(T) = 0;$$

$$\dot{\sigma}_{22} = -\frac{\sigma_{22}{}^2}{\alpha+1}, \quad \sigma_{22}(T) = 2\alpha.$$

The $\sigma_{22}(t)$ is defined by formula:

$$\sigma_{22}(t) = \frac{\alpha+1}{t - T + \frac{\alpha+1}{2\alpha}}.$$

Consider two cases: $T < 1$, $T > 1$. If $T < 1$ and $\alpha = 1$ then $\sigma(t)$ exists in all segment $[0, T]$. Hence, $u^I(t) = 0$ is the point of minimum of the functional I; $\sigma(t)$ exists in a part of the segment $[0, T]$ only and the control $u^I(t) = 0$ may be improved by this algorithm with $T > 1$ and $\alpha = 1$. Take $\alpha = \frac{1}{2T-1}$. Then $\delta y_2(t) = \epsilon t$, where $\epsilon \neq 0$ is a constant, $z^{II}(t) = \epsilon$, and approximation of control is contained in the class of admissible controls.

$$u = \begin{cases} \epsilon s, & t \in [0, \frac{1}{s}), \\ 0, & t \in [\frac{1}{s}, T]. \end{cases}$$

ACKNOWLEDGMENTS

These research are supported by the RFBR, Grants Nos. 97-01-960002, 97-01-96005 and 97-01-00047.

REFERENCES

Dykhta, V.A.(1994). Variational maximum principle and quadratic optimality conditions for impulsive and singular control. *Sib. Math. J.* **Vol. 35**, p. 70-82.(in Russian)

Goh, B.S.(1966). Necessary conditions for singular extremals involving multyple control variables.*SIAM J.Control,* **Vol.4**, N4. p.716-731.

Goh, B.S. (1967). Optimal singular Control for Multi-Input linear systems. *Journal of Math. Anal. and Appl.*, **Vol.20,** p.534-539.

Gurman, V.I.(1977). *Singular optimal control problems.* Nauka, Moscow.(in Russian)

Gurman, V.I., V.A. Baturin, I.V. Rasina (1983). *Approximating solution methods of the optimal control problems.* Irkutsk State University,Irkutsk.

Gurman, V.I. (1987). *New improving methods for controlled processes.* Nauka, Novosibirsk. (in Russian)

Gurman, V.I., V.A. Baturin, at all. (1987) *New methods of improving of controllable processes.* Nauka, Siberian Branch, Novosibirsk.

Kazakov, V.A., V.F. Krotov (1995). Iteration method of construction for discontinuous of optimal control problems. *Avtomatika i Telemekhanika* N1, p 29–43 (in Russian).

Kolokolnikova, G.A. (1992). Investigation of generalizated optimal control problems with linear infinite controls by multiple transformations. *Differenz. Uravnenia,* **Vol.28**, N11, pp 1919–1932 (in Russian).

Krotov, V.F., V.I.Gurman (1969). *Methods and problems of the optimal control.* Nauka, Moscow (in Russian).

ON AN OPTIMAL STOPPING TIME FORMULATION OF ADAPTIVE SIGNAL FILTERING

Italo Capuzzo Dolcetta* Roberto Ferretti**

* *Dip. di Matematica, Università di Roma "La Sapienza", P.le A. Moro, 2, 00185 Roma, Italy*
** *Dip. di Matematica, Università di Roma "Tor Vergata", v. Fontanile di Carcaricola, 00133 Roma, Italy*

Abstract: This paper presents an approach to signal filtering, based on an optimal stopping time problem for the evolution equation describing the filtering kernel. Convergence of semi-discrete approximate control problems is proved and a numerical test is presented and discussed.

Keywords: Signal processing, Optimal filtering, Distributed parameter systems, Switching times.

1. INTRODUCTION

This paper considers an application of optimal stopping time problems for parabolic linear evolution equations. It will follow the model problem

$$\begin{cases} y'(t) = \Delta y(t) \\ y(0) = y_0 \end{cases} \tag{1}$$

posed in the Hilbert space $H = L^2$, along with its numerical approximation. An optimal stopping time problem for equation (1) (see also (Bensoussan and Lions, 1984)) consists in minimizing (for a given initial state y_0) a cost in the integral form

$$J(y_0, t) = \int_0^t g(s, y(s))ds + \Phi(t, y(t)). \tag{2}$$

over all $t \geq 0$. A numerical discretization of (1), (2) requires a finite dimensional (semi-discrete) approximation of (1), that is

$$\begin{cases} y_n'(t) = A_n y_n(t) \\ y_n(0) = P_n y_0 \end{cases} \tag{3}$$

and a corresponding discretization for (2), that is

$$J_n(P_n y_0, t) = \int_0^t g(s, y_n(s))ds + \Phi(t, y_n(t)). \tag{4}$$

A general framework for semi-discrete approximations of a large class of non-quadratic control problems (including optimal stopping time problems) has been studied in (Ferretti, 1997*b*) and (Ferretti, 1997*a*). This approach provides minimizing sequences of approximate optimal solutions.

On the other hand, the linear filtering of a noisy data $y_0(x)$ (with $x \in R$ for signals, $x \in R^2$ for images) may be posed (see e. g. (Papoulis, 1987)) as the convolution

$$y(t, x) = y_0(x) * h(t, x) \tag{5}$$

where t heuristically plays the role of a scale factor of the filter, inversely related to the bandwidth. Assume for example that the kernel h has the gaussian structure

$$h(t, x) = \frac{1}{\sqrt{4\pi t}} e^{\frac{|x|^2}{4t}}. \tag{6}$$

Here, $2t$ is the second moment of the gaussian function, proportional to the inverse of the bandwidth. Thus, fixing t amounts to the choice of a particular bandwidth. However, $h(t,x)$ is also the fundamental solution of the heat equation (1) at time t, so that the choice of a particular bandwidth of the filter in order to optimize its performances (in a sense to be made clear) may be posed in the form of an optimal stopping time problem for (1). Of course, different filter structures might be taken into consideration, but in this paper only the heat kernel (6) will be treated (all results presented still hold for general linear parabolic equations).

The outline of the paper is the following. Section 2 states the problem in higher detail and gives a general result for the convergence of discrete approximations. In section 3 an example of application is given and the results of some numerical test are discussed, along with some possible extentions of this approach.

2. STATEMENT OF THE PROBLEM AND BASIC ASSUMPTIONS

Assume a state equation in the form (1), where y_0 and $y(t)$ belong to the Hilbert space $H = L^2(\Omega)$. It is well known that, for any initial state y_0, (1) has a unique continuous global solution $y \in C([0,+\infty[; H)$. A semi-discrete approximation of (1) is in the form (3) (see (Gottlieb and Orszag, 1977), (Lasiecka and Triggiani, 1991), (Raviart and Thomas, 1983)). Let $H_n \subset H$ be a sequence of vector spaces, and $P_n : H \to H_n$ be a sequence of projections, and assume that $\dim H_n = k_n$ (with $k_n \to \infty$), and that for any $\bar{y} \in H$:

$$\lim_{n\to\infty} \|\bar{y} - P_n\bar{y}\| = 0 \qquad (7)$$

(here and in the sequel, the norm $\|\cdot\|$ refers to the space H). Approximation (3) is assumed to be convergent and uniformly stable, in the sense that, for all $t > 0$ and $y_0 \in H$:

$$\|y(t) - y_n(t)\|_H \le C(y_0, n)(1+t)^{-\alpha} \qquad (8)$$

with $\lim_{n\to\infty} C(y_0, n) = 0$, and $\alpha > 1$.

Given the evolution equation (1) and the initial state y_0, the optimal control problem into consideration is to find a stopping time $t^* \in [0,+\infty]$ minimizing the cost (2), assuming (7), (8), and the (uniform in t) local Lipschitz continuity of g and Φ:

$$|g(t,y_1) - g(t,y_2)| \le L_g\|y_1 - y_2\| \qquad (9)$$

$$|\Phi(t,y_1) - \Phi(t,y_2)| \le L_\Phi\|y_1 - y_2\| \qquad (10)$$

for any $y_1, y_2 \in H$, $t \in R^+$ and with $L_g = L_g(\|y_1\|, \|y_2\|)$, $L_\Phi = L_\Phi(\|y_1\|, \|y_2\|)$.

In the approximate version of this problem, given the evolution equation (3) with initial state $P_n y_0$, one looks for a stopping time $t_n^* \in [0,+\infty]$ minimizing the cost (4) (note that the depedence of the optimal stopping times t^*, t_n^* on the initial state is dropped to simplify notation).

The value functions for both the original and the approximate problem are defined as:

$$v(x) := \inf_{t\ge 0} J(x,t), \quad v_n(x_n) := \inf_{t\ge 0} J_n(x_n,t). \qquad (11)$$

Such value functions will be used in the proof of the following main convergence result.

Theorem 1. Assume (7)–(10). Then, for any $y_0 \in H$, $|J(y_0, t_n^*) - v(y_0)| \to 0$ as $n \to \infty$.

It suffices to give a sketch of the proof which is a slight adaptation of the results in (Ferretti, 1997*a*). By the definition of v, v_n, for $y_0 \in H$ and any given $\varepsilon > 0$, it is possible to find two finite stopping times t^ε, t_n^ε (depending on ε) such that:

$$v(y_0) \le J(y_0, t^\varepsilon) \le v(y_0) + \varepsilon \qquad (12)$$

$$v_n(P_n y_0) \le J_n(P_n y_0, t_n^\varepsilon) \le v_n(P_n y_0) + \varepsilon. \qquad (13)$$

Step 1: $\limsup_n v_n(P_n y_0) \le v(y_0)$. To prove this step, note that by the definition of v_n one has:

$$v_n(P_n y_0) \le \int_0^{t^\varepsilon} g(y_n(s))ds + \Phi(y_n(t^\varepsilon)). \qquad (14)$$

Adding the terms $\pm J(y_0, t^\varepsilon)$, and using (9), (10) and (12), one obtains:

$$v_n(P_n y_0) \le \int_0^{t^\varepsilon} L_g\|y(s) - y_n(s)\|ds +$$

$$+ L_\Phi\|y(t^\varepsilon) - y_n(t^\varepsilon)\| + v(y_0) + \varepsilon \qquad (15)$$

whence, using the convergence of the scheme and the dominated convergence theorem, the step is proved.

Step 2: $\liminf_n v_n(P_n y_0) \ge v(y_0)$. By the definition of v,

$$v(y_0) \le \int_0^{t_n^\varepsilon} g(y(s))ds + \Phi(y(t_n^\varepsilon)). \qquad (16)$$

Operating as before, one has:

$$v(y_0) \le \int_0^{t_n^\varepsilon} L_g\|y(s) - y_n(s)\|ds +$$

$$+ L_\Phi\|y(t_n^\varepsilon) - y_n(t_n^\varepsilon)\| + v_n(P_n y_0) + \varepsilon. \qquad (17)$$

One cannot conclude as in (15) since the sequence t_n^ε may not be bounded as $n \to \infty$. However, since (8) holds, it is possible again to apply the dominated convergence theorem to both the first and the second term in the right-hand side of (17). This completes the step and ensures that $\lim_n v_n(P_n y_0) = v(y_0)$.

Step 3: $\lim_n |J(y_0, t_n^*) - v(y_0)| = 0$. It suffices to note that

$$|J(y_0, t_n^*) - v(y_0)| \leq |J(y_0, t_n^*) - J_n(P_n y_0, t_n^*)| + \\ + |v_n(P_n y_0) - v(y_0)| \quad (18)$$

and to apply the same arguments of step 2, noting that $v_n(P_n y_0) \to v(y_0)$.

3. APPLICATION TO SIGNAL PROCESSING AND NUMERICAL TESTS

Consider the heat equation on R^1, with Neumann boundary conditions:

$$\begin{cases} y_t(t,x) = y_{xx}(t,x) \\ y_x(t,0) = y_x(t,1) = 0 \\ y(0,x) = y_0(x). \end{cases} \quad (19)$$

Here, $x \in]0,1[$ denotes the space variable, and $H = L^2([0,1])$. Note that Neumann conditions are used to ensure that the mean value of y_0 is preserved (this is a reasonable requirement in signal processing). The cost has the form

$$J(y_0, t) = \Phi(t, y(t)) = \\ = c_1 t + c_2 \frac{\|y_0\|^2_{L^2}}{t^{1/2}} + \|y(t) - y_0\|^2_{L^2}. \quad (20)$$

In the general form of J, the first term accounts for the cost of computing time (and also causes J to be coercive), the second for the asymptotic behaviour of the energy of filtered noise and the third for the distance to the initial data (heavy deformation of the original noisy data should be avoided). It is trivial to prove the following

Proposition 2. Problem (19), (20) admits an optimal stopping time $t^* > 0$.

A semi-discrete finite element discretization of (19) has the form (see (Raviart and Thomas, 1983)):

$$\mathbf{M}\mathbf{Y}'(t) = \mathbf{A}\mathbf{Y}(t) \quad (21)$$

where $\mathbf{Y} \in R^{k_n}$ corresponds to $y \in H_n$ by $y_n(t,x) = \sum_k \mathbf{Y}_k(t)\phi_k(x)$, and $\mathbf{M}$ and $\mathbf{A}$ are respectively the mass and stiffness matrices whose entries are defined by

$$m_{ij} = \int_0^1 \phi_i(x)\phi_j(x)dx ,$$

$$a_{ij} = \int_0^1 \frac{\partial}{\partial x}\phi_i(x)\frac{\partial}{\partial x}\phi_j(x)dx. \quad (22)$$

The discretized version of the cost (20) is

$$J_n(P_n y_0, t) = \Phi(t, y_n(t)) = \\ = c_1 t + c_2 \frac{\|P_n y_0\|^2_{L^2}}{t^{1/2}} + \|y_n(t) - P_n y_0\|^2_{L^2}. \quad (23)$$

which may be easily computed taking into account that $\|y_n\|^2_{L^2} = \mathbf{Y}^t\mathbf{M}\mathbf{Y}$ ($(\cdot)^t$ denoting the transpose of a vector).

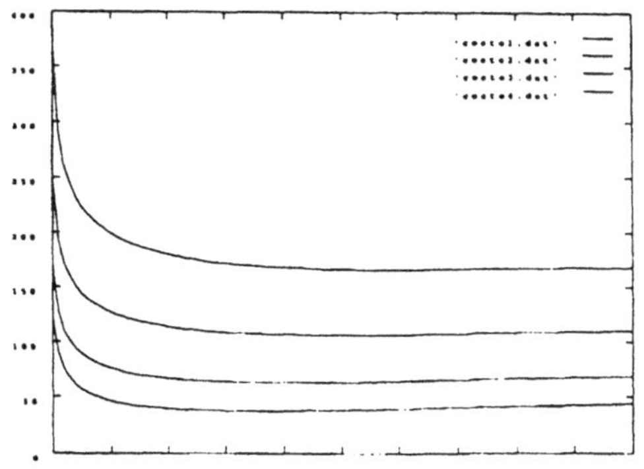

Fig. 1. Plots of $J_n(P_n y_0, t)$ for different noise levels.

The numerical tests have been performed using $y_0(x) = s(x) + n(x)$, where $s(x)$ is a square wave and $n(x)$ is a random noise of varying amplitude. A careful trimming of the coeficients c_1, c_2 is necessary to obtain the desired adaptive behaviour of the filter. Figure 1 compares values of J_n as a function of t for different amplitudes of the noise $n(x)$ (higher amplitudes for the upper plots), whereas figure 2 shows $s(x)$, $y_0(x)$ and $y(t_n^*, x)$ for the corresponding noise levels. The adaptive behaviour is shown by the increase in the abscissa of the minimum point for J_n at the increase of the noise level, although it is apparent from figure 2 that the stopping time is too large (that is, the bandwidth is too small) with this choice of the parameters.

4. CONCLUSIONS

The proposed technique, although at a very preliminary stage, seems a viable approach to adaptive filtering of noisy signals and images. However, the efficiency of the cost functional (20) still heavily depends on the choice of the constants c_1 and c_2, so that a first research direction might be to determine a more flexible criterion. A further direction is to investigate the performance of filters described by *nonlinear* evolution equations. One such filter, to be used for image processing (to which this research is ultimately directed), is described for example in (Alvarez *et al.*, 1992).

5. REFERENCES

Alvarez, L., P. L. Lions and J. M. Morel (1992). Image selective smoothing and edge detection by nonlinear diffusion. *SIAM J. Num. Anal.* **29**, 845-866.

Bensoussan, A. and J. L. Lions (1984). *Impulse control and quasi-variational inequalities*. Gauthier-Villars. Paris.

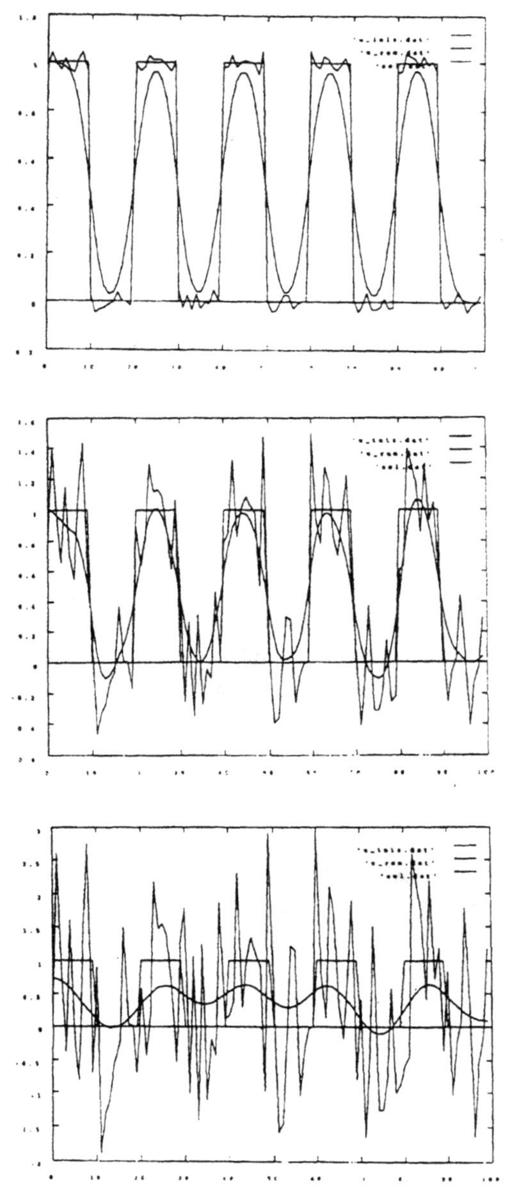

Fig. 2. Result of filtering for different noise levels

Ferretti, R. (1997*a*). Dynamic programming techniques in the approximation of optimal stopping time problems in hilbert spaces. In: *Partial differential equation methods in control and shape analysis* (G. Da Prato and J. P. Zolesio, Eds.). pp. 153–162. M. Dekker. New York.

Ferretti, R. (1997*b*). Internal approximation schemes for optimal control problems in hilbert spaces. *Journ. Math. Syst. Est. Contr.* 7, 115–118.

Gottlieb, D. and S. A. Orszag (1977). *Numerical analysis of spectral methods.* SIAM. Philadelphia.

Lasiecka, I. and R. Triggiani (1991). *Differential and algebraic Riccati equations with applications to boundary/point control problems: continuous theory and approximation theory.* Springer–Verlag. New York.

Papoulis, A. (1987). *Signal Analysis.* McGraw–Hill. Auckland.

Raviart, P. A. and J. M. Thomas (1983). *Introduction a l'analyse numérique des équations aux derivées partielles.* Masson. Paris.

UNIVERSAL MECHANISM SOFTWARE AND SIMULATION OF MECHANICAL CONTROL SYSTEMS

Dmitri Pogorelov, Evgeny Selensky

Bryansk State Technical University
241035 Bryansk, Russia
E-mail: pogorelov@bitmcnit.bryansk.su

Abstract: A description of the Universal Mechanism software is given. The human arm manipulator control problem is considered. Two kinematical models of the manipulator are discussed. Arm manipulator control algorithms for the models are developed. A hierarchical sequence of manipulator operations combining simplicity and flexibility of programming is created. The problem of obtaining the optimal in the sense of minimal length path in the manipulator configuration space is solved.

Keywords: human arm manipulator, control algorithms.

1. INTRODUCTION

The interest in the modelling of various human actions is motivated by the necessity to predict the mechanical behaviour of the human body. The applications of the human body behaviour simulation range from the prosthesis design and the reliability assessment for different man-operated control systems to the optimisation of sports equipment and the estimation of operating conditions comfort (Silva, *et al.*, 1997). The present paper deals with the problems of the computer-aided control of a human arm manipulator. The synthesis of the manipulator kinematical models as well as the motion simulation and control have been carried out with the help of the programme package Universal Mechanism.

2. UNIVERSAL MECHANISM SOFTWARE

The programme package Universal Mechanism (UM) has been developed at the Bryansk State Technical University (Pogorelov, 1997). It is aimed at automating the synthesis and analysis of the kinematics and dynamics of complex planar and spatial mechanical systems. Any object is considered by the software a system of rigid bodies connected with each other by means of kinematical pairs and force elements of different types.

UM is a useful tool for the computer-aided modelling of various kinds of systems: plane and spatial mechanisms, complex aerospace structures, robot systems, transport vehicles (such as locomotives, railway carriages, cars, planes and vessels), cables, truss structures, etc. A model of the object synthesised by UM may contain up to a few thousand elements. The equations of motion are generated automatically.

UM offers the designer a wide range of motion animation capabilities (the carcass, contour and surface built-in graphics). The process of motion as well as any characteristic or function of interest can be displayed. In the process of modelling the designer can vary the object parameters: inertial and geometric characteristics, parameters of control, etc.

The area of application of the programme package is not restricted by just simulating the operation of large multibody systems. One of the possible ways to implement the package might also be creating computer trainers in various areas of engineering to model emergencies. With the help of the software one can optimise principal structural parameters, acquire skills in controlling complex objects by means of maintaining the dialogue with the programme, test the reliability of adaptive control systems and so forth.

3. HUMAN ARM MANIPULATOR

An example of various systems capable of being treated by UM is a human arm manipulator.

3.1 Kinematical models of manipulator

Two different kinematical models have been considered for the manipulator (Pogorelov, *et al.*, 1996). The first one is aimed at creating and improving arm motion control algorithms while the second serves for planning manipulator motions for the case of obstacles in the manipulator workspace.

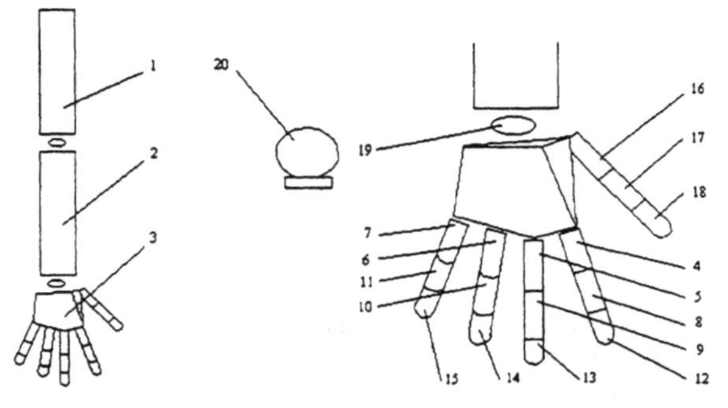

Fig. 1. Human arm manipulator kinematical model.

The first model (fig. 1) consists of 20 bodies. The model includes the shoulder spherical joint, that is fixed relative to the global system of co-ordinates, shoulder bone 1, the elbow rotational joint, forearm bone 2, the wrist spherical joint, hand 3 and the five fingers (bodies 4..18) attached to the hand by the joints with two rotational degrees of freedom (except the spherical joint between the hand and the thumb), and, finally, object 20 for transportation (a ball). Body 19 is fictitious and is introduced to ensure that the wrist Cartesian co-ordinates are explicit time functions. Each finger has three phalanxes. The phalanxes are connected consecutively with one another by the rotational joints.

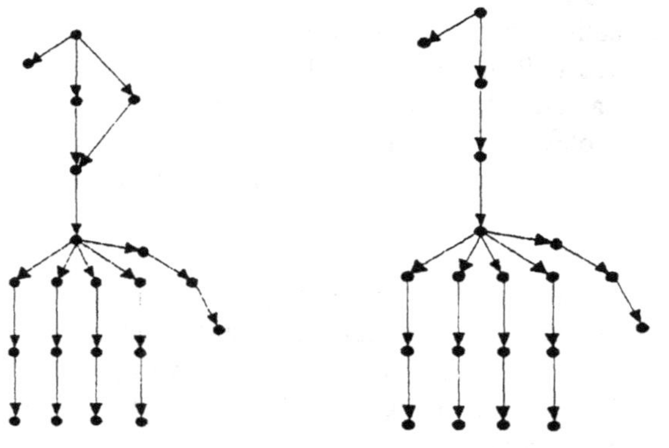

a) b)

Fig. 2. Oriented graphs of the 1st (a) and the 2nd (b) human arm manipulator kinematical models. The vertices of the graph represent the bodies of the manipulator, the arcs - the joints.

Thus, there are 20 bodies in the model that are connected with the basic body (the global inertial system of co-ordinates) and each other by means of 21 joints. Therefore the model contains one closed loop. The oriented graph of the system is shown in fig. 2a. The second model has no closed loops, i.e. its graph is a tree (fig. 2b).

3.2 Principles of human arm manipulator computer-aided control

Environment. The human arm manipulator computer-aided control is carried out in the UM environment in Borland Pascal.

The key principles of control. The computer-aided control is based on the similitude of motions of the actual human arm and the manipulator when performing the same tasks. As to the first model, the arm manipulator motion is controlled by the known time functions setting the values of the wrist co-ordinates, the fingers and the arm plane motions, whereas the angles in the shoulder and the elbow joints are considered unknown. In the second model the manipulator motion is still set by the time functions. However, the values of the angles in the shoulder and the elbow joints as well as the orientation of the arm plane are obtained through solving the problem of optimal trajectory in the manipulator configuration space. The obstacles in the workspace are taken into account. Here the problem of mapping the obstacle into the manipulator configuration space is dealt with.

Task. The manipulator carries out the following operations: stretch the arm towards the object, grip the object, lift and carry the object and, finally, release it.

Hierarchy of operations. In the control file the process of performing the task is split into separate operations. A hierarchical structure of operations is introduced. There exist the global and local sequences or chains of operations. If necessary, the hierarchy of operations can be developed both by adding separate operations to the structure at the existing levels and introducing new hierarchical levels. Such planning of the manipulator operations combines necessary flexibility and simplicity. The local operations within a global one are carried out simultaneously, whereas the global operations are done consecutively.

Each element of the local and the global chains is a record containing necessary parameters of a particular operation (duration, minimal and maximal values of local joint co-ordinates, etc.) as well as pointers to the previous and the next elements of the chain. The hierarchy of operations is shown in fig. 3.

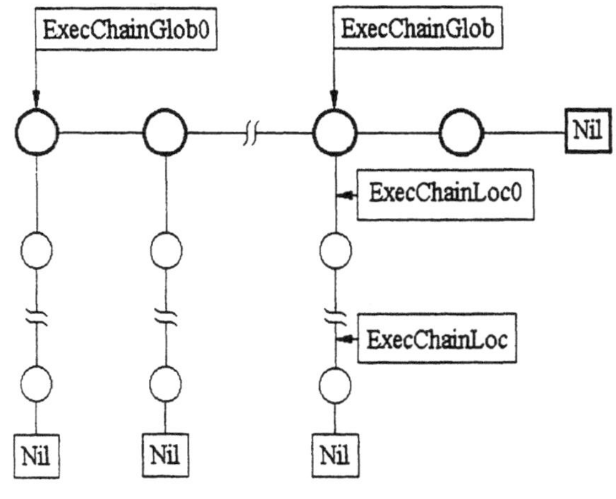

Fig. 3. Hierarchy of operations. *ExecChainGlob0* and *ExecChainLoc0* are the pointers to the beginning of the global and a local chain, respectively, whereas *ExecChainGlob* and *ExecChainLoc* are the pointers to the current elements of the corresponding chains. The horizontal chain is global. The vertical ones are local. The circles correspond to the particular operations.

Stretching the manipulator arm. The first manipulator kinematical model helps to generate the programme-controlled motion of the manipulator arm towards the object by means of representing the global Cartesian co-ordinates of the wrist as explicit functions of time *XWrist(t), YWrist(t), ZWrist(t)*. The operation is controlled as described by K.S. Fu *et al.* (1987). However, unlike the majority of industrial robots discussed there, the manipulator has one redundant degree of freedom as to stretching of the arm. The number of generalised co-ordinates in the shoulder and elbow joints (the angles in these joints are not controlled) equals four, whereas there are only three aforementioned explicit time functions determining the wrist motion, which can be used for determining the unknown angles. So to ensure unambiguity of the manipulator motion up to the motion of the hand it is necessary to set one more constraint to control any parameter or combination of parameters.

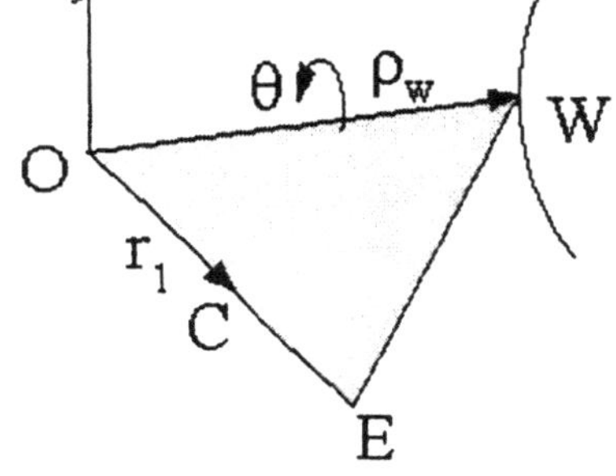

Fig. 4. Arm manipulator plane.

Here, it seems most natural to control the angle θ of rotation of the arm plane shaded in fig. 4. The plane is generated by the vector ρ_w directed from the origin O to the wrist W and the vector $\mathbf{r}_1$ directed from the origin O to the mass centre C of the shoulder bone. The plane rotates about ρ_w.

Given the dependence $\theta(t)$, the corresponding constraint equation is as follows:

$$\mathbf{n}_v{}^T\mathbf{n}_a - \cos\theta(t)n_v n_a = 0 ,$$

where $\mathbf{n}_v$ is the normal to the vertical plane generated by the global axis Oz and ρ_w; $\mathbf{n}_a$ is the normal to the arm plane; $\mathbf{n}_v{}^T\mathbf{n}_a$ is their scalar product, whereas $n_v = \|\mathbf{n}_v\|$ and $n_a = \|\mathbf{n}_a\|$.

Since

$$\mathbf{n}_v = \tilde{\rho}_w\mathbf{e}_3 , \qquad \mathbf{n}_a = \tilde{\mathbf{r}}_1\rho_w ,$$

where $\mathbf{e}_3$ is the unit vector of the axis Oz, $\tilde{\rho}_w$ and $\tilde{\mathbf{r}}_1$ are antisymmetrical matrices (Wittenburg, 1977), the constraint equation becomes

$$(\mathbf{r}_1{}^T\rho_w)(\rho_w{}^T\mathbf{e}_3)-\rho^2(\mathbf{r}_1{}^T\mathbf{e}_3)- c\sqrt{\left(\rho^2-(\rho_w{}^T\mathbf{e}_3)^2\right)\left(r^2\rho^2-(\rho_w{}^T\mathbf{r}_1)^2\right)} = 0 , \qquad (1)$$

where $\rho = \|\rho_w\|$, $r = \|\mathbf{r}_1\|$ and $c = \cos\theta(t)$. On differentiating Eq. 1 with respect to time it is found that

$$\mathbf{a}^T\mathbf{V}_1 + \mathbf{b}^T\mathbf{V}_w + s\,\theta^{\cdot}(t)\, n_v n_a = 0, \qquad (2)$$

where

$$\mathbf{a} = (\rho_w{}^T\mathbf{e}_3)\rho_w - \rho^2\mathbf{e}_3 - c\gamma\left(\rho^2\mathbf{r}_1 - (\mathbf{r}_1{}^T\rho_w)\rho_w\right),$$

$$\mathbf{b} = (\rho_w{}^T\mathbf{e}_3)\mathbf{r}_1 + (\mathbf{r}_1{}^T\rho_w)\mathbf{e}_3 - 2(\mathbf{r}_1{}^T\mathbf{e}_3)\rho_w - c\gamma^{-1}\left(\rho_w - (\rho_w{}^T\mathbf{e}_3)\mathbf{e}_3\right) - c\gamma\left(r^2\rho_w - (\mathbf{r}_1{}^T\rho_w)\mathbf{r}_1\right),$$

$$\mathbf{V}_1 = \mathbf{r}_1^{\cdot} ,$$

$$s = \sin\theta(t),$$

$$\gamma = n_v / n_a .$$

The vector $\mathbf{V}_w = (V_{wx}, V_{wy}, V_{wz})$ and $\theta(t)$ are the given time functions. The latter must have at least one continuous time derivative.

Eq. 2 can be written in a different form. Because $\mathbf{r}_1$

is not an explicit time function,

$$\mathbf{V}_l = \frac{\partial \mathbf{r}_l}{\partial \mathbf{q}}\mathbf{q}^{\cdot} + \frac{\partial \mathbf{r}_l}{\partial t} = \frac{\partial \mathbf{r}_l}{\partial \mathbf{q}}\mathbf{q}^{\cdot} = \mathbf{D}_l\mathbf{q}^{\cdot}.$$

Here $\mathbf{q}$ is the $(n \times 1)$ vector of generalised co-ordinates, $\mathbf{D}_l$ - the $(3 \times n)$ automatically generated matrix corresponding to body 1 (the shoulder bone). The $(1 \times n)$ product

$$\mathbf{G}\mathbf{D}_l^{\cdot} = \mathbf{J}$$

is a row of the Jacobi matrix for the current problem. It corresponds to the considered constraint equation and is necessary for the numerical solution of the non-linear algebraic constraint equations. Thus,

$$\mathbf{J}\mathbf{q}^{\cdot} + \mathbf{b}^T\mathbf{V}_w = 0, \qquad (3)$$

where $\mathbf{J}=\mathbf{a}^T\mathbf{D}_l$.

Gripping the object. On stretching the arm towards the object the gripping of the object is carried out. Each finger has a number of tactile points. If a tactile point is in contact with the object, there appears the elastic force of interaction. The palm is also given a set of such points. The friction force $\mathbf{F}$ is assumed to comply with the Coulomb law:

$$\mathbf{F} = -\mu N \frac{\mathbf{v}_r}{\|\mathbf{v}_r\|},$$

where $N = Cu$ is the normal reaction in the finger-tip and is equal to the product of the contact stiffness C and the deformation of the finger-tip u; μ is the friction coefficient in the contact area; $\mathbf{v}_r$ is the relative velocity of sliding. Gripping is controlled by the time functions setting the desired changing in time of the angles between the phalanxes. Gripping is done when any three fingers contact the object and the friction forces reach the values necessary for transportation.

Object transportation. As an alternative to modelling the process of transportation by means of friction forces it is worthwhile to impose extra constraints on the orientation of the object and the position of its mass centre. This makes the integration rate more than twice as fast.

After the object is gripped by the fingers and if the grip is stiff, the position of the mass centre and the orientation of the object in the hand-fixed system of co-ordinates are constant. The constraint equation corresponding to the constancy of the object mass centre co-ordinates relative to the hand-fixed system can be written in this system

$$\mathbf{A}_{k0}(\rho_b^{(0)} - \rho_k^{(0)}) - \rho_r^{(k)} = 0, \qquad (4)$$

where ρ_r is the vector directed from the origin of the hand-fixed system of co-ordinates to the object (fig. 5). The superscript $^{(k)}$ corresponds to the hand-fixed system of co-ordinates, the superscript $^{(0)}$ - to the global reference frame. The subscript $_b$ stands for the object, and finally, the subscript $_k$ - for the hand.

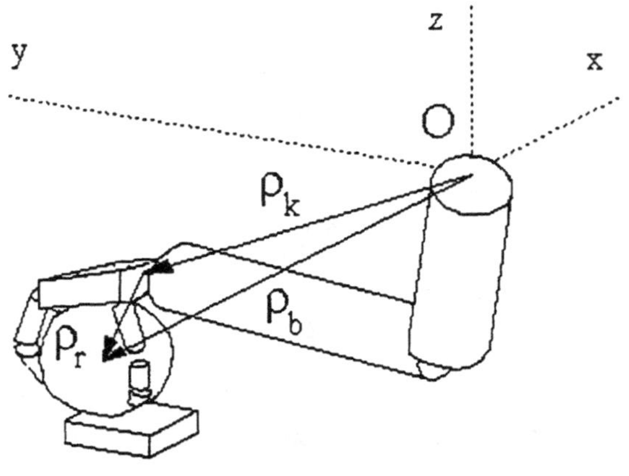

Fig. 5. Object transportation.

The vector $\rho_r^{(k)} = const$ is stored at the outset of transportation, whereas $\rho_k^{(0)} = \rho_k^{(0)}(t,q)$ and $\rho_b^{(0)} = \rho_b^{(0)}(t,q)$ are variable. $\mathbf{A}_{k0}$ is the matrix of rotation of the hand-fixed system relative to the global one.

Since $\dot{\rho}_k^{(k)} = 0$, after differentiating Eq. 4 with respect to time it is readily verified that

$$\mathbf{A}_{k0}\left[\left(\tilde{\rho}_b^{(0)} - \tilde{\rho}_k^{(0)}\right)\omega_k^{(0)} + \mathbf{V}_b^{(0)} - \mathbf{V}_k^{(0)}\right] = 0.$$

Bearing in mind that

$$\mathbf{V}_b^{(0)} = \mathbf{D}_b^{(0)}\mathbf{q}^{\cdot}, \quad \mathbf{V}_k^{(0)} = \mathbf{D}_k^{(0)}\mathbf{q}^{\cdot} + \mathbf{V}_k'^{(0)}$$

and

$$\omega_k^{(0)} = \mathbf{B}_k^{(0)}\mathbf{q}^{\cdot},$$

where $\omega_k^{(0)}$ is the angular velocity of the hand in the global system and $\mathbf{B}_k^{(0)}$ is the (3×1) automatically generated matrix corresponding to the hand, one can obtain the resulting equation

$$\mathbf{J}\mathbf{q}^{\cdot} + \mathbf{g} = 0,$$

where the matrix $\mathbf{J}$ corresponding to the imposed

constraints is

$$\mathbf{J} = (\tilde{\rho}_b^{(k)}\mathbf{B}_k^{(k)} + \mathbf{A}_{k0}\mathbf{D}_b^{(0)}) - (\tilde{\rho}_k^{(k)}\mathbf{B}_k^{(k)} + \mathbf{A}_{k0}\mathbf{D}_k^{(0)}),$$

whereas

$$\mathbf{g} = \mathbf{V}_k^{\prime(0)} = \frac{\partial \rho_k}{\partial t}.$$

In order to obtain the constraint equations corresponding to the constancy of the orientation of the object in the hand-fixed system and their 1st time derivatives the matrix

$$\mathbf{A}_{bk} = \mathbf{A}_{b0}\mathbf{A}_{0k} = \mathbf{A}_{b0}\mathbf{A}_{k0}^T$$

of transition from the object-fixed system of co-ordinates to the hand-fixed one must be obtained first. This matrix is stored at the beginning of the transportation phase.

At each step of integration during transportation the following constraint equation is generated:

$$\mathbf{A}^* = \mathbf{A}_{0b}\mathbf{A}_{bk}\mathbf{A}_{k0} = \mathbf{E},$$

where $\mathbf{E}$ is the identity matrix. The matrices $\mathbf{A}_{0b}$ and $\mathbf{A}_{k0}$ are variable.

The independent constraint equations corresponding to the constancy of the orientation of the object in the hand-fixed co-ordinate system after certain transformations become

$$\left.\begin{aligned} (a_{32}^* - a_{23}^*)/2 = 0 \\ (a_{13}^* - a_{31}^*)/2 = 0 \\ (a_{21}^* - a_{12}^*)/2 = 0 \end{aligned}\right\}, \tag{5}$$

where a_{ij}^* are the elements of $\mathbf{A}^*$.

On taking the 1st time derivative of Eqs. 5 it is found that

$$\omega_b^{(0)} - \omega_k^{(0)} = 0, \tag{6}$$

because the angular velocity of the object relative to the hand equals zero as a result of imposing the constraints. Here $\omega_b^{(0)}$ is the angular velocity of the object and $\omega_k^{(0)}$ is that of the hand in the global system of co-ordinates.

The Jacobi matrix corresponding to the imposed constraints is found to be

$$\mathbf{J} = \mathbf{A}_{0b}\mathbf{B}_b^{(b)} - \mathbf{A}_{0k}\mathbf{B}_k^{(k)}.$$

Returning to the initial position. This operation is carried out similarly to that of stretching the arm.

Stop phase. This operation is idle.

3.3 Obtaining the optimal path in the manipulator configuration space

The problem of the joint angles control capable of taking into account obstacles in the manipulator workspace is considered. The problem is solved with the help of the second kinematical model (see Sec. 3.1). The manipulator configuration space $\Re^4$, whose element is the vector

$$\mathbf{v} = (v_1, v_2, v_3, v_4)^T,$$

is discretized. Here v_1, v_2, v_3, v_4 are the angles of rotation in the shoulder spherical joint (about the global axes x, y, z) and the elbow rotational joint (about the local axis x), respectively. The rest degrees of freedom are left out of account. Thus, any point $K \in \Re^4$ determines the manipulator configuration in Euclidean space except for the position and orientation of the hand. The manipulator motion is set by time functions. The values of the angles in the shoulder and the elbow joints are obtained through solving the problem of optimal trajectory in the manipulator configuration space, given the initial and final manipulator configurations. The final configuration is chosen to ensure the safe gripping of the object. So the path of minimal length in the manipulator configuration space from the initial to the final point is obtained with the help of a simple wave algorithm. The algorithm is of complexity n, where n is the number of points in the configuration space mesh. The basic assumption of the algorithm is that the paths between adjacent finite elements of the configuration space have the same weight. It provides simplicity and a high integration rate.

The problem of mapping obstacles given in Euclidean space into the manipulator configuration space arises here. The obstacle discretization is introduced. The manipulator is supplied with a number of reference points. To obtain certain prohibited combinations of the controlled angles a reference point is placed into a finite element of the obstacle. The arm plane rotates about the vector ρ_w (fig. 4) while the reference point remains within the obstacle. The obtained values of v_1, v_2, v_3, v_4 are stored.

To optimise the process of mapping different obstacles the following method is offered. A finite element mesh is introduced in the manipulator Euclidean workspace. For each node referred to as an obstacle the corresponding mapping area is stored in a packed form. To generate the mapping of an obstacle the latter is represented as a set of finite elements. Its mapping is constructed as a logical unification of the mappings of separate elements which are obtained from the stored database.

In fig. 6 the manipulator can be seen at the moment of moving around the cylindrical obstacle.

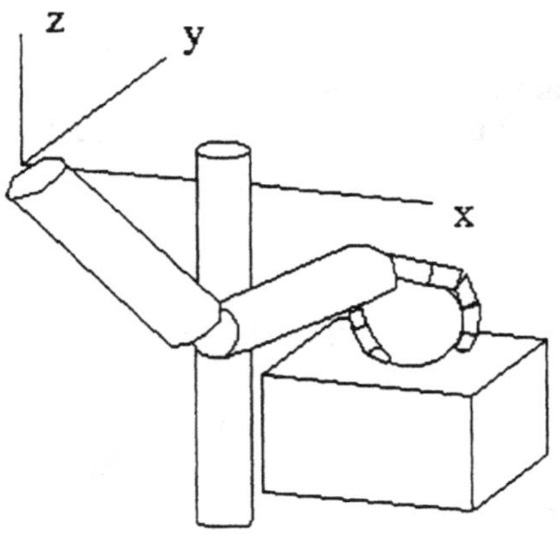

Fig. 6. The motion of the manipulator around the obstacle.

4. CONCLUSIONS

The human arm manipulator computer-aided control problem is considered. The problem is solved with the help of the programme package Universal Mechanism. Two kinematical models for the manipulator are offered. The presence of one closed loop in the graph of the first model facilitates the arm motion control. To simulate the actual motions of the human arm a simple and flexible hierarchical sequence of operations is introduced. The second model allows one to take into account obstacles in the workspace. A plain algorithm is developed to plan the manipulator motions in the workspace with obstacles. The problem of mapping the obstacle area given in Euclidean space into the configuration space is considered.

ACKNOWLEDGEMENTS

The authors are grateful to Dmitrochenko Oleg for his valuable assistance in the course of their work.

REFERENCES

Fu, K.S., R.C.Gonzalez and C.S.G. Lee (1987). Planning manipulator trajectories. In: *Robotics: Control, Sensing, Vision, and Intelligence*, pp. 168-221. McGraw-Hill, Inc., New York.

Pogorelov, D., Selensky E. and Sichkov E. (1996). Generation of symbolic motion equations for large multibody systems. In: *New computer technologies in control systems*, Proceedings of International Workshop (Dmitriev M.G., Gerdt V.P., Ed.), p. 50. IPS RAS, Pereslavl-Zalessky, Russia.

Pogorelov, D. (1997). *An introduction to multibody system dynamics modelling*. BGTU Publishers, Bryansk (in Russian).

Silva, M.P.T., J.A.C. Ambrosio and M.S. Pereira (1997). Biomechanical Model with Joint Resistance for Impact Simulation. *Multib. Syst. Dyn.* (W. Schielen, J.A.C. Ambrosio, Ed.), **1**, 65-84.

Wittenburg, J. (1977). *Dynamics of systems of rigid bodies*. B.G. Teubner, Stuttgart.

SECOND ORDER NECESSARY OPTIMALITY CONDITIONS FOR IMPULSE CONTROL PROBLEM AND MULTIPROCESSES

Vladimir Dykhta

Irkutsk State Economic Academy,
11, Lenin Str., Irkutsk, Russia
e-mail : dykhta@iinh.irkutsk.su

Abstract. We consider a class impulse optimal control problems in which the controls are represented by vector measures, trajectories are functions of bounded variation and are interpreted as robust solutions of the nonlinear dynamic system, and terminal constraints are contained. For such problems a Maximum Principle and second order necessary optimality conditions are presented. The examined measure is discontinuous at a finite number of times and has no the continuous singular component. The approach of investigation based on an interpretation of the impulse optimal control problem as a specific optimal multiprocesse problem.

Keywords. optimal control, impulse control, Maximum Principle, second order conditions.

1. INTRODUCTION

The aim of this paper is to give second order necessary optimality conditions for an optimal control problem (we label it (P)) in which control variables are represented by vector measures ("impulsive" control variables) and state trajectories are functions of bounded variation. Specifically, we consider the following problem:

$$\text{Minimize } J(t_0, t_1, x(t_0), x(t_1)) \tag{1}$$

$$\text{subject to } l(t_0, t_1, x(t_0), x(t_1)) \leq 0 \, , \tag{2}$$

$$k(t_0, t_1, x(t_0), x(t_1)) = 0 \, , \tag{3}$$

$$dx(t) = f(t, x(t))dt + G(t, x(t))\nu(dt) \, , \tag{4}$$

$$\nu(A) \in V \text{ for all } A \in B(T) \, . \tag{5}$$

Here the time interval $T = [t_0, t_1]$ is free, the dimensions of l, k, f are $d(l), d(k), d(x)$ respectively, $B(T)$ denotes the Borel subsets of T , V is closed convex cone in $R^{d(\nu)}$, the "impulsive" control ν is V-valued regular measure on $B(T)$, and G is $d(x) \times d(\nu)$ matrix.

The problem (P) is a relaxational extension of a conventional optimal control problem described by the ordinary differential system

$$\dot{x} = f(t, x) + G(t, x)u \, , \quad u(t) \in V \tag{6}$$

with measurable bounded controls $u(\cdot)$ instead (4), (5). Such problem is "degenerate" in the sense that, as a rule, it has no optimal solution in a traditional class functions $x(\cdot), u(\cdot)$ because the set V is unbounded and the system is linear with respect to control. In this degenerate case trajectories of a minimizing sequences converge to a discontinuous function and controls have a delta-like constituents and time's intervals of singular extremals. Such properties often arise in applications to robototechnique, economic, lazer influence, flight dynamics (one of the examples is the famous Lawden's problem). Hence, degenerate optimal control problem must be extended by means passage to the impulsive controls instead conventional one. In this way we obtain the problem (P).

We derive a second-order necessary conditions for the problem (P) under the main assumption that dynamic equation (4) is robust (or stable) respectively to approximations of a generalized control (vector measure) by conventional controls (Miller, 1978; Zavalischin and Sesekin, 1991; Bressan and

Rampazzo, 1991). Robustness of equation (4) is equivalently to so-called Frobenius condition for the matrix G, i.e. for any $i, k \in \{1, \dots, d(\nu)\}$

$$G_{ix}(t,x)G_k(t,x) - G_{kx}(t,x)G_i(t,x) = 0 \ , \quad (7)$$

where G_i is ith column of G. If Frobenius condition is hold then for any given measure ν on $[t_0, t_1]$ and initial position $x(t_0-) = x_0$ there exists unique corresponding trajectory as a right continuous on $(t_0, t_1]$ function of bounded variation such that

$$x(t) = x_0 + \int_{t_0}^{t} f(\theta, x(\theta))d\theta +$$

$$+ \int_{t_0}^{t} G(\theta, x(\theta))\nu_c(d\theta) + \quad (8)$$

$$+ \sum_{s_i \le t} (z(1; s_i, c^i) - x(s_i-)) \ .$$

Here ν_c represents the continuous part of ν,

$$\nu_a(dt) := \sum c^i \delta_{s_i}$$

represents the atomic part, i.e. the points $s_i \in [t_0, t_1]$ are atoms of ν (times of impulses), δ_s denote Dirac measure at s and the vectors $c^i \in V$ are jumps of ν; the function $z^i(\tau) := z(\tau; s_i, c^i)$ are a solution of the so-called limiting system for equation (4)

$$\frac{dz}{d\tau} = G(s_i, z)c^i \ , \quad z(0) = x(s_i-) \ . \quad (9)$$

It follows from (8),(9) that $x(s_i) = x(s_i+) = z^i(1)$ for any time of impulse; this conditions define an admissible jumps of the robust solution of equation (4). Any robust solution described by equation (8) is certain generalized solution for ordinary controlled equation (6).

In the common case, i.e. if the condition (4) does not hold, the description of a discontinuous generalized solutions become more complicated since for given (T, ν, x_0) there exist an uncountable set of generalized solutions. Every such solution depends on a choosing of sequence conventional controls, approximating ν. It means that a corresponding dynamic optimization problem is not an impulse optimal control problem and may be naturally interpreted as certain problem on a set of sequences with specific properties. First optimality conditions for the such "unstable" dynamic optimization problem with discontinuous trajectories have been obtained in works (Vinter and Pereira, 1988; Miller, 1992; Motta and Rampazza, 1996). Undoubtedly, an investigation of the optimization problem with discontinuous trajectories without Frobenius condition is very important. However, there are many interesting theoretical questions related to the problem (P) with robust condition (7). One of them is finding a pair of close second-order necessary and sufficient conditions, what is similar to conditions for regular extremals (Osmolovskii, 1995) and for singular extremals (Dmitruk, 1994) in the conventional optimal control problem. Recall that recently there is no any second-order condition for the impulse optimal control with one-side constraint on control type (5) even under Frobenius condition (for the case $V = R^{d(\nu)}$ close pair optimality conditions have been obtained in the work (Dykhta, 1994)). Meanwhile, quadratic optimality condition must be rather effective namely in impulse control problem since impulsive extremals are singular, i.e. Pontryagin function H has nonstrong maximum in controls. It is the reason for our attention to second-order tests for the problem (P) over processes with robust solutions. Note that variants of Maximum principles for the problem (P) under Frobenius condition and fixed t_0, t_1 have been obtained in (Miller, 1982; Dykhta, 1996).

Brief about the method of proof. We assume that remained impulsive process (T_*, x_*, ν_*) has finite number of discontinuous times and measure ν_* has no continuous singular component. Under the assumption we may interpret the restrictions of admissible trajectories to any interval between jumps as component trajectories in the auxiliary optimal multiprocesses problem with intermediate state constraints (Clarke and Vinter, 1989). This auxiliary problem admits imbedding in some abstract smooth optimization problem on the convex set. Optimality conditions for original problem (P) have been obtained by means deciphering known necessary second-order test for abstract optimization problem (Baturin et al., 1990).

2. BASIC ASSUMPTIONS. A S-MINIMUM

Let us describe assumptions for the problem (P).

H1. The functions J, k, l, f, G are twice continuously differentiable.

H2. V is closed convex cone in $R^{d(\nu)}$ with nonempty interior.

H3. The matrix G satisfies Frobenius condition (7).

H4. The remained admissible process (T_*, x_*, ν_*) such that

$$\nu_*(dt) = u_*(t)dt + \sum_{s_i \in S_a} c^i \delta_{s_i} \ , \quad (10)$$

where $S_a \subset T_* := [t_{0*}, t_{1*}]$ is the finite set of atoms ν_*, the function $u_* : T_* \to V$ is piecewise continuous and its restriction on every continuity interval is smooth function (hence there exist limits $u_*(s\pm), \dot{u}_*(s\pm)$ at any discontinuity point).

Speaking of optimality, we should point out a type of local minimum under consideration. The type of minimum that will be defined now is based on assumption **H4** and on the conception of the graph convergence for functions of bounded variation (Marques, 1993).

Let $\phi : T \to R^k$ be right continuous function of bounded variation from interval $T = [t_0, t_1]$. We define the filled-in graph $\overline{gr}\phi$ by adding, if necessary, some line segments to the graph of ϕ in such a way that all its gaps, i.e.

$$\overline{gr}\phi = \{(t, w)\ R^1 \times R^k \mid t_0 \le t \le t_1 \text{ and } w \in [\phi(t-), \phi(t)]\} ,$$

where $[a, b]$ stands for the line segment between two points in R^k. If ϕ, g are two functions of bounded variation, we define

$$\rho(\phi, g) := \rho(\overline{gr}\phi, \overline{gr}g) ,$$

where $h(A, B)$ is Hausdorff distance between two nonempty compact sets.

Now for any admissible process (T, x, ν) denote by $w(t; \nu)$ the distribution function of ν and put $S = S_a \cup S_c$, where $S_c \subset (t_{0*}, t_{1*})$ is the set of jump times for u_* (it is possible that $S_a \cap S_c \neq \emptyset$).

Definition. We say that (T_*, x_*, ν_*) is a S-minimum if there exists no admissible sequence $\{T_n, x_n, \nu_n\}$ such that

$$\rho(x_n, x_*) \to 0 , \quad \rho(w(\nu_n), w(\nu_*)) \to 0 ,$$
$$J(T_n, x_n, \nu_n) < J(T_*, x_*, \nu_*) .$$

We suggest first and second-order necessary conditions for a S-minimum.

3. FIRST-ORDER NECESSARY CONDITIONS: A MAXIMUM PRINCIPLE

Let

$$F(t, x, u) = f(t, x) + G(t, x)u ,$$
$$H(t, x, \psi, u) = < \psi, F(t, x, u) > + \psi_t ,$$
$$H_0(t, x, \psi) = < \psi, f(t, x) > + \psi_t ,$$
$$L(b) = \alpha_0 J(b) + < \alpha, l(b) > + < \beta, k(b) > ,$$

where $b := (t_0, t_1, x_0, x_1)$ and $\lambda := (\alpha_0, \alpha, \beta, \psi(\cdot), \psi_t(\cdot))$ is tuple of Lagrange multipliers.

For the process (T_*, x_*, ν_*) denote by Λ the set of tuples λ satisfying the following conditions:

a)

$$\alpha_0 \ge 0 , \quad \alpha \ge 0 , \quad < \alpha, l(b_*) > = 0 ,$$
$$\sum_0^{d(l)} \alpha_i + \mid \beta \mid = 1 ;$$

b)

$$\dot{\psi} = -H_x(t, x_*(t), \psi(t), u_*(t)) \text{ on } T_* \setminus S_a ,$$
$$\dot{\psi}_t = -H_t(t, x_*(t), \psi(t), u_*(t)) , \text{ on } T_* \setminus S_a ,$$
$$\psi(t_{0*}-) = L_{x_0}(b_*) , \quad \psi(t_{1*}+) = -L_{x_1}(b_*) ,$$
$$\psi_t(t_{0*}-) = L_{t_0}(b_*) , \quad \psi_t(t_{1*}+) = -L_{t_1}(b_*) ;$$

c) at every point $s_i \in S_a$

$$\psi(s_i+) = p(1; s_i, c_*^i) , \quad \psi_t(s_i+) = q(1; s_i, c_*^i) ,$$

where the functions $p(\tau; \ldots), q(\tau; \ldots)$ are a solutions of the limiting systems for costate equations, i.e

$$\frac{dp}{d\tau} = -H_{xu}(s_i, z_*^i(\tau), p)c_*^i , \quad p(0) = \psi(s_i-) ,$$
$$\frac{dq}{d\tau} = -H_{tu}(s_i, z_*^i(\tau), p(\tau))c_*^i, q(0) = \psi_t(s_i-),$$

$z_*^i(\tau) := z(\tau; s_i, c_*^i)$ is the corresponding solution of the system (9);

d) the switching function $g(t) := H_u(t, x_*(t), \psi(t))$ is absolutely continuous on T_* and

$$< g(t), v > \le 0 \text{ for all } (t, v) \in T_* \times V ,$$
$$< g(t), u_*(t) > = 0 \text{ on } T_* \setminus S_a ,$$
$$< g(t), c_i^* > = 0 \ \forall s_i \in S_a ;$$

e) $H_0(t, x_*(t), \psi(t)) \equiv 0$ on (t_{0*}, t_{1*}) and if ν_* has no atom at the point t_{0*} (or t_{1*}), then $H_0 \mid_{t=t_{0*}} = 0$ (respectively $H_0 \mid_{t=t_{1*}} = 0$).

The next theorem gives the maximum principle (MP) for the problem (P).

Theorem 1. If (T_*, x_*, ν_*) is a S-minimum, then the set Λ is nonempty.

As usually, fulfillment of the MP does not quarantee the optimality of the given process. Therefore, other optimality conditions, strengthening MP, may be obtained via investigations of the second order.

4. SECOND ORDER NECESSARY CONDITIONS

To state the quadratic necessary conditions, we need to introduce, in addition to Λ, the critical cone of variation K and a quadratic form Ω^λ, defined on this cone.

Let $S = \{s_0, \dots, s_N\}$, where $t_{0*} \leq s_0 < s_1 < \dots < s_N \leq t_{1*}$. Without loss of generality we may suppose that $S_0 = t_{0*}, s_N = t_{1*}$, putting $c_i^* = 0$ for $S \setminus S_a$. Denote by η tuple of variation, i.e.

$$\eta := (\delta s_0, \dots, \delta s_N, \delta x(\cdot), \delta u(\cdot),$$
$$\delta z^0(\cdot), \dots, \delta z^N(\cdot), \delta c^0, \dots, \delta c^N) ,$$

where $\delta x(\cdot)$ is piecewise continuous function absolutely continuous on each interval of the set $(t_{0*}, t_{1*}) \setminus S$, $\delta u(\cdot) \in L_\infty$ on T_*, $\delta z^i(\tau) \in C^1$ on $[0,1]$, $\delta c^i \in R^{d(\nu)}$ for $i = 0, \dots, N$. Critical cone K consists of all tuples η such that (recall that $con(V - v_*) := \{\gamma(v - v_*) \mid v \in V , \gamma > 0\}$):

i)

$$\delta \dot{x} = F_x(t, x_*(t), u_*(t))\delta u +$$
$$+F_u(t, x(t))\delta u \quad \text{on } T_* \setminus S ,$$
$$\delta u(t) \in con(V - u_*(t)) ;$$
$$\frac{d\delta z^i}{d\tau} = \phi_z(s_i, z_*^i, c_*^i)\delta z^i + \phi_c(s_i, z_*^i, c_*^i)\delta c^i +$$
$$+\phi_t(s_i, z_*^i, c_*^i)\delta s_i \quad \text{for } \tau \in [0,1] ,$$
$$\delta c^i \in cone(V - c_*^i) , \; i = 0, \dots, N ,$$

where $\phi(t, z, c) := G(t, z)c$;

ii)

$$\delta x^i(s_i-) + \dot{x}_*(s_i-)\delta s_i = \delta z^i(0) ,$$
$$\delta x^i(s_i+) + \dot{x}_*(s_i+)\delta s_i =$$
$$= \delta z^i(1) , \; i = 0, \dots, N ,$$
$$\delta x(s_0-) := \delta z^0(0) , \; \delta x(s_N+) := \delta z^N(1) ;$$

iii)

$$< J_b(b_*), \delta b > \leq 0 , \; k_b(b_*)\delta b = 0 ,$$
$$< l_{jb}(b_*), \delta b > \leq 0 \text{ for } j \in \{j \mid l_j(b_*) = 0\} ,$$

where

$$\delta b := (\delta t_0, \delta t_1, \delta x(t_{0*}-), \delta x(t_{1*})) =$$
$$= (\delta s_0, \delta s_N, \delta x(s_0-), \delta x(s_N+)) .$$

It is clear that the cone K may be obtained by certain linearization of the dynamic equation, the limiting system, the jump conditions and other data of the problem (P).

Now we introduce the quadratic form Ω^λ. For any $\lambda \in \Lambda$ and $\eta \in K$, let

$$\Omega^\lambda(\eta) =< L_{bb}(b_*)\delta b, \delta b > -$$
$$- \int\limits_{T_*} d^2 H(t, x_*, \psi, u_*) -$$
$$- \sum_0^N (\int_0^1 d^2 h(s_i, z_*^i, p^i, c_*^i) d\tau +$$
$$+2[< \dot{\psi}, \delta x >]_{s_i} \delta s_i + [< \psi, \ddot{x}_* >]_{s_i} \delta s_i^2) ,$$

where

$$d^2 H = (H_{xx}\delta x, \delta x) + 2(H_{xu}\delta u, \delta x) ,$$
$$h(t, z, p, c) := (p, \phi(t, z, c)) ,$$

the second differential $d^2 h$ taken with respect to z, c, t and depends on $\delta z^i, \delta c^i, \delta s_i$, $p^i(\tau) := p(\tau; s_i, c_*^i)$, the symbol $[w]_{s_i}$ denotes the jump of the function w at s_i, i.e. $[w]_{s_i} := w(s_i+) - w(s_i-)$. Note that two first terms in Ω^λ coincide with conventional the second variation of Lagrange function; other terms correspond to jumps of the "scattered" control $u_*(\cdot)$ and to points of impulses.

Define the functional

$$\omega(\eta) = \max_{\lambda \in \Lambda} \Omega^\lambda(\eta) .$$

Theorem 2. If (T_*, x_*, ν_*) is a S-minimum, then the set Λ is nonempty and the functional $\omega(\eta)$ is nonnegative on cone K.

Thus, a necessary condition for S-minimum is determined by the behavior of the maximum of quadratic forms Ω^λ over the compact set Λ on the critical cone K. This test is certain analogue of the classical Jacoby conditions in calculus variation in the form $\delta^2 J \geq 0$.

The following questions of interest are:

1) Which necessary conditions, having a pointwise character (similiar to classical Legengre test), succeed from theorem 2?

2) What strengthening of the necessary conditions leads to sufficient conditions for S-minimum?

These questions may be the subjects of a future studies.

Aknowledgements. This research was partially supported by RFFI grant 96-01-01739 and Gos. Com. Vuz. grant 95-0-1.9-58.

REFERENCES

Baturin, V.A., V.A.Dykhta, A.I.Moscalenko, et al. (1990). *Methods of solving control theory problems based on extension principle*, pp.5-48. Nauka, Novosibirsk. (In Russian).

Bressan, A. and F.Rampazzo (1991). Impulsive control system with commutative vector fields. *J.Optim.Theory Appl.* **Vol.53, No 5**, pp.50-58.

Clarke, F.H. and R.B.Vinter (1989). Optimal multiprocesses. *SIAM J. Control Optim.* **Vol. 27, No 5**, pp.1072-1094.

Dmitruk, A.V. (1994). Second order optimality conditions for singular extremals. In: *Computational Optimal Control* (R.Bulirsch and D. Kraft (Ed.)), **Vol.115**, pp.71-81, ISMN.

Dykhta, V.A. (1994). A variational maximum principle and quadratic conditions of optimality for impulsive and singular processes. *Sib.Math.J.* **Vol.35, No 1**, pp. 70-82. (In Russian).

Dykhta, V.A. (1996). Optimality conditions for impulsive processes and it's applications. *13th World Congress of IFAC. San Francisco, CA, USA, June 30 - July 5, 1996. Preprints. Vol.D: Control design II, Optimization*, pp. 345-350.

Marques Manuel D.P. (1993). *Differential inclusion in nonsmooth mechanical problems*, 182 p. Birkhauser. Basel-Boston-Berlin.

Miller, B.M. (1978). Stability of solutions of ordinary differential equations with measure. *Uspekhi Mat.Nauk.* **Vol.198, No 2**, p.198. (In Russian).

Miller, B.M. (1982). Optimality conditions in a problem of control system discribed by differential equation with measure. *Automat. Remote Control.* **Vol.43, No 5**, pp. 60-72. (In Russian).

Miller, B.M. (1992). Optimality conditions in problems of generalized control. *Automat. Remote Control.* **Vol.53, No 5**, pp.50-58.

Motta, M. and F.Rampazzo. Dynamic programming for nonlinear systems driven by ordinary and impulsive controls. *SIAM J.Control Optim.* **Vol.34, No 1**, pp.199-225.

Osmolovskii, N.P. (1995). Quadratic conditions for nonsingular extremal in optimal control (a theoretical treatment). *Russian J. Math. Physics.* **Vol.2, No 4**, pp.487-516.

Vinter, R.B. and F.L.Pereira (1988). A Maximum principle for optimal processes with discontinuous trajectories. *SIAM J.Control Optim.* **Vol.26, No 1**, pp.205-229.

Zavalischin, S.T. and A.N. Sesecin (1991). *Impulsive processes. Models and Applications*, 256 p. Nauka, Moscow. (In Russian).

OPTIMIZATION OF IMPULSE CONTROL IN ONE PROBLEM OF GUIDANCE WITH INCOMPLETE INFORMATION

Dmitry D. Emelyanov*

* *Institute of Control Sciences, 65 Profsoyuznaya Str., Moscow 117806, Russia, E-mail: emelian@ipu.rssi.ru*

Abstract: The planar problem of guidance with incomplete information is considered. The impulse-controlled pursuer is guided to the target which moves with constant unknown acceleration. The aim of guidance is to minimize distance between pursuer and target to ensure target impact. During the guidance passive angle-type measurements are performed. The closing velocity is constant, both pursuer's control and target's acceleration are normal to the closing velocity vector. Guidance algorithm is developed, using a description of "target-pursuer" position in terms of information sets. The algorithm is numerically tested, simulation results are presented.

Keywords: Impulses, Minimax techniques, Optimization problems, Monte Carlo simulation.

1. INTRODUCTION

The promising approach to the analysis of many problems of control with incomplete information has been developed in recent years. This approach is concerned with describing positions of the controlled system by so-called information set (IS), all points of which are compatible with the system's dynamics, control history, measurement model and observation history (Kurzhanskii, 1977), (Chernou'sko and Melikyan, 1978). In the framework of this approach the original control problem is reduced to a problem of control of the IS evolution. Usually, the aim of such control is to minimize some performance index as a function of information set.

Using this approach the planar problem of guidance with incomplete information is investigated. The impulse-controlled pursuer **P** is guided to the target **E** which moves with constant unknown acceleration. During the guidance the pursuer performs passive angle-type measurements corrupted by additive uniformly distributed noise. An important feature of the problem is the fact that the closing velocity is constant and independent of the pursuer's control, because both pursuer's impulse control and target's acceleration are normal to the closing velocity vector. The aim of guidance is to minimize a distance between **P** and **E** to ensure target impact. Similar problems were examined in works (Kumkov and Patsko, 1995; Kumkov and Patsko, 1996) in the context of minimax filtration and theory of differential games.

As the criterion for the IS evolution control the miss averaged over IS is taken. The pursuer's control is chosen in the class of so-called "conditionally programmed (CP) controls", i.e. programmed controls which are updated after each measurement. To update the control the auxiliary optimal impulse control problem is solved. In this problem a performance index has the meaning of expense of impulse control over time-to-go, and a terminal condition means zero average miss. The auxiliary problem is transformated to the quadratic mathematical programming problem which is solved using Kuhn-Tucker optimality conditions.

To demonstrate the efficiency of the constructed CP control the guidance process was simulated on a computer. The CP control method was tested and compared with the "strategy of symmetric miss" (SM) (Kumkov and Patsko, 1995; Kumkov and Patsko, 1996). The results shows that the CP control has an advantage over the strategy of SM both in the sense of accuracy of guidance and impulse contol expense.

2. PROBLEM FORMULATION

The guidance geometry is shown in Fig.1 in axes fixed to the target **E**.

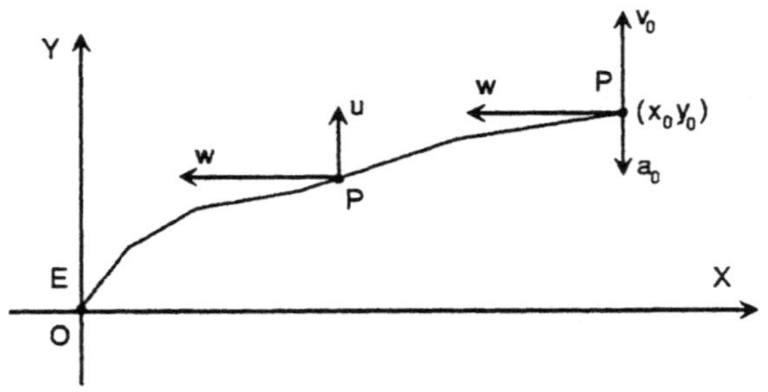

Fig. 1. Problem geometry

The relational motion of the pursuer **P** and the target **E** in the Cartesian reference system **OXY** is described by the equations:

$$x(t) = x_0 + w(t - t_0), \tag{1}$$

$$y(t) = y_0 + \int_{t_0}^{t} v(s)\, ds, \tag{2}$$

$$v(t) = v_0 - a_0(t - t_0) + \sum_{t_i \le t} u(t_i), \tag{3}$$

where t_0 – the initial instant of time, x, y – coordinates of the pursuer, v – it's velocity along **OY** axis, a_0 – target's acceleration, w – given constant velocity, $w < 0$, $u = u(t)$ – pursuer's impulse control represented by a totality of impulses $u_i = u(t_i)$ applied at fixed instants $t_i = t_0 + i\Delta$, $i = 0, 1, \ldots$, Δ – given time interval.

The initial parameters x_0, y_0, v_0, a_0 are assumed to be unknown but restricted by the inequalities

$$\begin{aligned} 0 < x_1 \le x_0 \le x_2, \quad & y_1 \le y_0 \le y_2, \\ v_1 \le v_0 \le v_2, \quad & a_1 \le a_0 \le a_2, \end{aligned} \tag{4}$$

where x_1, x_2, y_1, y_2, v_1, v_2, a_1, a_2 – given constants.

During the guidance the pursuer performs measurements $\zeta_i = \zeta(t_i)$ described by the equation

$$\zeta_i = \frac{y(t_i)}{x(t_i)} + \eta_i, \qquad i = 0, 1, \ldots, \tag{5}$$

where η_i – uniformly distributed measurement noise:

$$|\eta_i| \le c, \tag{6}$$

c – given constant. The observation process and control process are terminated when

$$x(t_f) < \varepsilon, \tag{7}$$

where ε – given small distance, $\varepsilon > 0$.

The problem of guidance is to construct (using the measurements $\{\zeta_i\}, t_i \le t$) the impulse control $u(t) = \{u_j\}, t \le t_j < t_f$ which minimizes the miss $|y(\theta)|$ at the instant θ when $x(\theta) = 0$.

3. THE AUXILIARY IMPULSE CONTROL PROBLEM

To solve the problem stated above an approach concerned with construction of information sets is taken. The dynamics of the problem is given by (1)-(3), and the nonlinear measurement function is described by (5). In order to obtain linear form of the measurement function and facilitate the construction of information set, the following change of coordinates is made:

$$\{x, y, v, a\} \longrightarrow \{x, \varphi, \omega, a\}, \tag{8}$$

where

$$\varphi(t) = \frac{y(t)}{x(t)}, \qquad \omega(t) = \frac{d\varphi}{dt}.$$

In terms of variables (x, φ, ω, a) the equations of the relational motion of **P** and **E** (1)-(3) take the form

$$x(t) = x_0 + w(t - t_0), \tag{9}$$

$$\varphi(t) = \varphi_0 + \int_{t_0}^{t} \omega(s)\, ds, \tag{10}$$

$$\omega(t) = \omega_0 - \int_{t_0}^{t} \frac{2w\omega(s) + a(s)}{x(s)}\, ds + \sum_{t_i \le t} \frac{u_i}{x(t_i)}, \tag{11}$$

$$a(t) = a_0, \tag{12}$$

where the initial parameters x_0, φ_0, ω_0, a_0 are restricted by the inequalities which are obtained directly from (4) by changing the variables (8).

$$\begin{aligned} x_1 \le x_0 \le x_2, \quad & \varphi_1(x_0) \le \varphi_0 \le \varphi_2(x_0), \\ a_1 \le a_0 \le a_2, \quad & \omega_1(x_0) \le \omega_0 \le \omega_2(x_0), \end{aligned} \tag{13}$$

where

$$\varphi_1(x) = \frac{y_1}{x}, \quad \omega_1(x) = \frac{v_1 x - w y_1}{x^2}$$

$$\varphi_2(x) = \frac{y_2}{x}, \quad \omega_2(x) = \frac{v_2 x - w y_2}{x^2}.$$

The linear measurement model is given by

$$\zeta_i = \varphi(t_i) + \eta_i, \qquad i = 0, 1, \ldots, \qquad (14)$$

Let's denote the information set associated with the dynamics (9)-(12), the restrictions (13) and the model (14), (6), (7) as I. In this study the informational set is defined in a recursive way by describing it's evolution. The initial information set I_0 before processing any measurements is given by the inequalities (13). Let's denote information set after processing the measurements ζ_i, $i = 0, \ldots, j-1$ and before processing the measurement ζ_j as I_j. By definition, the information set consists of points which are compatible with the measurement model and the observation history. It follows from (14), (6), (7) that any point (x, φ, ω, a) compatible with the measurement ζ_j should satisfy the conditions

$$\zeta_j - c \le \varphi \le \zeta_j + c, \qquad x \ge \varepsilon.$$

Thus, after processing the measurement ζ_j, the set I_j takes the form

$$\widehat{I}_j = I_j \cap H_j, \qquad (15)$$

where set H_j is given as

$$H_j = \{\varphi \in [\zeta_j - c,\, \zeta_j + c],\, x \ge \varepsilon\}.$$

The evolution of the information set on the interval $[t_j, t_{j+1}[$ between measurements is given by the equation

$$I_{j+1} = \mathcal{L}(u_j)\,\widehat{I}_j, \qquad (16)$$

where $\mathcal{L}(\cdot)$ – one-step extrapolation operator, which can be found from (9)-(12) by integrating over $[t_i, t_i + \Delta]$ with some control u_i.

$$\mathcal{L}(u_i): \begin{cases} x' &= x + w\Delta, \\ \varphi' &= \varphi + (\omega + u_i/x)q\Delta - \\ & \quad -a\Delta(1-q)/(2w), \\ \omega' &= (\omega + u_i/x)q^2 - \\ & \quad -a(1-q^2)/(2w), \\ a' &= a, \end{cases} \qquad (17)$$

where $(x', \varphi', \omega', a') \in I_{i+1}$, $(x, \varphi, \omega, a) \in \widehat{I}_i$, and

$$q = \frac{x}{x + w\Delta}.$$

Note that, the dependence of $\mathcal{L}(u)$ on the control variable u is linear. Thus, the evolution of the information set I is defined completely by the equations (15),(16) and the initial conditions (13).

To formulate a problem of information set control let's introduce: t_n – the instant of control process termination, $\widetilde{I}_n(u)$ – information set extrapolated to the instant t_n, and $\Pi_n(u)$ – terminal miss averaged over set $\widetilde{I}_n(u)$. By definition,

$$t_n = t_k + \Delta \min_{A \in \widehat{I}_k} \left[\frac{\varepsilon - x_A}{w\Delta}\right], \qquad (18)$$

$$\widetilde{I}_n(u) = \mathcal{L}(u_n)\,\mathcal{L}(u_{n-1}) \ldots \mathcal{L}(u_k)\,\widehat{I}_k, \qquad (19)$$

$$\Pi_n(u) = \frac{1}{Q_n} \int\limits_{\widetilde{I}_n(u)} \pi \, dx\, d\varphi\, d\omega\, da, \qquad (20)$$

where t_k – an instant taken as initial, x_A – x-coordinate of point A, square brakes denote an integer part of an expression, $u = \{u_i\}$, $i = k, \ldots, n$, Q_n – phase volume of $\widetilde{I}_n$ assumed to be non-zero, π – terminal miss function given by the equation

$$\pi = \pi(A) = -a\frac{x^2}{2w^2} - \omega\frac{x^2}{w}, \qquad (21)$$

where $A = (x, \varphi, \omega, a)$ – a point of the information set. In terms of t_k, t_n and $\Pi_n(\cdot)$, the auxiliary problem of the impulse control of the informational set evolution can be formulated as follows.

Problem 1. Let t_k be an initial instant, $\widehat{I}_k$ be an informational set after processing measurement ζ_k and t_n be an instant of control process termination given by (18). It is required to choose the impulse control $u = \{u_i\}$, $i = k, \ldots, n$ in (19) in order to minimize the performance index

$$G_n(u) = \sum_{i=k}^{n} u_i^2 \longrightarrow \min_u, \qquad (22)$$

and satisfy the terminal condition

$$\Pi_n(u) = 0. \qquad (23)$$

4. SOLUTION OF THE AUXILIARY CONTROL PROBLEM

It follows from (17) that the information set extrapolation (19) can be described by the following equations

$$x' = x + l_{nk},$$

$$\varphi' = \varphi + \frac{\Delta(n-k)}{x + l_{nk}} \left(\omega x - \frac{a\Delta}{2}(n-k) + \right.$$

$$+\sum_{i=k}^{n}\frac{n-i}{n-k}u_i\Bigg),$$

$$\omega'=\frac{1}{(x+l_{nk})^2}\left(\omega x^2+\sum_{i=k}^{n}(x+l_{ik})\,u_i-\right.$$

$$\left.-\frac{a}{2w}l_{nk}(2x+l_{nk})\right),$$

$$a'=a,$$

where $(x',\varphi',\omega',a')\in\widetilde{I}_n(u)$, $(x,\varphi,\omega,a)\in\widehat{I}_k$, $l_{ik}=w\Delta(i-k)$, $i=k,\ldots,n$. These equations define a change of variables in the integral (20). After changing variables, the terminal condition (23) takes the form

$$F_m+\sum_{i=k}^{n}F_{ik}u_i=0, \qquad (24)$$

where

$$F_m=\int_{\widehat{I}_k}\left(\omega+\frac{a}{2w}\right)\frac{x^4\,dx\,d\varphi\,d\omega\,da}{(x+l_{nk})^2}, \qquad (25)$$

$$F_{ik}=\int_{\widehat{I}_k}\frac{x^3+x^2l_{ik}}{(x+l_{nk})^2}\,dx\,d\varphi\,d\omega\,da. \qquad (26)$$

Thus, the Problem 1 is transformed to the quadratic programming problem with the pay-off function (22) and linear condition (24). To solve this problem, let's introduce the following Lagrangian function

$$L(u,\lambda)=\sum_{i=k}^{n}u_i^2+\lambda\left(F_m+\sum_{i=k}^{n}F_{ik}\,u_i\right),$$

where λ is a Lagrange multiplier, $\lambda\geq 0$. It follows from Kuhn-Tucker theorem that the optimality condition for the control is

$$\frac{\partial L(u,\lambda)}{\partial u_i}=0, \qquad i=k,\ldots,n. \qquad (27)$$

From (27) and (24) the optimal control u_i^*, $i=k,\ldots,n$ and λ are obtained:

$$u_i^*=-\frac{F_{ik}\,F_m}{S_{nk}}, \qquad \lambda=\frac{2\,F_m}{S_{nk}},$$

where $S_{nk}=\sum_{i=k}^{n}F_{ik}^2$. The optimal impulse u_k^* is given as

$$u_k^*=-\frac{F_{kk}\,F_m}{S_{nk}}. \qquad (28)$$

5. GUIDANCE ALGORITHM

Using the solution (28) of the Problem 1, the pursuer's control can be constructed in the class of so-called conditionally programmed (CP) controls, i.e. programmed controls which are updated after each measurement. The CP guidance algorithm is formulated as follows. The algorithm consists of four steps.

1) Measurement ζ_k processing (15).

2) Calculation of instant t_n, number n and integrals (25), (26).

3) Calculation of optimal impulse u_k^*, using (28).

4) One-step extrapolation (16) of the information set with optimal control u_k^*.

This updating procedure given by steps 1)–4) is then repeated until the observation process is terminated.

6. NUMERICAL RESULTS

The guidance process (1)-(7) was simulated on a computer with the pursuer employing two different control algorithms. First algorithm was the conditionally programmed control (CP) described in previous section and second one was the strategy of symmetric miss (SM) developed in (Kumkov and Patsko, 1995; Kumkov and Patsko, 1996). For the problem under study the SM impulse control is calculated as follows. At each instant t_k two points A^-, A^+ in the information set $\widehat{I}_k$ are chosen in order to minimize and maximize the terminal miss function π given by (21). The control impulse u_k is found from the condition of symmetric miss:

$$u_k=-w\frac{\pi^++\pi^-}{x^++x^-},$$

where x^+, x^- are x-coordinates of points A^+, A^-, and

$$\pi^+=\max_{A\in\widehat{I}_k}\pi(A), \qquad \pi^-=\min_{A\in\widehat{I}_k}\pi(A).$$

For the purpose of simulation, the informational set is implemented numerically by using it's approximation by set of polyhedrons P_l, $l=0,\ldots,N$, constructed in the space $\{\varphi,\omega,a\}$:

$$I\approx\bigcup_{l=0}^{N}\{x=x_l,\{\varphi,\omega,a\}\in P_l\},$$

where $N+1$ is number of polyhedrons, $x_l=x_l(t)$ is x-coordinate corresponding to the polyhedron P_l. The initial set $\{P_l\}$ is defined by (13) with discrete parameter $x_0=x_l(t_0)$ given by the equation

$$x_l(t_0)=x_1+\frac{x_2-x_1}{N}l, \qquad l=0,\ldots,N.$$

During the guidance the evolution of each polyhedron is described by equations (15), (16), where $x(t) = x_l(t)$.

The simulation starts with the following initial conditions:

$$w = -4000 \text{ m/sec}, \ \Delta = 0.1 \text{ sec},$$
$$\varepsilon = 500 \text{ m}, \qquad c = 0.002 \text{ rad}.$$

The initial uncertainties in the relational coordinates x_0, y_0, velocity v_0 and the target's acceleration a_0 are defined by the inequalities (4) with the following parameters:

$$x_1 = 75000 \text{ m}, \quad y_1 = -1000 \text{ m},$$
$$x_2 = 85000 \text{ m}, \quad y_2 = 1000 \text{ m},$$
$$v_1 = -100 \text{ m/sec}, \ a_1 = -5 \text{ m/sec}^2,$$
$$v_2 = 100 \text{ m/sec}, \quad a_2 = 5 \text{ m/sec}^2.$$

The initial parameters x_0, y_0, v_0 and a_0 are assumed to be random variables uniformly distributed in the following intervals

$$x_0 \in [78000, 82000] \text{ m}, \ y_0 \in [-200, 200] \text{ m},$$
$$v_0 \in [-20, 20] \text{ m/sec}, \quad a_0 \in [-2, 2] \text{ m/sec}^2.$$

The performance of both CP and SM control algorithms is evaluated by using Monte-Carlo technique. For each realization of the random variables x_0, y_0, v_0, a_0 and random noise $\{\eta_i\}$ the guidance process is simulated with the pursuer employing SM and CP control algorithms. The number of Monte-Carlo trials for each control algorithm is 1000. After each trial two values are calculated - the terminal miss $M = |y(\theta)|$ and the total expense of impulse control r given by the equation.

$$r = \sum_{t_i \leq \theta} |u(t_i)|.$$

The simulation results are presented in Fig. 2, 3 and summarized in Table 1.

Table 1. Simulation results.

Control algorithm	M, m	r, m/sec
CP control	0.00 ± 0.45	55.4 ± 13.6
SM strategy	0.00 ± 0.51	258.8 ± 7.7

The obtained results show the following.

1) The developed CP guidance algorithm has an advantage over SM strategy in the sense of accuracy of guidance, because the r.m.s. deviation of the terminal miss generated by CP algorithm is lower than that for the SM algorithm.

2) An important result is that the total expense of impulse control for CP guidance algorithm is lower by a factor of 4-5 than that for SM algorithm.

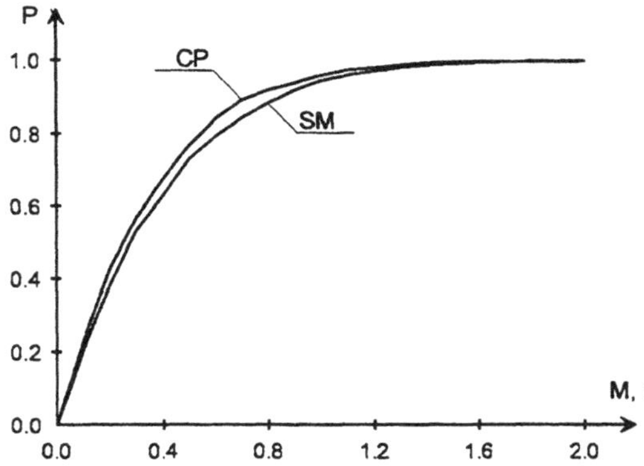

Fig. 2. Terminal miss distribution.

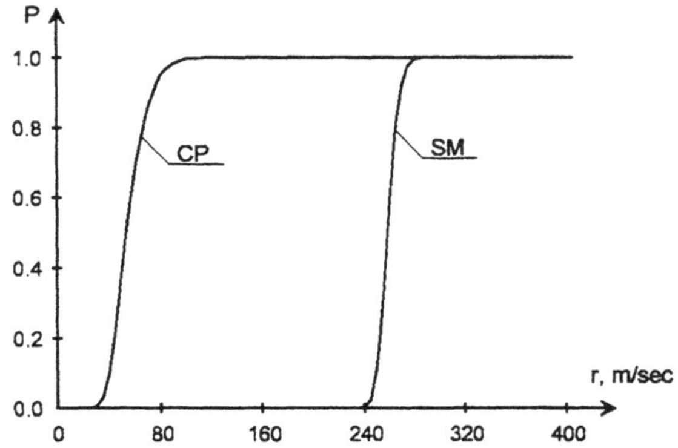

Fig. 3. Impulse control expense distribution.

7. CONCLUSIONS

The problem of guidance with incomplete information has been considered. To solve the problem an approach concerned with a construction of information sets (IS) has been used. In terms of IS the original problem has been reduced to an auxiliary impulse control problem of the IS evolution. The auxiliary control problem has been solved, and the guidance algorithm has been developed. The performance of the developed algorithm has been shown to be superior to that of the strategy of symmetric miss, especially in the sense of impulse control expense.

8. REFERENCES

Chernou'sko, F.L. and A.A. Melikyan (1978). *Game problems of control and searching.* Nauka. Moscow.

Kumkov, S.I. and V.S. Patsko (1995). Optimal strategies in a pursuit problem with incomplete information. *J. Appl. Math. Mech.* **59**, 75–85.

Kumkov, S.I. and V.S. Patsko (1996). A model problem of space vehicle homing. In: *Proceedings of the Seventh International Symposium on Dynamic Games and Applications* (J.A. Filar, Ed.). Vol. I. Dec. 16-18, 1996, Kanagawa, Japan. pp. 547–556.

Kurzhanskii, A.B. (1977). *Control and observation in uncertainty conditions.* Nauka. Moscow.

CONSTRAINED FEEDBACK CONTROL OF IMPERFECTLY KNOWN, LINEAR, TIME-DELAY SYSTEMS OF NEUTRAL TYPE

D.P. Goodall

Control Theory and Applications Centre, Coventry University, Priory Street, Coventry CV1 5FB, U.K.

Abstract: A class of robust continuous feedback controls, with memory, is synthesized and a delay-free stability criterion is presented for a class of imperfectly known time-delay systems of the neutral type, when there are constraints on the control input. The uncertain systems are modelled as nonlinear perturbations to a known linear neutral time-delay control system and, as well as being time and state dependent, the nonlinear perturbations are assumed to be state delay-dependent and input dependent. Moreover, prior information on the bound of the system uncertainty is required. A deterministic approach is taken and, using a Lyapunov-Krasovskiĭ functional, a global uniform asymptotic stability property is investigated for the class of systems.

Keywords: Control constraints; feedback stabilization; time-delay systems of the neutral type; uncertain dynamic systems.

1 INTRODUCTION

For real systems, mathematical models always contain uncertain elements which model the designer's lack of knowledge about parameter values, disturbances and the imprecision in the model. In addition, real dynamical control systems often incorporate time-delay elements. Uncertainties, arising from imperfect knowledge of system inputs and inaccuracies in the mathematical model, and time-delay components may contribute to performance degradation of the feedback control system. Problems due to uncertainties and time-delays in the system need to be addressed. Two approaches have been widely used to obtain desired system response, namely stochastic and deterministic. In this paper, a deterministic approach is taken, where the imperfectly known systems are modelled as a class of nonlinear perturbations influencing a known nominal linear time-delay system of the neutral type.

The main objective of this investigation is that of designing robust feedback controls to (asymptotically) stabilize a class of imperfectly known, linear time-delay systems of the neutral type, subject to control constraints. Neutral time-delay systems have been investigated, for the linear case, in (Byrnes, *et al.*, 1984; Fliess, *et al.*, 1995), amongst others, and, for the nonlinear case, in (Bellen, *et al.*, 1995; Kuang and Feldstein, 1991; Kolmanovskii and Nosov, 1979; Tchangani, *et al.*, 1996). Uncertain systems with control constraints have been studied by a number of researchers, including (Blanchini, 1991; Corless and Leitmann, 1993; Sussmann, *et al.*, 1994).

Continuous controllers, with possibly high gains, have been used to stabilize the imperfectly known, delay-free systems; however, early work with such controllers only ensured global uniform ultimate boundedness, which is a weaker property than global uniform asymptotic stability. To overcome this problem, discontinuous controllers have been synthesized which guarantee global uniform asymptotic stability of the zero state, as shown, for example, in (Goodall and Ryan, 1988). Recently, a new type of continuous controller, described in (Wu and Mizukami, 1993) and applied to time-delay problems in (Wu and Mizukami, 1996), has been used to obtain global uniform asymptotic stability results. Using a slightly modified controller, with the same structure as those described in (Wu and Mizukami, 1993), global uniform asymptotic stability can be achieved for a class of uncertain time-delay systems of the neutral type, subject to control constraints.

The uncertainty term, which additively perturbs the nominal model, is assumed to be both state delay-dependent and input-dependent and may include time-dependent or nonlinear elements. *A priori* bounding knowledge on the system uncertainties is also assumed. A deterministic approach, based on Lyapunov theory and the use of a Lyapunov–Krasovskiĭ functional, is adopted and is analogous to that used in (Goodall, 1995), where the imperfectly known linear systems are of the retarded type, *viz.* time-delays only occurring in the state vector. Here, a delay-independent stability criterion is presented for the class of uncertain linear time-delay systems of the neutral type with control constraints.

2 PRELIMINARIES AND PROBLEM STATEMENT

First some mathematical preliminaries are introduced. Define $\mathbb{R} := (-\infty,\ \infty)$, $\mathbb{R}^+ := (0,\ \infty)$ and $\mathbb{R}_0^+ := [0,\ \infty)$. Let the Euclidean inner product (on $\mathbb{R}^n$ or $\mathbb{R}^m$ as appropriate) and the induced norm be denoted by $\langle\cdot\,,\,\cdot\rangle$ and $\|\cdot\|$, respectively. Define, for a linear map L, $\|L\| := \{\max\sigma(L^{\mathrm{T}}L)\}^{\frac{1}{2}}$, where σ denotes spectrum. Let $C(\mathbb{R}^p;\mathbb{R}^q)$ denote the space of all continuous functions mapping $\mathbb{R}^p \to \mathbb{R}^q$. Given $T > 0$, the notation $x_t = x_t(r) := x(t+r)$ $(r \in [-T,\ 0])$ is introduced, which denotes the restriction of $x(\cdot)$ to the interval $[t-T,\ t]$. Note that if $x_t \in C([-T,\ 0];\mathbb{R}^p)$, then $\|x_t\|_T := \sup_{r\in[-T,\ 0]} \|x(t+r)\|$. In addition, a set of bounded functions in $C([-T,\ 0];\mathbb{R}^p)$ is defined by $\mathcal{Q}_A := \{q \in C([-T,\ 0];\mathbb{R}^p) : \|q\|_T < A\}$, with $0 < A < \infty$. Finally, $\mathbb{R}^{p\times p}$ represents the set of real square matrices of order $p \times p$ and, if $P \in \mathbb{R}^{p\times p}$, $P > 0$ denotes that P is positive definite.

2.1 Nominal time-delay system

The **nominal** time-delay system, considered here, is a linear system of the neutral type described by

$$\dot{x}(t) = G\dot{x}(t-T_d) + A_0x(t) + A_1x(t-T_d) + Bu(t)\ ,\quad T_d > 0\ ,\quad t \in (t_0,\ \infty), \tag{1}$$

$$x_{t_0}(\theta) = \phi(\theta),\ \theta \in [-T_d,\ 0] \text{ with } \phi(0) = x^0, \tag{2}$$

where $x(t) \in \mathbb{R}^n$ is the state vector, $u(t) \in \mathbb{R}^m$ $(1 \le m \le n)$ is the control (or input) vector, the constant matrices A_0, A_1, $G \in \mathbb{R}^{n\times n}$ are known, $B \in \mathbb{R}^{n\times m}$ is full rank, T_d is a known bounded delay which affects the system, and $\phi \in C([-T_d,\ 0];\mathbb{R}^n)$ is a given function specifying the initial condition. For this nominal system the following hypotheses are assumed to hold.

H1:

(a) $0 < \|G\| < 1$.

(b) Given symmetric $P_0, P_1, Q \in \mathbb{R}^{n\times n}$, with $P_0, P_1, Q > 0$, there exists a unique symmetric $K > 0$ which satisfies the condition

$$\tilde{S} := P_0 - G^{\mathrm{T}}KA_1 - A_1^{\mathrm{T}}KG + A_1^{\mathrm{T}}KP_1^{-1}KA_1 > 0 \tag{3}$$

and the algebraic Riccati-type matrix equation

$$KA_0 + A_0^{\mathrm{T}}K + A_0^{\mathrm{T}}KGP_0^{-1}G^{\mathrm{T}}KA_0 + Q + P_1 + \tilde{S} = O\ . \tag{4}$$

Remarks:

(*i*) If $P_0 - G^{\mathrm{T}}P_1G > 0$, then condition (3) is satisfied since

$$\tilde{S} = P_0 - G^{\mathrm{T}}P_1G + (P_1^{-\frac{1}{2}}KA_1 - P_1^{\frac{1}{2}}G)^{\mathrm{T}}(P_1^{-\frac{1}{2}}KA_1 - P_1^{\frac{1}{2}}G) \ge P_0 - G^{\mathrm{T}}P_1G\ .$$

(*ii*) If hypotheses H1 hold, then the existence of a symmetric $K > 0$ satisfying (3)-(4) guarantees that the open-loop time-delay system

$$\dot{x}(t) = G\dot{x}(t-T_d) + A_0x(t) + A_1x(t-T_d)$$

is asymptotically stable in the sense of Definition 2.1 (see Hale and Lunel (1993)), given below.

2.2 Class of uncertain neutral delay systems

The class of uncertain time-delay systems to be investigated consists of the nominal linear time-delay system (1) augmented by an unknown function, that is

$$\dot{x}(t) = G\dot{x}(t-T_d) + A_0x(t) + A_1x(t-T_d) + B\left[u(t) + g(t, x(t), x(t-T_d), u(t))\right] \tag{5}$$

$$x_{t_0}(\theta) = \phi(\theta),\ \theta \in [-T_d,\ 0] \text{ with } \phi(0) = x^0, \tag{6}$$

where the unknown function $g : \mathbb{R} \times \mathbb{R}^n \times \mathbb{R}^n \times \mathbb{R}^m \to \mathbb{R}^m$ belongs to a known nonempty class which comprises all possible uncertainties in the system description, including any known time-dependent or nonlinear elements, and satisfies the following boundedness assumption.

H2: For all $(t,p,q,r) \in \mathbb{R} \times \mathbb{R}^n \times \mathbb{R}^n \times \mathbb{R}^m$, there exist real constants $\beta, \gamma \geq 0$, $\xi \in [0,\ 1)$ and a real continuous function $\alpha : \mathbb{R}_0^+ \to [0,\ \bar{\alpha}]$, with $0 < \bar{\alpha} < \infty$, such that

$$|g(t,p,q,r)| \leq \alpha(t) + \beta \|p\| + \gamma \|q\| + \xi \|r\| .$$

2.3 Stability properties

Consider the system

$$\frac{\mathrm{d}}{\mathrm{d}t} a(t, x_t) = b(t, x_t), \quad t > t_0 \geq 0 , \tag{7}$$

$$x_{t_0}(\theta) = \phi(\theta) \in C([-T_d,\ 0]; \mathbb{R}^p),\ \phi(0) = x^0 \tag{8}$$

where $a(t,\theta) = \theta(0) - c(t,\theta)$. Here, $b, c : [t_0,\ \infty) \times \mathcal{Q}_A \to \mathbb{R}^p$ are continuous and satisfy $b(t,0) \equiv c(t,0) \equiv 0$, for all $t \in \mathbb{R}$. In addition, it is assumed that there exists $k > 0$ such that $\|b(t,\theta)\| < k$ for all $t \geq t_0$ and $\theta \in \mathcal{Q}_A$. A solution to (7)-(8), if it exists, is denoted by $x(\cdot, t_0, \phi)$. By solution, it is meant that $x(\cdot, t_0, \phi)$, defined on some interval $\mathcal{T}$, is an absolutely continuous function that satisfies (7)-(8) for almost all $t \in \mathcal{T}$.

Definition 2.1 *The solution $x = 0$ is* ***uniformly asymptotically stable*** *for system* (7)-(8) *if the following holds :*

> (*i*) existence of solutions: *for each $(t_0, \phi, x^0) \in \mathbb{R}_0^+ \times \mathcal{Q}_A \times \mathbb{R}^p$, there exists a solution $x(\cdot, t_0, \phi)$ defined on $[t_0 - T_d,\ \infty)$.*
>
> (*ii*) uniform stability: *for each $\varepsilon > 0$ and for every $t_0 \in \mathbb{R}_0^+$, there exists $\delta = \delta(\varepsilon) > 0$, which is independent of t_0, such that if $x^0 \in \mathbb{B}_p(\delta)$ then $x(t, t_0, \phi) \in \mathbb{B}_p(\varepsilon)$ for all $t \geq t_0$ on every solution $x(\cdot, t_0, \phi)$ of* (7)-(8).
>
> (*iii*) uniform attractivity: *for each $\varepsilon > 0$, $t_0 \in \mathbb{R}_0^+$ and $x^0 \in \mathbb{R}^p$, there exists $\delta > 0$, which is independent of t_0 and ε, and a real number $T_\varepsilon(x^0) \geq 0$, independent of t_0, such that $x(t, t_0, \phi) \in \mathbb{B}_p(\varepsilon)$ for all $t \geq t_0 + T_\varepsilon(x^0)$ on every solution $x(\cdot, t_0, \phi)$ of* (7)-(8) *whenever $x^0 \in \mathbb{B}_p(\delta)$.*

Moreover, if the above hold with δ arbitrarily large, then $x = 0$ is said to be ***globally uniformly asymptotically stable*** *with respect to system* (7)-(8).

2.4 Statement of the problem

It is assumed that the constraint

$$\|u\| \leq \rho \tag{9}$$

is imposed on the control input for system (5)-(6), where $\rho > 0$ is known. Given an initial state x^0 and ϕ satisfying (6), the main objective is to design a class of continuous state-feedbacks $\mathcal{F}$, satisfying the constraint (9), which globally stabilize system (5)-(6). More precisely, for each initial state x^0 and ϕ satisfying (6), every feedback control $u = f(x_t) \in \mathcal{F}$, with $\|u\| \leq \rho$ and ρ prescribed, guarantees that $x = 0$ is globally uniformly asymptotically stable under the dynamics of (5).

3 FEEDBACK CONTROLS WITH MEMORY

The class of feedback controls consists of nonlinear controls $u(t) = f(x_t) \in \mathcal{F}$, with memory, where each control consists of two parts; one component is a saturation-type function which stabilizes the nominal linear time-delay system (1)-(2) and the second component, loosely speaking, is used to counteract the uncertainty in the system. The feedback function is given by

$$f(x_t) := -\Bigg[\mu \mathrm{sat}\Big(B^{\mathrm{T}}K(x(t) - Gx_t(-T_d))\Big) + \frac{(\rho-\mu)B^{\mathrm{T}}K(x(t) - Gx_t(-T_d))}{\|B^{\mathrm{T}}K(x(t) - Gx_t(-T_d))\| + \nu\|x(t)\|^2}\Bigg], \tag{10}$$

where $\mu,\ \nu > 0$ are design parameters and

$$z \mapsto \mathrm{sat}(z) := \begin{cases} z\,, & z \in \overline{\mathbb{B}}_m(1) \\ z/\|z\|\,, & \text{otherwise} \end{cases} .$$

Suppose μ is chosen to satisfy $\mu < \rho$, then $u = f(x_t)$ satisfies the constraint (9) and, by definition, such feedback functions are continuous, as shown in (Wu and Mixukami, 1993). To guarantee stabilizability of the system using constrained feedbacks, a feasibility hypothesis is required, namely:

H3: $\rho > (1-\xi)^{-1} \times \left[\bar{\alpha} + \frac{1}{4}(\beta+\gamma)\left(\|B^{\mathrm{T}}K\| + \|B^{\mathrm{T}}KG\|\right)^{-1}\right]$.

4 SYSTEM STABILIZATION

System (5)-(6), with $u(t) = f(x_t)$, can be expressed in the form

$$\dot{x}(t) = G\dot{x}(t - T_d) + h(t, x_t)\ ; \quad t \geq t_0 \tag{11}$$

$$x_{t_0}(\theta) = \phi(\theta),\ \theta \in [-T_d,\ 0],\ \phi(0) = x^0, \tag{12}$$

where

$$h(t, x_t) := A_0 x(t) + A_1 x(t - T_d) + B\left[f(x_t) + g(t, x(t), x(t - T_d), f(x_t))\right] .$$

It is assumed that $h : [t_0,\ \infty) \times \mathcal{Q}_A \to \mathbb{R}^n$ is continuous and $\|h(t,\theta)\| < k$, $0 < k < \infty$, for all

$t \geq t_0$ and $\theta \in \mathcal{Q}_A$. The above conditions on h are sufficient for the existence of solutions (see Kolmanovskii and Myshkis (1992)).

To obtain the required stability property for system (11)-(12), with f given by (10), the following lemma will be invoked (see Theorem 1.4, Chapter 4, in (Kolmanovskii and Myshkis, 1992). Let Ω denote the set of scalar nondecreasing functions $\omega \in C([-T_d,\ 0];\mathbb{R})$ such that $\omega(r) > 0$ for $r > 0$ and $\omega(0) = 0$.

Lemma 4.1 *Consider system* (7)-(8) *and suppose there exists* $\omega_1 \in \Omega$ *and real* $0 < \zeta < 1$ *such that*

$$\|c(t+r,\psi) - c(t,\chi)\| \leq \omega_1(r) + \zeta \|\psi - \chi\|_{T_d}\ , \tag{13}$$

with $\psi,\ \chi \in \mathcal{Q}_A$, $t \geq t_0$ *and* $r \geq 0$. *If there exist* $\omega_2,\ \omega_3,\ \omega_4 \in \Omega$ *and a continuous functional* $V(t,\psi,\theta)$ *such that, for all* $(t,\psi,\theta) \in [t_0,\ \infty) \times \mathcal{Q}_A \times \mathcal{Q}_A$,

(a)

$$\omega_2(\|a(t,\psi)\|_{T_d}) \leq V(t,\psi,a(t,\psi)) \leq \omega_3(\psi)$$

(b) *along solutions to* (7)-(8),

$$\dot{V}(t,x_t,a(t,x_t)) + \omega_4(\|x(t)\|) \leq 0\ ,$$

then $x = 0$ *is uniformly asymptotically stable.*

Invoking Lemma 4.1 and utilizing a Lyapunov-Krasovskii functional candidate of the form:

$$v(\psi) := \langle \psi(0) - G\psi(-T_d),\ K(\psi(0) - G\psi(-T_d))\rangle + \int_{-T_d}^{0} \langle \psi(\theta),\ S\psi(\theta)\rangle\ d\theta\ , \tag{14}$$

where $\psi \in C([-T_d,\ 0];\mathbb{R}^n)$, $S = \tilde{S} + \lambda Q > 0$ and $0 \leq \lambda < 1$, one may deduce the following result. Let $\sigma_{\min}(\cdot)$ denote the minimum eigenvalue of a symmetric positive definite matrix.

Theorem 4.1 *Suppose* H1-H3 *hold. If*

$$\sigma_{\min}(Q) > 2(\beta+\gamma)\left(\|B^{\mathrm{T}}K\| + \|B^{\mathrm{T}}KG\|\right)\ , \tag{15}$$

then, with $u(t) = f(x_t)$ *and the design parameters* μ *and* ν *chosen to satisfy*

$$\frac{(\beta+\gamma)(1-\xi)^{-1}}{4\left(\|B^{\mathrm{T}}K\| + \|B^{\mathrm{T}}KG\|\right)} < \mu \leq \rho - \bar{\alpha}(1-\xi)^{-1} \tag{16}$$

and

$$\nu < \tfrac{1}{2}\Big[\sigma_{\min}(Q) - 2(\beta+\gamma) \times \left(\|B^{\mathrm{T}}K\| + \|B^{\mathrm{T}}KG\|\right)\Big]\bar{\alpha}^{-1}\ ,$$

$x = 0$ *is globally uniformly asymptotically stable for the class of systems modelled by* (5)-(6).

Proof: Identifying system (5)-(6) with (7)-(8), it is seen that $c(t,\theta) = G\theta$ and, hence, (13) holds with $\omega_1 \in \Omega$ arbitrary and $\zeta = \|G\|$.
To verify conditions (a) and (b) of Lemma 4.1 hold, an analysis is performed using the candidate Lyapunov-Krasovskii functional (14) with $S = \tilde{S}+\lambda Q$ and λ at present unassigned. Straightforward calculations show that $v(x_t)$ satisfies the inequality

$$\sigma_{\min}(K)\,\|x(t) - Gx_t(-T_d)\|^2_{T_d} \leq v(x_t) \leq \sigma_{\max}(K)\,\|x(t) - Gx_t(-T_d)\|^2_{T_d} + T_d\sigma_{\max}(S)\,\|x_t\|^2_{T_d}$$

and, therefore, property (a) of Lemma 4.1 is valid. As a consequence of H1 and applying the inequality

$$2|\langle v,\ w\rangle| \leq \langle v,\ Pv\rangle + \langle w,\ P^{-1}w\rangle\ , \text{ where } P > 0,$$

it can be shown that, along solutions to (11)-(12) with f defined by (10) and for almost all t,

$$\begin{aligned}\dot{v}(x_t) \leq\ & -(1-\lambda)\langle x(t),\ Qx(t)\rangle \\ & - \lambda\langle x_t(-T_d),\ Qx_t(-T_d)\rangle \\ & + 2\Big\langle B\big[f(x_t) + g(t,x(t),x_t(-T_d),f(x_t))\big]\ , \\ & \qquad K(x(t) - Gx_t(-T_d))\Big\rangle\ .\end{aligned}$$

Suppose $\left\|B^{\mathrm{T}}K(x(t) - Gx_t(-T_d))\right\| > 1$.
Then, in view of H2 and H3, it can be shown that

$$\begin{aligned}&\Big\langle B\big[f(x_t) + g(t,x(t),x_t(-T_d),f(x_t))\big]\ , \\ &K(x(t) - Gx_t(-T_d))\Big\rangle \leq \nu\alpha(t)\,\|x(t)\|^2 \\ &+ \left\|B^{\mathrm{T}}K(x(t) - Gx_t(-T_d))\right\| \times \\ &[\beta\,\|x(t)\| + \gamma\,\|x_t(-T_d)\| - \mu(1-\xi)]\ .\end{aligned} \tag{17}$$

Assume, for the present, that (A) $\beta+\gamma \neq 0$ (B) $2\gamma\left\|B^{\mathrm{T}}KG\right\|[\sigma_{\min}(Q)]^{-1} < \lambda < 1$, then along solutions to (11) and utilizing (17)

$$\dot{v}(x_t) \leq -\Big[(1-\lambda)\sigma_{\min}(Q) - 2\left(\beta\left\|B^{\mathrm{T}}K\right\| + \nu\alpha(t)\right) + \frac{\left(\beta\left\|B^{\mathrm{T}}KG\right\| + \gamma\left\|B^{\mathrm{T}}K\right\|\right)^2}{\lambda\sigma_{\min}(Q) - 2\gamma\left\|B^{\mathrm{T}}KG\right\|}\Big]\,\|x(t)\|^2\ ,$$

almost everywhere. The constant λ is chosen to be

$$\lambda = [\sigma_{\min}(Q)]^{-1}\Big((\beta+2\gamma)\left\|B^{\mathrm{T}}KG\right\| + \gamma\left\|B^{\mathrm{T}}K\right\|\Big)\ , \tag{18}$$

which minimizes

$$\lambda\sigma_{\min}(Q) + \frac{\left(\beta\left\|B^{\mathrm{T}}KG\right\| + \gamma\left\|B^{\mathrm{T}}K\right\|\right)^2}{\lambda\sigma_{\min}(Q) - 2\gamma\left\|B^{\mathrm{T}}KG\right\|}\ .$$

In view of (18), $\lambda > 2\gamma\left\|B^{\mathrm{T}}KG\right\|[\sigma_{\min}(Q)]^{-1}$, whilst, in view of (15),

$$\lambda\sigma_{\min}(Q) \leq 2(\beta+\gamma)\left(\left\|B^{\mathrm{T}}K\right\| + \left\|B^{\mathrm{T}}KG\right\|\right) < \sigma_{\min}(Q)\ ,$$

which confirms the supposition that $\lambda < 1$. Thus, along solutions to (11) and for almost all t,

$$\dot{v}(x_t) \le -\Big[\sigma_{\min}(Q) - 2(\beta+\gamma) \times \big(\|B^{\mathrm{T}}K\| + \|B^{\mathrm{T}}KG\|\big) - 2\nu\alpha(t)\Big] \|x(t)\|^2 \;;$$

whilst, if $\beta + \gamma = 0$, then, with $\lambda = 0$,

$$\dot{v}(x_t) \le -\Big[\sigma_{\min}(Q) - 2\nu\alpha(t)\Big] \|x(t)\|^2 \;.$$

Define

$$\varepsilon := \sigma_{\min}(Q) - 2(\beta+\gamma)\big(\|B^{\mathrm{T}}K\| + \|B^{\mathrm{T}}KG\|\big) > 0 \,. \tag{19}$$

Then, along solutions to (11), with f given in (10) and λ given by (18) and for all β and γ,

$$\dot{v}(x_t) \le -(\varepsilon - 2\nu\bar{\alpha}) \|x(t)\|^2 \;,$$

almost everywhere. Select $\nu < \frac{1}{2}\varepsilon\bar{\alpha}^{-1}$, then there exists $\delta > 0$ such that

$$\dot{v}(x_t) \le -\delta \|x(t)\|^2 \;,$$

and, hence, every solution will eventually (in finite time) satisfy the inequality

$$\big\|B^{\mathrm{T}}K(x(t) - Gx_t(-T_d))\big\| \le 1 \,.$$

The case $\big\|B^{\mathrm{T}}K(x(t) - Gx_t(-T_d))\big\| \le 1$ is now investigated. In view of H3 and assuming (C) $\beta\gamma \neq 0$ (D) $\gamma^2\left[2\mu(1-\xi)\sigma_{\min}(Q)\right]^{-1} < \lambda < 1$, it can be shown that, along solutions to (11) and for almost all t,

$$\dot{v}(x_t) \le -\left[(1-\lambda)\sigma_{\min}(Q) - 2\nu\alpha(t)) - \frac{\beta^2}{2\mu(1-\xi)} - \frac{(\beta\gamma[\mu(1-\xi)]^{-1})^2}{4\left(\lambda\sigma_{\min}(Q) - \frac{\gamma^2}{2\mu(1-\xi)}\right)}\right] \|x(t)\|^2 \,.$$

Now

$$\lambda\sigma_{\min}(Q) + \frac{\beta^2\gamma^2}{4\mu^2(1-\xi)^2}\left(\lambda\sigma_{\min}(Q) - \frac{\gamma^2}{2\mu(1-\xi)}\right)^{-1}$$

has a local minimum when

$$\lambda = \frac{\gamma(\beta+\gamma)}{2\mu(1-\xi)}[\sigma_{\min}(Q)]^{-1} \,. \tag{20}$$

Thus, with λ given by (20), $\lambda > \dfrac{\gamma^2[\sigma_{\min}(Q)]^{-1}}{2\mu(1-\xi)}$ and as a consequence of (16) and (19),

$$\lambda\sigma_{\min}(Q) = \frac{\gamma(\beta+\gamma)}{2\mu(1-\xi)} \le \sigma_{\min}(Q) - \varepsilon \,.$$

This implies $\lambda < 1$ and so condition (D) holds. Moreover, since $\nu < \frac{1}{2}\varepsilon\bar{\alpha}^{-1}$,

$$\dot{v}(x_t) \le -(\beta+\gamma)\left[2\big(\|B^{\mathrm{T}}K\| + \|B^{\mathrm{T}}KG\|\big) - \frac{(\beta+\gamma)}{2\mu(1-\xi)}\right] \|x(t)\|^2 \;,$$

along solutions to (5). The choice of design parameter satisfying (16) ensures that

$$\frac{(\beta+\gamma)}{2\mu(1-\xi)} < 2\big(\|B^{\mathrm{T}}K\| + \|B^{\mathrm{T}}KG\|\big)$$

and Hypothesis H3 guarantees that

$$\frac{\beta+\gamma}{4\left(\|B^{\mathrm{T}}K\| + \|B^{\mathrm{T}}KG\|\right)} < \rho(1-\xi) - \bar{\alpha} \,.$$

If $\beta\gamma = 0$, then, with $\lambda = 0$ and along solutions to (11),

$$\dot{v}(x_t) \le -\left[\sigma_{\min}(Q) - \varepsilon - \frac{\beta^2}{2\mu(1-\xi)}\right] \|x(t)\|^2 \;,$$

almost everywhere. Therefore, for all β and γ, almost all t and along solutions to (11), there exists $\delta > 0$ such that

$$\dot{v}(x_t) \le -\delta \|x(t)\|^2 \,.$$

Invoking Lemma 4.1 and since x^0 (specified in (12)) is arbitrary, the result readily follows.

□

5 EXAMPLE

Consider an imperfectly known dynamical system with time-delay modelled by

$$\dot{x}(t) = G\dot{x}(t-\tau) + A_0x(t) + A_1x(t-\tau) + B[u(t) + g(t, x(t), x(t-\tau), u(t))] \,, \tag{21}$$

where $x(t) = [x_1(t)\ x_2(t)]^{\mathrm{T}} \in \mathbb{R}^2$, $u(t) \in \mathbb{R}$, with known time-delay τ,

$$G = \begin{bmatrix} \frac{1}{2} & 0 \\ 0 & \frac{1}{2} \end{bmatrix}, \quad A_0 = \begin{bmatrix} -8 & 0 \\ 0 & -4 \end{bmatrix},$$

$$A_1 = \begin{bmatrix} 0 & 1 \\ -1 & 0 \end{bmatrix} \quad \text{and} \quad B = \begin{bmatrix} 0 \\ 1 \end{bmatrix}.$$

It is assumed that the constraint $\|u\| \le 2.5$ is imposed on the control input. Clearly H1(a) holds with $\|G\| = \frac{1}{2}$. Let I denote the 2×2 identity matrix, then, with $P_0 = 2I$ and $P_1 = \frac{1}{2}I$, condition (3) of H1(b) is satisfied (see the Remark (*i*) immediately following H1). In addition, with $K = \frac{1}{2}I$, condition (4) of assumption H1(b) is satisfied with $Q = \begin{bmatrix} 3 & 0 \\ 0 & \frac{1}{2} \end{bmatrix}$.

It is assumed that, for all $(t, p, q, r) \in \mathbb{R} \times \mathbb{R}^2 \times \mathbb{R}^2 \times \mathbb{R}$,

$$|g(t, p, q, r)| \leq \alpha(t) + \beta \|p\| + \gamma \|q\| + \xi |r| ,$$

where $t \mapsto \alpha(t) : \mathbb{R}_0^+ \to [0,\ 1]$, $\beta = 0.2$, $\gamma = 0.1$ and $\xi = 0.5$. Thus, assumption H2 is valid.
Now $\|B^{\mathrm{T}} K\| = 0.5$ and $\|B^{\mathrm{T}} KG\| = 0.25$. Therefore, inequality (13) holds, since

$$\sigma_{\min}(Q) > 2(\beta+\gamma)(\|B^{\mathrm{T}}K\| + \|B^{\mathrm{T}}KG\|) = 0.45 .$$

In addition, H3 is satified since

$$\rho > (1-\xi)^{-1}\Big[\bar{\alpha} + \tfrac{1}{4}(\beta+\gamma)\left(\|B^{\mathrm{T}}K\| + \|B^{\mathrm{T}}KG\|\right)^{-1}\Big] = 2.2 .$$

With

$$0.2 = \frac{(\beta+\gamma)(1-\xi)^{-1}}{4(\|B^{\mathrm{T}}K\| + \|B^{\mathrm{T}}KG\|)} < \mu \leq \rho - \bar{\alpha}(1-\xi)^{-1} = 0.5$$

and

$$\nu < \tfrac{1}{2}\varepsilon\bar{\alpha}^{-1} = 0.025 ,$$

the stabilizing control is designed to be

$$u(t) = \frac{-(2.5-\mu)z(t)}{|z(t)| + \nu[x_1^2(t) + x_2^2(t)]} - \mu z(t)\begin{cases} 1 , & |z(t)| \leq 1, \\ |z(t)|^{-1} , & \text{otherwise,} \end{cases}$$

where

$$z(t) := B^{\mathrm{T}}K(x(t) - Gx_t(-\tau)) = \tfrac{1}{2}x_2(t) - \tfrac{1}{4}x_2(t-\tau) .$$

Invoking Theorem 4.1, $x = 0$ is globally uniformly asymptotically stable under the dynamics of system (21).

REFERENCES

Bellen, A., V.B. Kolmanovskii, L. Torelli, R. Vermiglio (1995). About stability of some functional-differential equations of neutral type. textitJ. Mathematical Analysis and Applications, **189**, 59–84.

Blanchini, F. (1991). Constrained control for uncertain linear systems. *J. Optimization Theory and Applications*, **71**, 465–484.

Byrnes, C.I., M.W. Spong and T.J. Tarn (1984). A several complex variables approach to feedback stabilization of linear neutral delay differential systems. *Math. Systems Theory*, **17**, 97–133.

Corless, M. and G. Leitmann (1993). Bounded controllers for robust exponential convergence. *J. Optimization Theory and Applications*, **76**, 1–12.

Fliess, M., H. Mounier, P. Rouchon and J. Rudolph (1995). Controllability and motion planning for linear delay systems with an application to a flexible rod. *Proc. 34th IEEE Conference on Decision and Control*, New Orleans, Louisiana, U.S.A., 2046–2051.

Goodall, D.P. (1995). Comments on a Razumikhin-type condition for feedback stabilization of uncertain dynamical time-delay systems. *Proc. 3rd European Control Conference ECC95*, Rome, 3342–3347.

Goodall, D.P. and E.P. Ryan (1988). Feedback controlled differential inclusions and stabilization of uncertain dynamical systems. *SIAM J. Control and Optimization*, **26**, 1431-1441.

Hale, J.K. and S.M. Verduyn Lunel (1993). *Introduction To Functional Differential Equations.* Springer-Verlag, New York, U.S.A..

Kuang, Y. and A. Feldstein (1991). Boundedness of solutions of a nonlinear nonautonomous neutral delay equation. *J. Mathematical Analysis and Applications*, **156**, 293-304.

Kolmanovskii, V. and A. Myshkis (1992). *Applied Theory Of Functional Differential Equations.* Kluwer Academic Publishers, Boston, U.S.A..

Kolmanovskii, V.B. and V.R. Nosov (1979). Stability of neutral systems with a deviating argument. *PMM*, **43**, 209–218.

Sussmann, H.J., E.D. Sontag and Y. Yang (1994). A general result on the stabilization of linear systems using bounded controls. *IEEE Trans. Automatic Control*, **39**, 2411–2425.

Tchangani, A.P., M. Dambrine and J.P. Richard (1996). Stability and stabilization of neutral systems. *Proc. Symposium on Modelling, Analysis and Simulation, IMACS/IEEE-SMC Multiconference on Computational Engineering in Systems Applications CESA'96*, Lille, France, **1**, 812–815.

Wu, W. and K. Mizukami (1993). Exponential stability of a class of nonlinear dynamical systems with uncertainties. *Systems and Control Letters*, **21**, 307-313.

Wu, W. and K. Mizukami (1996). Linear and nonlinear stabilizing continuous controllers of uncertain dynamical systems including state delay. *IEEE Trans. Automatic Control*, **41**, 116-121.

SYNTHESIS OF IMPULSE FEEDBACKS FOR STABILIZATION OF DYNAMIC SYSTEMS

F.M. Kirillova, N.V. Balashevich

Institute of Mathematics
National Academy of Sciences
Surganov str. 11, 220072 Minsk
Belarus
imanb@imanb.belpak.minsk.by

Abstract: The stabilization problem for dynamic systems by impulse controls is investigated. A method of realizing bounded stabilizing feedbacks is proposed. The basis for the method is a special procedure of correcting optimal open loop controls for an auxiliary problem of optimal control in real time. For effective calculating a realization of bounded impulse stabilizing feedback the dual method of linear programming is used. Three types of stabilizers depending on the control criterion of an auxiliary optimal control problem are presented.

Keywords: impulse feedback, real time, algorithm, stabilization, optimal control

1. INTRODUCTION

In the paper the stabilization problem for dynamic systems in the case when the behaviour of a dynamic system is described by ordinary differential equations and stabilizing signals are impulse, i.e. are analogous to known Dirac δ-functions is considered. For the case of usual standard controls, the stabilization problem is deeply investigated in the classical and the modern theories of control (Tsien, 1954; Kwakernaak and Sivan, 1972). As it is one of the central problems in the theory of automatic regulation, so various methods were developed for its solving. The majority of these methods lean upon the fundamental results on the theory of stability. In doing so the structure (linear) of stabilizing feedback was set beforehand and the stabilization problem was reduced to choosing parameters of feedback which provided the sufficient conditions of stability for the closed system. Besides, constraints on values of stabilizing signals were not taken into account. In the modern theory of stabilization linear feedbacks were complemented by bang-bang feedbacks with linear switching surfaces that caused the creation of stable systems with variable structure (Emelyanov and Utkin, 1964; Barbashin, 1967).

A new stage in the theory of feedbacks came when the theory of optimal control appeared. At this stage it has become possible not to set the structure of feedback beforehand but to seek it on the base of optimality of certain control criterion. The first result on stabilization problem for dynamic systems obtained by methods of optimal control belongs to Lyotov (1960) and Kalman (1961). Using a linear-quadratic problem of optimal control with infinite horizon Kalman constructed a linear stabilizer for a controllable dynamic system. The results of Lyotov and Kalman have found wide use and have formed the basis of various investigations on the problem of stabilization (Kwon and Pearson, 1977; Michalska and Mayne, 1991). However in all these works constraints on values of stabilizing signals were not taken into account.

In the paper we use another approach to the stabilization problem based on methods of optimal control. Since 70th investigations on constructive methods of extremal problems were carried out in Minsk (Belarus). A number of algorithms for solving linear, nonlinear programming problems and optimal control problems had been elaborated and tested by computer (Gabasov and Kirillova, 1984). On the base of the algorithms mentioned an approach to constructing optimal feedback controls was elaborated that used a procedure of correction of optimal open loop control in real time (Gabasov, *et al.*, 1992, 1995). This approach together with the idea of sliding horizon (Kwon and Pearson, 1977) made it possible to construct bounded stabilising feedbacks in the class of piecewise continuous controls (Balashevich, *et al.*, 1994). In the paper this approach is used for constructing bounded stabilizing feedback in the class of impulse controls. Taking into account specifics of impulse functions the realization of stabilizing feedback is constructed by the dual method of linear programming. From a technical standpoint, the main achievement consists in the description of an algorithm of operating stabilizers which in real time generate values of stabilizing feedbacks during each concrete process of stabilization.

2. PROBLEM STATEMENT

Consider a control system

$$\dot{x} = Ax + bu \tag{1}$$

where x is an n-vector of state, u is a scalar control signal,

$$\text{rank}\{b, Ab, \ldots, A^{n-1}b\} = n. \tag{2}$$

The mathematical model (1) describing the local (differential) behaviour of the dynamic control system is convenient and universal with the use of standard control signals set by piecewise continuous (measurable) functions. When using impulse controls like Dirac δ-functions

$$u(t) = \delta(t - t_1), \quad t \geq 0, \quad t_1 \geq 0,$$

the model (1) is not very convenient as at the moment $t = t_1$ it does not give any information about the behaviour of the system although in this moment its behaviour essensially changes. In this connection in the paper the model (1) will be used only for formal description of system, for substantial description of local behaviour of control system under impulse controls an other mathematical model will be used. So let us define the class of used impulse functions.

A function $u(t), t \geq 0$, is said to be a finite-impulse open loop control if
1) on any bounded interval T it only on a finite set of points T_u differs from 0,
2) it takes at these points finite values $u(t), t \in T_u$,
3) it generates a trajectory $x(t), t \geq 0$, of the system (1) with initial state $x(-0) = x_0$ such that

a) at $t \notin T_u$ it is a continuous solution to the equation $\dot{x} = Ax$, b) at moments $t \in T_u$ it makes jumps according to the law

$$x(t + 0) = x(t - 0) + bu(t). \tag{3}$$

A function $u(x)$, $x \in R^n$, is said to be an impulse feedback control with parameter ν if

1) its realization $u(t) = u(x(t)), t \geq 0$, along each trajectory $x(t), t \geq 0$, of the closed system

$$\dot{x} = Ax + bu(x), \quad x(-0) = x_0, \tag{4}$$

is finite-impulse;

2) a trajectory of the closed system (4) on each interval $[(k-1)\nu, k\nu[, k = 1, 2, \ldots$ coincides with a solution to the equation $\dot{x} = Ax$, and at the moment $k\nu$ changes according to the law

$$x(k\nu + 0) = x(k\nu - 0) + bu(k\nu),$$

$$u(k\nu) = u(x(k\nu - 0)), \quad k = 0, 1, \ldots.$$

The impulse feedback control is said to be a bounded stabilizing feedback in a domain $G \subset R^n$, $0 \in G$, if

1) $|u(x)| \leq L, \quad x \in G$,

2) $u(0) = 0$, i.e. the closed system holds the state of equilibrium of the initial system,

3) the state of equilibrium of the closed system is asymptoticallly stable in the domain G.

The problem of stabilizing dynamic systems by bounded controls consists not only in constructing a function $u(x)$ for which the system (4) is asymptotically stable in some domain G where $|u(x)| \leq L$. The purpose is also in that for a given L the domain G of attraction of the asymptotically stable solution $x(t) \equiv 0$, $t \geq 0$, be the biggest, i.e. be little different from the set of controllability of the system (4). If the problem is understood not so, the latter is solved simply because an arbitrary stabilizing feedback $u(x)$ is a bounded stabilizing feedback in any domain $G = \{x : |u(x)| \leq L\}$ while the trajectory of the closed system does not leave the set G.

Construction of a bounded stabilizing feedback in an explicit form is an extraordinary difficult problem. To make the essence of the proposed approach to the stabilization problem clear, let us analyze the use of stabilizing feedback during a concrete process of stabilization. Denote by $x^*(t)$, $t \geq 0$, a trajectory of the stabilized system (1). From the above definition of stabilizing feedback it is obvious that during each process of stabilization the feedback $u(x)$, $x \in G$, is not used entirely, all one need to do is to know its values along the stabilized trajectory $x^*(t)$, $t \geq 0$, of the dynamic system and moreover the values $u^*(k\nu) = u(x^*(k\nu - 0))$, $k = 0, 1, \ldots$, are not needed to be known beforehand, it is enough to know them only at a current moment $k\nu$.

A function $u^*(k\nu) = u(x^*(k\nu - 0))$, $k = 0, 1, \ldots$, is said to be a realization of the impulse bounded stabilizing feedback in a concrete process. Any device which for a chosen $\nu > 0$ is able to calculate the values $u^*(k\nu)$, $k = 0, 1, \ldots$, during stabilization in real time is said to be an impulse stabilizer.

Thus the stabilization problem for dynamic system in the class of impulse controls is reduced to constructing an algorithm of operating an impulse stabilizer. In the paper this problem is solved by positional solution to an auxiliary problem of optimal control (APOC). For definiteness, we restrict our consideration to three types of optimal control problems. In Section 3 the optimal control problem presents the problem of damping dynamic system by controls of minimal intensity. In Section 4 the sum of intensities of input impulses is taken as a control criterion of APOC. In Section 5 for stabilization of the dynamic system the time-optimal problem is used as APOC. Every time it is shown, that the positional solution to APOC possesses a stabilizing property and a method of realizing positional solutions to APOC in real time is proposed.

3. THE INTENSITY STABILIZER

On the basis of dynamic system (1), after choosing an integer number N, $n < N < \infty$, formulate the following APOC

$$\rho(z) = \min \rho,$$

$$\dot{x} = Ax, \quad x(-0) = z, \tag{5}$$

$$x(k\nu + 0) = x(k\nu - 0) + bu(k\nu), \quad x(N\nu + 0) = 0,$$

$$|u(k\nu)| \leq \rho, \quad k = 0, \ldots, N.$$

We shall further assume that the parameters A, b, ν of the problem are such that the condition

$$\text{rank}\{b, Bb, B^2 b, \ldots B^{n-1} b\} = n \tag{6}$$

$$\Big(B = \exp(A\nu)\Big),$$

holds which under above assumption (2) will take place almost for any $\nu > 0$.

A function $u^0(k\nu | z)$, $k = \overline{0, N}$, is said to be an optimal open loop solution to the problem (5) if a generated trajectory $x^0(t|z)$, $t \in [0, N\nu + 0]$, satisfies the constraint $x^0(N\nu + 0|z) = 0$ and the control criterion reaches the minimal value.

A function $u^0(z) = u^0(0|z)$, $z \in R^n$, is said to be an optimal feedback start control.

Introduce a set $G_N = \{x \in R^n : |u^0(x)| \leq L\}$. Let S_L be the set of controllability of the system (1) by bounded controls, i.e. the set containing all states x_0 for which it is possible to transfer the system (1) to the origin of coordinates $x = 0$ by bounded impulse controls $|u(t)| \leq L$, $t \in [0, t^*]$ in a finite time t^*. For any $\varepsilon > 0$ one can indicate such a number $N < \infty$ that the ε-neighbourhood $[G_N]_\varepsilon$ of the set G_N contains S_L.

Let us show that the function $u^0(x)$, $x \in R^n$, is an impulse bounded stabilizing feedback.

Really, from the statement of APOC it follows, that
1) $u^0(0) = 0$;
2) for each bounded set $G \subset R^n$ there exists such a number $L > 0$ that $\rho(x) \leq L$ for all $x \in G$, so $|u^0(x)| \leq L$;
3) the trajectory $x^*(t)$, $t \geq 0$, on each interval $[k\nu + 0, (k+1)\nu - 0]$, $k = 0, 1, \ldots$, is defined as a solution to the equation $\dot{x} = Ax$ and from the jump condition $x(k\nu + 0) = x(k\nu - 0) + bu^0(x(k\nu - 0))$.

To prove that the closed system (4) ($u(x) = u^0(x)$) is asymptotically stable we consider an arbitrary current moment $\tau = l\nu$ and a state $x^*(\tau - 0|x_0)$ corresponding to an arbitrary initial state $x_0 \in G$. For the state $x^*(\tau - 0|x_0)$ the control criterion of the problem (5) takes the value $\rho(x^*(\tau - 0|x_0))$. It is calculated by solving the problem (5) with $z = x^*(\tau - 0|x^0)$.

It is easy to show, that the last problem is equivalent to the following linear programming problem

$$\rho \longrightarrow \min,$$

$$\sum_{i=0}^{N} F((N-i)\nu) b u_i = -F(N\nu) x^*(\tau - 0|x^0),$$

$$|u_i| \leq \rho, i = \overline{0, N}, \tag{7}$$

where $F(t), t \geq 0$, is a fundamental matrix of solutions to the system $\dot{x} = Ax$.

Denote by $u^0(\tau) = (u_i^0, i = \overline{0, N})$ the optimal feasible solution to the problem (7) and by $K^0(\tau)$ the optimal basis of the problem (7) (Dantzig, 1963; Gabasov, 1994).

At the moment $\tau + \nu$ the closed system (4) transfers to the state

$$x^*(\tau + \nu - 0|x_0) = F(\nu)\big(x^*(\tau - 0|x_0) +$$

$$+ bu^0(x^*(\tau - 0|x_0))\big).$$

Let us show that the value of the control criterion of the problem (5) for the new state $z = x^*(\tau +$

$\nu|x_0)$ satisfies the inequality $\rho(x^*(\tau+\nu|x_0)) \leq \rho(x^*(\tau-0|x_0))$. Really, the control $u(k\nu) = u^0((k+1)\nu|x^*(\tau-0|x_0)), k = \overline{0, N-1}, u(N\nu) = 0$ transfers the system (5) from the initial state $x^*(\tau+\nu-0|x_0)$ to the origin of coordinates and satisfies the inequality $|u(t)| \leq \rho(x^*(\tau-0|x_0)$. Hence, on the optimal control $u^0(k\nu|x^*(\tau+\nu-0|x_0)), K = \overline{0,N}$, also the inequality $\rho(x^*(\tau+\nu-0|x_0)) \leq \rho(x^*(\tau-0|x_0))$ will be fulfilled. It remains to show that the equality $\rho(x^*(\tau+\nu-0|x_0)) = \rho(x^*(\tau-0|x_0))$ can be fulfilled no more than over N steps.

Assume that during stabilization the equalities

$$\rho(x^*(\tau-0|x_0)) = \rho(x^*(\tau+i\nu-0|x_0)), \quad i=\overline{1,N}.$$

take place.

At the moment $\tau + N\nu$ to calculate the control $u^0(x^*(\tau+N\nu-0|x_0))$ the stabilizer solves the problem (5) with $z = x^*(\tau+N\nu-0|x_0)$. This problem is equivalent to the following linear programming problem

$$\rho \longrightarrow \min,$$

$$\sum_{i=0}^{N} F((N-i)\nu)bu_i + \sum_{i=N+1}^{2N} F((N-i)\nu)bu_i =$$

$$- F(N\nu)x^*(\tau-\nu-0|x_0), \tag{8}$$

$$u^0(\tau+i\nu|x^*(\tau-0|x_0)) \leq u_i \leq$$

$$\leq u^0(\tau+i\nu|x^*(\tau-0|x_0)), \quad i=\overline{0,N},$$

$$|u_i| \leq \rho, \quad i=\overline{N+1,2N}.$$

The set $\bar{u} = (\bar{u}_i, i=\overline{0,2N}), \bar{u}_i = u^0(\tau+i\nu|x^*(\tau-0|x_0)), i=\overline{0,N}, \bar{u}_i = 0, i=\overline{N+1,2N}$, is a feasible solution to the problem (8) on which the control criterion takes the value $\rho(x^*(\tau-0|x_0))$.

We replace the basis $K^0(\tau)$ by a new basis K composed of elements of the set $\{N+1,\ldots,2N\}$. The basic feasible solution $\{\bar{u}, K\}$ is nondegenerate, i.e $|\bar{u}_i| < \rho, i \in K$. Among estimates $\Delta_i = y'F((N-i)\nu)b, i=\overline{N+1,2N}$, calculated by the basis K where y is corresponding to the basis K the optimal n-vector of the Lagrange multipliers, nonzeroes are necessarily found because the identity $y'F((N-i)\nu)b \equiv 0, i=\overline{N+1,2N}$ contradicts to (6). So after solving the problem (8) we obtain

$$\rho(x^*(\tau+N\nu-0|x_0)) < \rho(x^*(\tau-0|x_0)). \tag{9}$$

Using the inequality (9), by reasoning typical for the method of the Lyapunov functions it is easy to show that $\rho(x^*(k\nu-0|x_0)) \longrightarrow 0, \quad ||x^*(k\nu|x_0)|| \longrightarrow 0, k \longrightarrow \infty$. By virtue of specific character of impulse controls, from here it follows that also $x^*(t|x_0) \longrightarrow 0, t \longrightarrow \infty$, as was to be proved.

Now pass to describing the algorithm of operating the impulse stabilizer.

Assume that the stabilizer has operated at moments of $0, \nu, \ldots, l\nu$. At the moment $\tau = l\nu$ to generate the control $u^*(\tau)$ it uses the value of the open loop optimal control $u^0(0|x^*(\tau-0|x_0))$ of the problem (5) for $z = x^*(\tau-0|x_0)$. This problem is equivalent to the LP problem (7). To generate the control $u^*(\tau+\nu)$ the stabilizer has to know the solution to the problem (5) with the initial state $z = x^*(\tau+\nu-0|x_0)$. This problem has the form

$$\rho \longrightarrow \min,$$

$$\sum_{i=0}^{N} F((N-i)\nu)bu_i = -F(N\nu)x^*(\tau+\nu-0|x^0),$$

$$|u_i| \leq \rho, \quad i=\overline{0,N}. \tag{10}$$

The new problem differs from the problem (7) only by a right side vector. In such situation natural and the most effective method of the construction of the solution to the problem (10) is the dual method of linear programming (Dantzig, 1963; Gabasov, 1994) which for solving the problem (10) takes as initial approximation the optimal basis $K^0(\tau)$ of the problem (7). For small ν the solution to the new problem is constructed by a little number of iterations. In case of large ν to reduce a number of iterations of the dual method, the interval $[k\nu,(k+1)\nu]$ can be divided into a number of enough small subintervals. Then after solving the problem (7) by the dual method on these subintervals, to calculate the optimal control $u^0(i\nu|x^*((k+1)\nu|x_0), I=\overline{0,N}$, at the moment $(k+1)\nu$ a little number of iterations are required.

If the time it takes for calculating the control $u^*(k\nu)$ does not exceed ν then one may say that the stabilizer constructs a realization of a stabilizing control in real time. It is seen from the described algorithm of operating the stabilizer that for every system (1) one can select a computer device able to perform all necessary calculations in a time not exceed ν. And on the other hand, for every computer device one can indicate systems for which this device can calculate stabilizing signals according to described above scheme.

4. THE FUEL STABILIZER

The algorithm of operating this stabilizer is based on the solution to the following APOC

$$\mu(z) = \min \sum_{i=0}^{N} |u(i\nu)|,$$

$$\dot{x} = Ax, \quad x(-0) = z, \tag{11}$$
$$x(k\nu + 0) = x(k\nu - 0) + bu(k\nu), \quad x(N\nu + 0) = 0,$$
$$|u(k\nu)| \leq L, k = 0, \dots, N.$$

As in the previous consideration denote by $u^0(k\nu|z)$, $k = \overline{0, N}$, an optimal open loop control of the problem (11). Let G be a set of all $z \in R^n$ for which the problem (11) has a solution.

A function

$$u^0(z) = u^0(0|z), \quad z \in G, \tag{12}$$

is said to be an optimal feedback start control. Let us show that the function $u^0(x)$, $x \in G$, is an impulse bounded stabilizing feedback.

Prove the stabilizing property of the new feedback. As the Lyapunov function we again consider the optimal value of the control criterion of the problem (11).

It is clear that $\mu(x)$, $x \in G$, is a continuous function having the property $\mu(0) = 0$, $\mu(x) > 0$, $x \neq 0$.

Let at a moment $\tau = l\nu$ in the problem (11) with $z = x^*(\tau - 0|x_0)$ the optimal open loop control $u^0(k\nu|x^*(\tau - 0|x_0)), k = \overline{0, N}$, is constructed on which the control criterion gets the value $\mu(x^*(\tau - 0|x_0)) > 0$. At the moment $\tau + \nu$ the system transfers to the state

$$x^*(\tau + \nu - 0|x_0) = F(\nu)(x^*(\tau - 0|x_0) +$$
$$+ bu^0(\tau|x^*(\tau - 0|x_0)).$$

In the problem (11) with $z = x^*(\tau + \nu - 0|x_0)$ the control

$$u(k\nu) = u^0((k+1)\nu|x^*(\tau - 0|x_0)), \quad k = \overline{0, N-1}, \tag{13}$$
$$u(N\nu) = 0$$

is admissible and $\sum_{i=0}^{N} |u(i\nu)| = \mu(x^*(\tau - 0|x_0)) - u^0(0|x^*(\tau - 0|x_0))$. It is clear that if $u^0(0|x^*(\tau - 0|x_0)) \neq 0$ then $\mu(x^*(\tau + \nu - 0|x_0)) < \mu(x^*(\tau - 0|x_0))$. If $u^0(0|x^*(\tau - 0|x_0)) = 0$ and $\mu(x^*(\tau + \nu - 0|x_0)) = \mu(x^*(\tau - 0|x_0))$, the control (13) is the optimal open loop control of (11) with $z = x^*(\tau + \nu - 0|x_0)$. By virtue of the fact that $\mu(x^*(\tau - 0|x_0)) > 0$, such a number $k^* < N$ is found that $u^0(k^*\nu|x^*(\tau - 0|x_0)) \neq 0$, and this provides the inequality $\mu(x^*(\tau + (k^* + 1)\nu - 0|x_0)) < \mu(x^*(\tau - 0|x_0))$.

Thus the Lyapunov function $\mu(x)$, $x \in G$, monotonically decreases along the sequence $x(k\nu)$, $k = 1, 2, \dots$. By the Lyapunov functions method it can be shown that the limit can be only the zero value : $\lim \mu(x(k\nu)) = 0$, $k \longrightarrow \infty$. From here it follows that $||x(k\nu)|| \longrightarrow 0$, $k \longrightarrow \infty$, $||x(t)|| \longrightarrow 0$, $t \longrightarrow \infty$. The stabilizing property of the feedback (12) is proved.

As $\mu(x^*(\tau + \nu - 0|x_0)) \leq \mu(x^*(\tau - 0|x_0)) - |u^*(\tau)|$ so the realization $u^*(k\nu)$, $k = 0, 1, \dots$, of impulse bounded stabilizing feedback possesses the following property

$$\sum_{k=0}^{\infty} |u^*(k\nu)| \leq \sum_{k=0}^{N} |u^0(k\nu)|$$

where $u^0(k\nu)$, $k = \overline{0, N}$, is the optimal open loop control of the problem (11) for the initial state $z = x_0$. Thus for any $N > 0$ whole expenses for stabilization are not exceed expenses for optimal damping the system in the time $N\nu$ that can be regarded as a characteristic of the stabilizer.

The optimal control problem (11) is equivalent to the linear programming problem

$$\sum_{i=0}^{N} (v_i + w_i) \longrightarrow \min,$$

$$\sum_{i=0}^{N} F((N - i)\nu) b(v_i - w_i) = -F(N\nu)z, \tag{14}$$

$$v_i + w_i \leq L, \quad v_i \geq 0, \quad w_i \geq 0, \quad i = \overline{0, N}.$$

In every moment $\tau = k\nu$ the stabilizer constructs by the dual method a solution $(v_i^0, w_i^0, i = \overline{0, N})$ to the problem (14) for $z = x^*(\tau - \nu - 0|x_0)$ using as an initial basis the optimal basis of the problem (14) for $z = x^*(\tau - \nu - 0|x_0)$. At the moment τ the control $u^*(\tau) = v_0^0 - w_0^0$ is fed to the input of the system and the same procedure to be continued for the next moment.

5. THE TIME STABILIZER

In APOC used by the intensity and the fuel stabilizers parameter N determining the horizon of damping was set fixed. As the examples demonstrate, the use of variable parameter N can occur an effective tool for stabilization of systems under constantly acting disturbances.

In this Section as such auxiliary problem with variable parameter N the classical time-optimal problem in the class of impulse controls

$$N(z) = \min N,$$
$$\dot{x} = Ax, \quad x(-0) = z, \tag{15}$$
$$x(k\nu + 0) = x(k\nu - 0) + bu(k\nu), \quad x(N\nu + 0) = 0,$$
$$\sum_{i=0}^{N} |u(i\nu)| \leq \mu^*, \quad |u(k\nu)| \leq L, \quad k = 0, \dots, N,$$

is considered.

A function $u^0(k\nu|z)$, $k = \overline{0, N(z)}$, is called an optimal open toop control of the problem (15) if
1) it satisfies the constraint $|u(k\nu)| \leq L, k = 0, \ldots, N(z)$,
2) the corresponding trajectory $x^0(t|z), t \in [0, N(z)\nu + 0]$, satisfies the terminal constraint $x^0(N(z)\nu + 0|z) = 0$,
3) $\sum_{i=0}^{N(z)} |u^0(i\nu)| = \min_u \sum_{i=0}^{N(z)} |u(i\nu)|$ where the minimum is taken from controls satisfying first two conditions.

The optimal open loop control of the problem (15) is possible to be constructed by the method of solving the problem (11).

Let G be a set of all $z \in R^n$ for which the problem (15) has a solution.

A function

$$u^0(z) = u^0(0|z), \quad z \in G, \tag{16}$$

is said to be a feedback start control.

It is easy to show that the feedback (16) is a stabilizing one in the domain G. In doing so the property of asymptotical stability of the closed system (4) under the feedback (16) takes place in strict form: there exists such a moment t^* that $x(t) \equiv 0$ for $t \geq t^*$.

By virtue of mentioned connection of the problem (15) with the problem (11), the realization of the feedback (16) is analogous to the realization of the feedback (12). At practical applications it is reasonable to use the feedback (16) while $N(x^*(\tau - 0|x_0)) \geq N^*$ where $N^* > n$ is a given number and to finalize the process of the stabilization by the feedback (12).

Remark. So that during operating the fuel and the time stabilizers the closed system (4) without disturbances is not damped for finite time, one can pass to use of the intensity stabilizer at the finish of process of stabilization.

6. CONCLUSION

A method of constructing impulse bounded stabilizing feedbacks in real time is suggested which is based on the use of auxiliary problems of optimal control. The proposed method differs from many known methods in that it takes into account constraints on values of stabilizing signals, does not set the structure of feedback beforehand and produces a variety of types of stabilizers depending on the control criterion of an auxiliary optimal control problem.

The research was partially supported by Grant F96-011 from the Foundation for Basic Research of Belarus.

REFERENCES

Balashevich, N.V., R. Gabasov and F.M. Kirillova (1994). Optimal damper of dynamic systems. *Automation & Remote Control*, No. 5, 615–623.

Barbashin, E.A. (1967). *Introduction to Stability Theory.* Nauka, Moscow, (in Russian).

Dantzig, G.B. (1963). *Linear Programming and Extensions.* Princeton University Press, Princeton.

Emelyanov, S.V. and V.I. Utkin (1964). About stability of motion of one class of automatic control systems with variable structure. *Technical Cybernetics*, No. 2, 24–35, (in Russian).

Gabasov, R. (1994). Adaptive method of solving linear programming problems. *Preprint Series of University of Karlsruhe*, Institute for Statistics and Mathematics, Germany, Karlsruhe.

Gabasov, R. and F.M. Kirillova (1984). Consideration of optimal control problems specificity on generalizing mathematical programming algorithms. *Preprints of the 9th IFAC Congress on Automatic Control*, Hungary, Budapest, **9**, 264–269.

Gabasov, R., F.M. Kirillova and O.I. Kostyukova (1992). Construction of optimal controls of feedback type in a linear problem. *Soviet Math. Dokl.*, **44**, No. 2, 608–613.

Gabasov, R., F.M. Kirillova and S.V. Prischepova (1995). *Optimal Feedback Control.* Lecture Notes in Control an Information Sciences. (Thoma M. ed.) Springer. **207**.

Kalman, R. (1961). Control systems general theory. *Preprints of the 1st IFAC Congress on Automation Control*, USSR, Moscow, **1**, 521–547.

Kwakernaak, H. and R. Sivan (1972). *Linear Optimal Control Systems.* Wiley, New York.

Kwon, W.H. and A.E. Pearson (1977). A modified quadratic cost problem and feedback stabilization of a linear system. *IEEE Trans. Automat. Contr.*, **AC-22**, No. 5, 838.

Lyotov, A.M. (1960). Analytic construction of regulators. Pts. I–III. *Avtomatika i Telemekhanika*, **21**, 436–441; 561–568; 661–665.

Michalska, H. and D.Q. Mayne (1991). Receding horizon control of nonlinear systems without differentiability of the optimal value function. *Syst. Contr. Lett.*, **16**, No. 2, 123–130.

Tsien, H.S. (1954). *Engineering Cybernetics.* McGraw-Hill, New York.

OPTIMAL GENERALIZED TRAJECTORIES OF A DIFFERENTIAL INCLUSION

Kolokolnikova Galina

664015, Irkutsk, Lenin Street 11,
Irkutsk State Economic Academy, Russia
e-mail: kolok@iinh.irkutsk.su
tel.: (395-2)-24-28-19, (395-2)-33-43-59

Abstract. An optimization problem for a differential inclusion with an unbounded right-hand side is under consideration. Discontinuous trajectories are introduced and investigated. To regularizate the optimization problem, the transformation methods are used and developed. The transformations help to describe discontinuous trajectories and to get a first-order necessary conditions of their optimality.

Keywords. Optimization problems, differential inclusions, transformations, discontinuous trajectories, optimality, tests.

1. INTRODUCTION

Consider the following optimization problem P : to minimize a functional $J[x] = \Phi(x(T))$ on the set D of absolutely continuous trajectories $x \in AC^n[0,T]$ of a differential inclusion

$$\dot{x} \in F(t,x), \quad (1)$$

such that the terminal state restrictions are satisfied:

$$x(0) \in C_0, \quad x(T) \in C_T, \quad (2)$$

where C_0 and C_T are known compact sets from R^n,. In distinction from (Clarke, 1983; Blagodatskih and Filippov,1985), it is supposed here that the multifunction F transfers a point $(t,x) \in [0,T] \times R^n$ into unbounded subset of R^n. Such state of the problem is a natural generalization of the optimal control problems with an unbounded set of velocities (Gurman, 1977) which are widely spread in practice. Even the simplest examples permit to notice one of the main features of the considering optimization problem: as a rule, there is no the optimal trajectory among the absolutely continuous functions, satisfying the differential inclusion (1). A minimizing sequence is typically such that its trajectories converge to a discontinuous function.

For example, let $x_1(1)$ be minimized ($J[x] = x_1(1) \rightarrow \inf$) on the trajectories of the differential inclusion:

$$\dot{x}_1 \in F_1(t,x(t)) = x_2^2(t) + [0,2] \cap Q, \quad \dot{x}_2 \in [0,+\infty)$$

with the initial conditions $x_1(0) = 0, \ x_2(0) = -1$, where Q is the set of all rational numbers.

It is obvious that $\dot{x}_1(t) \geq 0$ and $x_1(1) \geq 0$.

The sequence of trajectories such that

$$x_2^k(t) = \begin{cases} -1 + kt, & t \in [0, 1/k), \\ 0, & t \in [1/k, 1], \end{cases}$$

and $x_1^k(t)$ is a solution of the differential equation $\dot{x}_1 = (x_2^k(t))^2$ under $x_1(0) = 0$, - is minimized and converges to the discontinuous vector-function x^* :

$$x_2^*(t) = \begin{cases} -1, & t = 0, \\ 0, & t \in (0,1]; \end{cases} \qquad x_1^*(t) \equiv 0.$$

It is natural to call x^* the generalized solution of the problem.

The known optimization tests, for example (Clarke, 1983; Blagodatskih and Filippov,1985), can't be applied to the problem P even for the case of a convex closed F (here the multifunction F may be convex or nonconvex, closed or not).

The aim of this paper is to study the methods of the problem P transformations to such an optimization problem, in which the known optimality tests are effective, and such that its an ordinary solution determines the generalized solution of the problem P. Here the generilized solution of the problem P is any function x^*, which may be discontinuous, such that there exists a sequence of absolutely continuous trajectories $\{x_s\} \subset D$: $\Phi(x_s(T)) \longrightarrow \inf_D J, \ \ x_s(t) \longrightarrow x^*(t) \ \ \forall\, t \in [0,T]$.

It will be said that $\{x_s\} \subset AC[0,T]$ is a sequence of quasi - trajectories of the differential inclusion (1) iff $\rho(\dot{x}_s(t), F(t, x_s(t))) \stackrel{s\to\infty}{\longrightarrow} 0 \quad \forall\, t \in [0,T]$, where $\rho(b, A) = \inf_{a\in A} |a-b|$.

Any function x will be called in this paper a generalized trajectory of the differential inclusion (1) iff there exists a sequence $\{x_s\} \subset D : x_s(t) \longrightarrow x(t) \ \ \forall\, t \in [0,T]$ or there exists a sequence $\{x_s\} \subset AC^n[0,T]$ of quasi-trajectories of the differential inclusuon (1) such that $x_s(t) \longrightarrow x(t) \ \forall\, t \in [0,T]$.

The problem P has a natural extension: to find out an optimal generalized trajectory.

Under definite assumption, the way of the generalized trajectories desciption will be given, and the necessary conditions of their optimality will be derived in the form of the variational maximum principle, analogeous to the known optimality conditions for the optimal control problems with impulsive controls (Dykhta, 1994; Kolokolnikova, 1996).

2. THE LIMIT CONE AND THE METHODS OF EXTENSION

Answer the question, what discontinuous function the sequence of $\{x_s\} \subset D$ may converge to, or where can one-sided limits of a discontinuous trajectory at the jump time moments lie. It will be shown that the set of accessibility of the limit differential inclusion helps answer the question.

Let $\overline{co}F(t,x)$ be a closed convex hull of an unbounded set $F(t,x)$. Consider the differential inclusion

$$\dot{x} \in \overline{co}F(t,x). \tag{3}$$

Use the result proved in the work (Kolokolnikova, 1997) that any absolutely continuous trajectory $\bar{x}$ of the weak differential inclusion (3) is a generalized trajectory of the differential inclusion (1) if the multifunction F is continuous on t and locally Lipschitzean on x (here $\bar{x}$ may be slide or a limit-slide trajectory). The right-hand side of (3) is unbounded so its limit cone may be introduced.

Let V be any unbounded set in R^n. The set

$$\mathcal{L}_V = \Big\{ g \in R^n \mid \exists \{v_k\} \subset V \ \ \exists \alpha \geq 0 \ \ (\alpha \in R^1) : \lim_{k\to\infty} |v_k| = +\infty, \quad \lim_{k\to\infty} \frac{\alpha v_k}{|v_k|} = g \Big\}$$

will be called a limit cone. Some properties of $\mathcal{L}_V$ were proved earlier (Kolokolnikova, 1997):

1) $\mathcal{L}_V \setminus \{0\} \neq \emptyset$ if and only if V is unbounded.
2) If V is a convex unbounded set then $\mathcal{L}_V$ coincides with a recessive cone of V.
3) $\overline{co}V + \mathcal{L}_{\overline{co}V} V = \overline{co}V$ where $\overline{co}V$ is a convex hull of the set V and an upper line is a sign of closure.

Under the additional assumption: $\mathcal{L} \stackrel{\text{def}}{=} \mathcal{L}_{\overline{co}V}$ is a nontrivial subspace in R^n, – the following properties are valid:

4) A projection $B(V)$ of the set $\overline{co}V$ on the annulator $\mathcal{L}^\perp \stackrel{\text{def}}{=} \{g : (l,g) = 0 \ \ l \in \mathcal{L}\}$ is a bounded set;
5) A decomposition $\overline{co}V = B(V) + \mathcal{L}$ is true;
6) $\sigma_{\overline{co}V}(l^*) = \sigma_V(l^*) = +\infty \quad \forall l^* \notin \mathcal{L}^\perp$, where $\sigma_V(l^*)$ is a support function of the set V.

Let $\mathcal{L}(t,x)$ be the limit cone of the set $\overline{co}F(t,x)$.

The differential inclusion

$$\frac{dx}{d\tau} \in \mathcal{L}(t,x) \tag{4}$$

(where t is a parameter), is called the limit differential inclusion for (1). Let $Q(\bar{t}, y)$ be the set of accessibility from the point y by the absolutely continuous trajectories of the limit differential inclusion (4) for $t = \bar{t}$. It will be proved below than any point from $Q(\bar{t}, y)$ may be achieved approximately by trajectories or quasi-trajectories of the original differential inclusion (1).

Theorem 1. If the multifunction F is continuous on t, x; $y_2 \in Q(\bar{t}, y_1)$, $y_2 \neq y_1$ then there exist both a number sequence $\{\Delta t_s\} : \Delta t_s \to 0$ and a sequence of quasi-trajectories on $[\bar{t}, \bar{t} + \Delta t_s]$ of the differential inclusion (1) $\{x_s\}$ such that $x_s(\bar{t}) = y_1$, $x_s(\bar{t} + \Delta t_s) \stackrel{s\to\infty}{\longrightarrow} y_2$.

In other words, it is possible to get from the point y_1 into any small neighbourhood of y_2 for any small time interval by the quasi-trajectories of (1)

Proof. As $y_2 \in Q(\bar{t}, y_1)$, there exists $\bar{x}(\tau) \in AC[0,1]$: $\bar{x}(0) = y_1$, $\bar{x}(1) = y_2$, $\dfrac{d\bar{x}}{d\tau} = \bar{l}(\bar{t}, \tau) \in \mathcal{L}(\bar{t}, \bar{x}(\tau))$ a.e [0, 1]. It is natural to suppose that $\bar{l}(\bar{t}, \tau) \neq 0$ as $y_1 \neq y_2$. Let $||\bar{l}||_{L_\infty} = k < \infty$. By the definition of the limit cone, there exist $\{v_s(\bar{t}, \tau)\} \subset F(\bar{t}, \bar{x}(\tau))$ and $\bar{\alpha}(\tau) > 0$ such that $|v_s(\bar{t}, \tau)| \stackrel{s\to\infty}{\longrightarrow} +\infty$,

$$\bar{\alpha}(\tau) \frac{v_s(\bar{t}, \tau)}{|v_s(\bar{t}, \tau)|} \stackrel{s\to\infty}{\longrightarrow} \bar{l}(\bar{t}, \tau) \tag{5}$$

It is easily to see, $\bar{\alpha}(\tau) = |\bar{l}(\bar{t}, \tau)|$ and $||\bar{\alpha}(\tau)||_{L_\infty} = k < \infty$.
Consider an auxiliary system:

$$\frac{dx_s}{d\tau} = \bar{\alpha}(\tau)\frac{v_s(\bar{t},\tau)}{|v_s(\bar{t},\tau)|},\ x_s(0) = y_1,$$
$$\frac{dt_s}{d\tau} = \bar{\alpha}(\tau)\frac{1}{v_s(\bar{t},\tau)},\quad t_s(0) = \bar{t}.$$

It is obvious that:

1) $t_s(1) = \bar{t} + \int\limits_0^1 \frac{\bar{\alpha}(\tau)}{|v_s(\bar{t},\tau)|}d\tau \overset{s\to\infty}{\longrightarrow} \bar{t};$

2) $x_s(\tau) \to \bar{x}(\tau)\quad \forall \tau \in [0,1]$ due to (5);

3) $\dot{x}_s = \frac{dx_s}{dt_s} = \frac{dx_s}{d\tau}/\frac{dt_s}{d\tau} = v_s(\bar{t},\tau_s(t))$, where $\tau_s(t)$ is a function, inverse to $t_s(\tau)$.

The inclusion $\dot{x}_s(t) \in F(\bar{t},\bar{x}(\tau_s(t))$ for $t \in [\bar{t}, t_s(1)]$ is hold due to 3). Then,

$$\rho(\dot{x}_s(t), F(t,x_s(t)) \le \rho(\dot{x}_s(t), F(\bar{t},\bar{x}(\tau_s(t)))+$$
$$+d(F(\bar{t},\bar{x}(\tau_s(t))), F(t,\bar{x}(\tau_s(t))))+d(F(t,\bar{x}(\tau_s(t))),$$
$$F(t,x_s(\tau_s(t)))) = d(F(\bar{t},\bar{x}(\tau_s(t))), F(t,\bar{x}(\tau_s(t))))+$$
$$+d(F(t,\bar{x}(\tau_s(t))), F(t,x_s(\tau_s(t)))),\qquad (6)$$

where

$$d(A,B) = \max\{\inf_{a\in A}\rho(a,B), \inf_{b\in B}\rho(A,b)\}.$$

The right-hand side of (6) tends to zero for $s \to \infty$. It follows from 1), 2) and an assumption that F is continuous on t, x. Thus, $\{x_s\}$ is a sequence of quasi-trajectories of the differential inclusion (1). It is necessary to mention that $t_s(1) \to \bar{t}$ and $x_s(t_s(1)) = x_s(1) \overset{s\to\infty}{\longrightarrow} \bar{x}(1) = y_2$. ■

The statement of the Theorem 1 may be strenghthened for the most typical case of the limit cone.

Theorem 2. Let $\mathcal{L}(t,x) = G(t,x)K\quad \forall (t,x) \in [0,T]\times R^n$, where K is a fixed k-dimensional cone, the $(n \times k)$ - matrix-function G is continuous on t, x. If $y_2 \in Q(\bar{t}, y_1)$, $y_2 \ne y_1$, then there exists a sequence of trajectories $\{x_s\} \subset D$ defined on $[\bar{t}, \bar{t}+\Delta t_s]$ respectively, and such that $x_s(\bar{t}) = y_1$, $\Delta t_s \to 0$, $x_s(\bar{t}+\Delta t_s) \to y_2$.

Here the assumption of the multifunction F continuity is absent but it is substituted by an continuity assumption on the matrix - function G.

Proof. As $y_2 \in Q(\bar{t}, y_1)$, there exists $\bar{x}(\tau) \in AC^n[0,1]:\ \bar{x}(0) = y_1,\ \bar{x}(1) = y_2,\ d\bar{x}/d\tau = G(\bar{t},\bar{x}(\tau))\zeta(\tau)$, where $\zeta(\tau) \in K$; $\zeta(\tau) \ne 0$ a.e [0,1] (as $y_2 \ne y_1$). Due to continuity of $\bar{x}$ on [0,1], there exists a compact set $A \subset R^n:\ \bar{x}(\tau) \in A\ \forall \tau \in [0,1]$. From the continuity of G, it is followed that $\exists B > 0:\ |||G(t,x)|| \le B\ \forall t \in [0,T]\ \forall x \in A$. Moreover, $G(t,x)\zeta(\tau) \in \mathcal{L}(t,x)\quad \forall (t,x),\ \forall \tau$, so there exist $\{v_s(t,x)\} \subset F(t,x)$ and $\alpha(\tau,t,x) > 0$ such that

$$\alpha(\tau,t,x)\frac{v_s(t,x)}{|v_s(t,x)|} \overset{s\to\infty}{\longrightarrow} G(t,x)\zeta(\tau)$$

and vraimax $\alpha(\tau,t,x) \le ||G(t,x)|| \cdot ||\zeta(\tau)||_{L^k_\infty} < B||\zeta||_{L_\infty} < \infty$. Consider an auxiliary system:

$$\frac{dx_s}{d\tau} = \alpha(\tau,t_s,x_s)\frac{v_s(t_s,x_s)}{|v_s(t_s,x_s)|},\ x_s(0) = y_1,$$
$$\frac{dt_s}{d\tau} = \alpha(\tau,t_s,x_s)\frac{1}{|v_s(t_s,x_s)|},\quad t_s(0) = \bar{t}.$$

It is obvious that $t_s(1) \to \bar{t}$; $dx_s/d\tau \to G(t,x_s(\tau))\zeta(\tau)\quad \forall\tau$ and $x_s(\tau) \to \bar{x}(\tau)\ \forall\tau$, so $x_s(1) \to \bar{x}(1) = y_2$.

Denote by $\tilde{x}_s(t)$ the function $x_s(\tau_s(t))$ where $\tau_s(t)$ is a function, inverse to $t_s(\tau)$. Then $\tilde{x}_s(t)$ is determined on $[\bar{t}, t_s(1)]$, $t_s(1) - \bar{t} \overset{s\to\infty}{\longrightarrow} 0$, $x_s(t_s(1)) = x_s(1) \to y_2$ and $\frac{d}{dt}\tilde{x}_s = \frac{dx_s}{d\tau}/\frac{dt_s}{d\tau} = v_s(x_s(t)) \in F(t,x_s(t))$, so $\{x_s\}$ is a sequence of trajectories of the differential equation (1). ■

Below the generalized trajectories are considered from the class of the functions of bounded variation with possible jumps along the sets $Q(t, x(t-0))$ at the moment t. To describe an absolutely continuous component of such generalized trajectory, we shall use the first integrals of the differantial inclusion (4) will be used.

Let $\eta_i(t,x)$, $i = \overline{1,r}$ be the independent first integrals of the differential inclusion (4). It means that $\forall i\ \eta_i(t,x)$ is differentiable on t, x; for any trajectory x of the differential inclusion (4) $\eta_i(t,x(\tau))$ is not depend on τ, or a constant $c(t)$ for all τ and $rank\ \eta_x(t,x) = r\ \forall (t,x) \in [0,T]\times R^n$. If the first integrals of (4) are exist, the original problem P may be transformed to the extended problem EP :

$$\dot{y}(t) \in \cup_{x\in Q(t,y(t))}\{\eta_t(t,x) + \eta_x(t,x)v \mid v \in$$
$$\in F(t,x)\} \overset{\text{def}}{=} Y(t,y(t)) \qquad (7)$$
$$y(0) = Y_0 = \{\eta(t,x_0)|x_0 \in C_0\} \qquad (8)$$
$$\tilde{\Phi}(y(T)) = \min\{\Phi(x)|x \in Q(T,y(T)) \cap C_T\} \quad (9)$$

It is easily to see that for every $x(\cdot) \in D$ (for every trajectory x such that (1) and (2) are hold) the function $y(t) = \eta(t,x(t))$ satisfies (7) - (8), so EP is an extension of the problem P. From another side, any jump along $Q(t,y(t))$ may be approximated by trajectories or quasi-trajectories of (1) due to the Theorems 1, 2. The question about the equaivalence of P and EP will be discussed more precisely below for the particular case of the limit cone. It is necessary to mention that an extension of the problem P to the problem (7) - (8) is usefull in practice and gives a possibility to avoid the unboundary inclusion (to reduce them to boundary case when the conditions (Clarke, 1983; Blagodatskih and Fillipov, 1985) are effective).

Example 1. To minimize $x_1(1)$ on the trajectories of the differential inclusion

$$\frac{dx_1}{dt} \in F_1(t, x_1, x_2) = \{(2 - x_2)^2 + \\ + x_1 + v | v \in [0, 20], v \geq x_2\},$$
$$\frac{dx_2}{dt} \in F_2 \equiv [0, +\infty),$$
$$x_1(0) = 0, \; x_2(0) = 1.$$

Here $F_2(t, x) \equiv [0, +\infty)$ is unbounded. The limit cone

$$\mathcal{L} = \left\{ \begin{pmatrix} 0 \\ \alpha \end{pmatrix} : \alpha \geq 0 \right\}$$

and the limit differential inclusion has a description: $\frac{dx_1}{d\tau} = 0$; $\frac{dx_2}{d\tau} \in \{\alpha | \alpha \in [0, +\infty)\}$. Obviously, $\eta(t, x) \equiv x_1$ is a first integral of the limit differential inclusion. As $x_2(0) = 1$ and $\frac{dx_2}{dt} \geq 0$, then $x_2(t) \geq 1$. Consider the reduced problem:

$$\begin{cases} \frac{dx_1}{dt} \in \cup_{x_2 \geq 1} \{(2 - x_2)^2 + x_1 + v | v \in [x_2, 20] \\ \quad \text{if } x_2 \leq 20\} \\ x_1(0) = 0, \; x(1) \to \min \end{cases}$$

The Hamiltonian, introduced in the item 3.2 of the (Clarke, 1983), is

$$\mathcal{H}(x_1, \psi) = \max_{x_2 \geq 1}(\psi(2 - x_2)^2 + \psi x_1 + \max_{v \in [x_2, 20]} \psi v),$$

where $-\dot{\psi} \in \partial_{x_1} \mathcal{H}$, $\psi(1) = -1$, so $\psi(t) = -e^{1-t}$.

Then $v^* = \arg \max_{v \in [x_2, 20]} \psi v = x_2$; $x_2^*(t) = \frac{3}{2}$, $t \in (0, 1]$. Taking into account the initial conditions, it is easy to find

$$x_2^*(t) = \begin{cases} 1, & t = 0, \\ \frac{3}{2}, & t \in (0, 1], \end{cases}$$

$$x_1^*(t) = \frac{7}{4}(e^t - 1), \quad J^* = x_1^*(1) = \frac{7}{4}(e - 1).$$

3. DESCRIPTION OF GENERALIZED TRAJECTORIES OF ASYMPTOTICALLY LINEAR DIFFERENTIAL INCLUSIONS

Consider the case when the limit cone $\mathcal{L}(t, x)$ is a subspace of the constant dimension k and has a description

$$\mathcal{L}(t, x) = \{G(t, x)v \mid v \in R^k\} \tag{10}$$

where $G(t, x)$ for every fixed (t, x) is a $(n \times k)$-matrix. The differential inclusion (1) will be called here unbounded, asymptotically linear iff the limit cone of the set $\overline{co}F(t, x)$ has a description (10).

Due to the properties 4) - 5) of the limit cone (see above 2), $\overline{co}F(t, x)$ allows a decomposition

$$\overline{co}F(t, x) = B(t, x) + G(t, x)v, \; v \in R^k,$$

where $B(t, x)$ is a projection of $\overline{co}F(t, x)$ on the annulator $\mathcal{L}^{\perp}(t, x) = \{g \in R^n | (g, l) = 0 \; \forall l \in \mathcal{L}(t, x)\}$ and $B(t, x)$ is a closed bounded and convex subset of R^n for any (t, x).

If $F(t, x)$ is continuous on t and locally Lipschitzean on x, the differential inclusion

$$\dot{x} \in B(t, x) + G(t, x)R^k \tag{11}$$

is equivalent to (1) in the limit sense (Kolokolnikova, 1997). The (11) can be partly parametrizated: $\dot{x} \in \{B(t, x) + G(t, x)v \mid v \in R^k\}$ and v plays a role of control, so a transformation method, analogeous to (Dykhta, 1981), is natural. It allow to give an analytical description of the generalized trajectories of (1).

Consider a system

$$\frac{\partial x}{\partial z} = G(t, x), \quad z \in R^k \tag{12}$$

and assume the matrix-function $G(t, x)$ to be continuous differentiable on t, x and such that the Frobenius condition is fulfillled:

$$\begin{aligned} &G_x^i(t, x)G^j(t, x) - G_x^j(t, x)G^i(t, x) = 0 \\ &\forall (t, x) \in [0, T] \times R^n, \end{aligned} \tag{13}$$

where G^i is a i-th column of the matrix G. Let $\xi(z; t, y)$ be a solution of the multi-dimensional system of differential equations (12) with an initial condition $\xi(0; t, y) = y$. Then the initial value y is the first integral of the system (12) $\eta(t, x, z)$ which has a dimension n. The integral satisfies a condition

$$\eta_z(t, x, z) + G'(t, x)\eta_x(t, x, z) = 0 \tag{14}$$

Reduce (11) to the differential inclusion

$$\dot{y}(t) \in \mathcal{F}(t, y(t)) \stackrel{\text{def}}{=} \{\eta_t(t, x, z) + \eta_x(t, x, z)v_1 \mid \\ v_1 \in B(t, x), \quad x = \xi(z; t, y(t)), \quad z \in Z\},$$

where Z is a domain of an unprolongable solution of the system (12), and consider an optimization problem RP (the reduced problem):

$$\begin{aligned} &\dot{y} \in \mathcal{F}(t, y), \quad y(0) \in C_0, \\ &y(T) \in Y_T \stackrel{\text{def}}{=} \{\eta(T, x, s) | x \in C_T, s \in Z\} \end{aligned} \tag{15}$$

$$\begin{aligned} &\bar{J}[y] = \bar{\Phi}(y(T)) = \inf_{s \in Z} \Phi(\xi(s; T, y(T)), \\ &\bar{J} \to \inf \end{aligned} \tag{16}$$

The function $x(t)$, $t \in [0, T]$ of bounded variation is said to be a function of the A_F class iff there exist an absolutly continuous function $y \in AC^n[0, T]$, a measurable, bounded function $z \in L_\infty^k[0, T]$ and a vector $s \in Z \subset R^k$, satisfying (15), such that

$$\begin{aligned} &x(t) = \xi(z(t); t, y(t)), t \in (0, T) \\ &x(0) = y(0), \quad x(T) = \xi(s; T, y(T)) \end{aligned} \tag{17}$$

Theorem 3. Any function $x \in A_F$ is a generalized trajectory of the differential inclusion (1) if 1) multifunction $F(t, x)$ is continuous on t, x and locally Lipschitzean on x; 2) the differential inclusion is unbounded, asymptotically linear with basic matrix - function $G(t, x)$, which is continuously differentiable on t, x and such that (13) is fulfilled.

A proof of the Theorem 3 is analogeous by the schere to one used in (Nikiforova, 1990), but it is supplemented by using the results of (Kolokolnikova, 1997) about the connection of the differential inclusions (1) and (3).

Apply to the problem RP the Clarke's conditions of optimality (Clarke, 1983). It will be assumed here, that $C_T = R^n$ (the terminal restriction is absent). Then, if $(y^*(\cdot), z^*(\cdot), s^*)$ is optimal in (15) - (16), there exists a Hamiltonian multiplier $(\psi(\cdot), \zeta)$, where $\psi(\cdot) \in AC^n[0,T]$, $\zeta \in R^n$ such that:

a) $-\dot{\psi}(t) \in \partial\mathcal{H}(t, y^*(t), \psi(t))$;

b) $\zeta \in \partial\bar{\Phi}(y^*(T))$;

c) for some $r > 0$ $\quad \psi(0) \in r\partial d_{C_0}(y^*(0))$;

d) $\psi(T) = -\zeta$.

Here $\mathcal{H}(t,y,\psi) \stackrel{\text{def}}{=} \max\{\langle\psi, v\rangle | v \in \mathcal{F}(t,y)\}$, $d_{C_0}(y)$ is a function of the Euclidian distance from the variable point y to the given set C_0, $\partial\varphi$ is a generalized gradient of the Clarke's type (see the chapter 2 of the work (Clarke, 1983)), of the function φ.

The conditions a) - d) may be defined more exactly as the sets $\mathcal{F}(t,y)$ are partly parametrizated (by z). Introduce an auxiliary function of Pontrjagin's type for (15):

$$\tilde{H}(t,y,z,\psi) \stackrel{\text{def}}{=} \Big[\langle\psi, \eta_t(t,x,z)\rangle + \\ + \max_{v_1 \in B(t,x)} \langle\psi, \eta_x(t,x,z)v_1\rangle\Big]_{x=\xi(z;t,y)}. \tag{18}$$

A statement, that $(y^*((\cdot), z^*(\cdot), s^*)$ is optimal in (15) - (16), implies that there exists $(\psi(\cdot), \zeta)$ such that (see the chapter 5 of the work (Clarke, 1983))

$$\begin{aligned} &-\dot{\psi}(t) \in \partial_y \tilde{H}(t, y^*(t), z^*(t), \psi(t)), \\ &\psi(T) + \zeta = 0, \end{aligned} \tag{19}$$

$$\begin{aligned} &\zeta \in \partial_y \Phi(\xi(s^*; T, y^*(T))), \\ &\psi(0) \in r d_{C_0}(y^*(0)) \end{aligned} \tag{20}$$

$$\begin{aligned} &\tilde{H}(t, y^*(t), z^*(t), \psi(t) = \\ &= \max_{z\in Z} \tilde{H}(t, y^*(t), z, \psi(t)) \quad a.e. \end{aligned} \tag{21}$$

It is possible to write the condition (19)-(21) in the terms of the original problem P. For the first step, introduce the Hamiltonian multiplier of the problem of a minimization of the original functional $J[x] = \Phi(x(T))$ on the set of trajectories satisfying (11) and (2). Let $(p(\cdot), \gamma) : p(\cdot) \in AC^n[0,T], \gamma \in R^n$ be such that

$$\begin{aligned} &-\dot{p}(t) \in \partial_x H(t, x^*(t), v^*(t), p), \\ &\gamma + p(T) = 0, \quad \gamma \in \partial_x \Phi(x^*(T)) \\ &p(0) \in r d_{C_0}(x^*(0)), \end{aligned} \tag{22}$$

where $H(t,x,p,v) = \max\limits_{v_1 \in B(t,x)} \langle p, v_1\rangle + \langle p, G(t,x)v\rangle$.

Lemma 1. The function $\psi(\cdot)$ from (19)-(21) and $p(\cdot)$ from (22) are interconnected:

$$\psi(t) = \xi_y(z^*(t); t, \eta(t, x^*(t), z^*(t))p(t),$$
$$p(t) = \eta_x(t, \xi(z^*(t); t, y^*(t)), z^*(t))\psi(t).$$

The result is analogeous to known one in the theory of optimal impulsive controls, and the scheme of the Lemma 1 proof is analogeous to lemma 3 from 2.2 of the work (Nikiforova, 1990).

Lemma 2. The condition (21) has a variational form

$$\int_0^1 \langle m(\tau, z), z - z^*(t)\rangle d\tau \le 0$$

$$\forall z \in Z,\ \forall m(\tau, z) \in \partial_z \tilde{H}(t, y^*(t), z^*(t) + \tau(z - \\ - z^*(t)), \psi(t)).$$

If $\tilde{H}$ is smooth on z then

$$\triangle_z \tilde{H} = \int_0^1 \langle \tilde{H}_z(z^* + \tau(z - z^*(t)), z - z^*(t)\rangle d\tau.$$

Here we use the Rademacher's theorem that any locally Lipschitzean function is differentiale almost everywhere, and the Clarke's theorem of the chapter 2.5 from the work (Clarke, 1983) that a generalized gradient is a convex hull of the limits of the usual gradients, namely

$$\partial\varphi(z) = co\{\lim_{i\to\infty} \nabla\varphi(z_i) : z_i \to z, z_i \notin \Omega_\varphi\},$$

where Ω_φ is a set of points, in which φ is nondifferentiable.

Lemma 3. Put

$$\chi(\tau) = \xi(z^*(t) + \tau(z - z^*(t)); t, y^*(t)),$$

$q(\tau) = \eta_x(t, \xi, z^* + \tau(z - z^*(t)); t, y^*(t)), z^*(t) + \tau(z - z^*(t))\psi(t)$.

Then

$$\begin{aligned} &\frac{d\chi}{d\tau} = G(t,\chi)(z - z^*(t)), \\ &\chi(0) = x^*(t), \\ &\frac{dq}{d\tau} = -p(t)G_x(t,\chi)(z - z^*(t)), \\ &q(0) = p(t). \end{aligned} \tag{23}$$

The result is analogeous to item 2.4 of (Nikiforova, 1980). It follows from the Lemma 1 and the definition of ξ and η.

Let $x^*(\cdot)$ be the given function of bound variation, having jumps at the points $\tau_1, \tau_2, \cdots$:

$$x^*(t) = x_a^*(t) + \sum_{\tau_i < t} h_i,$$

where $x_a^*(t)$ is an absolutely continuous component of x^*.

Put

$$v^*(t) = \sum_{\tau<t} h_i\delta(\tau_i), \qquad z^*(t) = \sum_{\tau_i<t} h_i \tag{24}$$

Theorem 4. If the assumptions of the Theorem 3 are fulfilled and $x^* \in A_F$ is optimal then there exist a function $p(\cdot)$ and a vector γ such that:

1) $-\dot{p}(t) \in \partial_x \mathcal{H}_B(t, p(t), x^*(t)) +$
$+p(t)' G_x(t, x^*(t)) v^*(t)$;

2) $p(0) \in r\partial d_{C_0}(x^*(0))$ for some $r > 0$;

3) $p(T) + \gamma = 0, \quad \gamma \in \partial\Phi(x^*(T))$;

4) $-\int_0^1 \langle \tilde{m}(\tau, z), z - z^*(t)\rangle d\tau \leq 0 \quad \forall z \in Z,$

$\forall \tilde{m}(\tau, z) \in [qG_t(t, x) + G_x(t, x)\mathcal{H}_B(t, q, x) +$
$+G(t, x)\partial_x \mathcal{H}_B(t, q, x)]_{x=\chi(\tau) \;\; q=q(\tau)}$;

5) $\gamma' G(t, x^*(t)) = 0$ if $Z = R^k$.

Here $\mathcal{H}_B(t, p, x) = \max\{(p, v) \mid v \in B(t, x)\}$; $\chi(\tau)$ and $q(\tau)$ are solution of (23); v^* and z^* are defined by (24); $d_{C_0}(x)$ is a function of the Euclidian distance from the variable point x to the set C_0; $\partial\varphi$ is a generalized (Clarke's) gradient of the function φ.

The statement of the Theorem 4 is a retranslation of (19) - (21) with account of (18), (17) and lemmas 1-3. It is necessary to mention that the condition 1) from the Theorem 4 has generalized sense: $\dot{p}(t)$ may have delta - component (as v^* has) at the points of x^* discontinuity.

The parts of the research are supported by the Russian Fund of Fundumental Researches, Grants 96-01-01739, 97-01-00109, and by the State Com. of Russ. Vuz., Grant 95-0-19-58.

REFERENCES

Blagodatskih,V.I, A.F.Filippov (1985). Differential inclusions and optimal control. *Trudy of Math. Inst AS USSR.* **Vol.169**, p.194-252. (in Russian)

Clarke,F.H. (1983). *Optimization and nonsmooth analysis.* John Wiley & Sons, New York.

Dykhta,V.A. (1981). Conditions of the local minimum for singular regimes in the systems with linear control. *Automatica and telemech.* **N 12**, p.5-10. (in Russian)

Dykhta,V.A. (1994). Variational maximum principle and quadratic optimality conditions for impulsive and singular processes. *Sib.Math.J.* **Vol. 35**, p.70-82. (in Russian)

Gurman,V.I. (1977). *Degenerated Problems of optimal control.* Nauka, Moscow. (in Russian)

Kolokolnikova,G.A. (1996). Discontinuous trajectories optimality in the nonlinear optimal control problems. *In: 13th World Congress of IFAC.* **vol.E**, p.353-357.

Kolokolnikova,G.A. (1997). The variational maximum principle for discontinuous trajectories of the unbounded, asymptotically linear controlled systems. *Differential equations.* **V.33**, N 6. (in Russian)

Nikiforova,I.A. (1990). *Quadratic conditions of the impulsive regimes optimality.* Diss of cand. phys.-math.science. Irkutsk State University. Irkursk.

OPTIMAL CONTROL OF STOCHASTIC LOGISTIC MODEL[1]

G.E. Kolosov, D.V. Nejemetdinova

*Cybernetics Dept., Moscow University
of Electronics and Mathematics,
B.Trekhsvjatitel'ski per., 3/12,
109028, Moscow, Russia*

Abstract: Stochastic version of a problem of optimal control with infinite horizon arising in some applications is considered. Optimal control function in the form of synthesis is obtained by solving the Bellman equation corresponding to the considered problem. This function has a "bang-bang" form with switch point derived from the condition of smooth sewing for the profit function.

Keywords: optimal control, stochastic control, dynamic programming, control system synthesis.

1. INTRODUCTION AND STATEMENT OF THE PROBLEM

The problem of optimal control of the population when population dynamics is described by one-dimensional logistic model is considered. As is known the simplest logistic model is defined by the differential equation

$$\dot{x} = r\Big(1 - \frac{x}{K}\Big)x \tag{1}$$

describing the dynamics of isolated population. In equation (1) $x = x(t)$ denotes the size (or density) of the population at time t; positive numbers r and K denote the rate of natural growth of population and the maximum value of population size (capacity of the medium), respectively. In the presence of external actions consisting, say, in removal of some specimens from the surroundings the following controlled logistic model

$$\dot{x} = r\Big(1 - \frac{x}{K}\Big)x - qux \tag{2}$$

is given, where $u = u(t)$ $(0 \leq u(t) \leq u_m)$ is the rate of catching process at time t, which is used as the control function and number $q > 0$ is the catchability coefficient. So that the value

$$Q = q\int_{t_1}^{t_2} u(\tau)x(\tau)\,d\tau$$

is equal to the quantity of removed specimens during the time interval $[\,t_1, t_2\,]$.

The functional

$$I(u) = \int_0^T \big(pqx(t) - c\big)u(t)\,dt \tag{3}$$

[1]This research was partially supported by Russian Foundation of Basic Researches Grant 95-01-00191

is equal to the total yield of the catching process being defined by the control function $u(t): 0 \leq t \leq T$ and the initial state $x(0) = x_0$ of the system (2). Thus the problem of optimal control considers to be solved when the optimal control function $u^*(t): 0 \leq t \leq T$ maximizing the value of the profit functional (3) is found (optimal control $u^*(t)$ is sought among the set of admissible control functions $u(t)$ consisting of all piecewise continuous bounded functions $u(t): 0 \leq t \leq T$ satisfying the restriction $0 \leq u(t) \leq u_m$).

It should be noted that more significant results of solution of the problem (2)-(3) may be obtained when the problem of optimal control with infinite horizon is considered. In this case the final time moment $T \to \infty$ and instead of functional (3) the optimality criterion of the form

$$I(u) = \int_0^\infty e^{-\delta t}(pqx(t) - c)u(t)\,dt \tag{4}$$

is used (the number $\delta > 0$ is given). Precisely this profit functional was used by Goh (1980) and Clark (1985) when solving the problems of optimal fisheries managements.

This paper is devoted to the solution of the optimal control problem which is stochastic generalization of the problem (2),(4). Namely, the behaviour of the population is supposed to be described by stochastic controlled logistic equation of the form

$$\begin{aligned} \dot{x} &= r\Big(1 - \frac{x}{K}\Big)x - qux + \sqrt{2B}x\,\xi(t), \\ t &> 0, x(0) = x_0, 0 \leq u(t) \leq u_m \end{aligned} \tag{5}$$

where $\xi(t)$ denotes Gaussian random process of white noise type of unit intensity and B is given positive number. Due to stochastic nature of the functions $x(t)$ and $u(t)$ the profit functional (4) in this case has to be replaced by the corresponding mean value and the optimal control problem under consideration may be written in the form

$$I(u) = M\Big[\int_0^\infty e^{-\delta t}(pqx(t) - c)u(t)\,dt\Big] \longrightarrow \sup_{\substack{u(t)\in[0,u_m],\\ t\geq 0}} \tag{6}$$

where $M[\,\cdot\,]$ denotes the mean value of $[\,\cdot\,]$.

2. SOLUTION OF THE PROBLEM (5)-(6)

The dynamic programming approach is used to solve the problem (5)-(6).Let

$$F(x) = \sup_{\substack{u(\tau)\in[0,u_m],\\ \tau\geq t}} M\Big[\int_t^\infty e^{-\delta(\tau-t)}(pqx(\tau) - c)u(\tau)\,d\tau \Big| x(t) = x\Big] \tag{7}$$

be the profit function for the problem under consideration. Here, $M[(\,\cdot\,)|x(t) = x]$ denotes the conditional mean value of $(\,\cdot\,)$. As it was proved by Stratonovich (1968) and Krylov (1980) the profit function (7) is twice continuously differentiable and satisfies the following Bellman equation $\Big(F_x = \frac{dF}{dx},\ F_{xx} = \frac{d^2F}{dx^2}\Big)$:

$$\sup_{0\leq u\leq u_m}\Big[\Big(r + B - qu - \frac{r}{K}x\Big)xF_x + Bx^2F_{xx} - \delta F + (pqx - c)\,u\Big] = 0. \tag{8}$$

Remark 1. When writing (8), a symmetric stochastic integral (Stratonovich, 1968) was used as more appropriate in applications.

Equation (8) provides a way to define the optimal control in the form of the function $u_*(x)$ of the current state of the system (5) (the synthesis function). To construct the function $u_*(x)$, let us note that according to (8) the set $R^+ = [0, \infty)$ of all admissible states of the system (5) is divided in two subsets: subset R_0, where $\varphi(x) = pqx - c - qxF_x < 0$ and $u_*(x) = 0$, and subset R_1, where $\varphi(x) > 0$ and $u_*(x) = u_m$. The boundary separating these two subsets is defined by the equality

$$pqx - c - qxF_x = 0. \tag{9}$$

As follows from the further calculations, there exists a unique point x_* satisfying the condition (9), and therefore the subsets R_0 and R_1 have the form of intervals $R_0 = [0, x_*)$ and $R_1 = [x_*, \infty)$. Note also that at the very point x_* the choice of control effort is not essential, and each admissible control $u \in [0, u_m]$ is suitable. Consequently the optimal control function $u_*(x)$ may be represented in the form

$$u_*(x) = \begin{cases} 0 & , \text{ if } 0 \leq x < x_* \\ u_m & , \text{ if } x \geq x_*. \end{cases} \tag{10}$$

By virtue of (10) the final solution of the synthesis problem amounts to determination of the switch point x_*. To calculate the value of x_*, it is necessary to solve the Bellman equation (8).

Remark 2. In the case when capacity of the medium $K \to \infty$ (stochastic controlled Malthus model) the problem (5)-(6) was solved by Kolosov (1996). In this special case the switch point $x_* = x^0$, where x^0 is defined by the explicit formula

$$x^0 = \frac{c(\delta - r - B + qu_m)}{pq(\delta - r - B + \mu qu_m)}, \tag{11}$$

$$\mu = \frac{\lambda_1 - 1}{\lambda_1 - \lambda_2};\ \lambda_i = a_i\left(\sqrt{1 + \frac{\delta}{Ba_i^2}} - 1\right),\ i = 1, 2;$$

$$a_1 = \frac{r}{2B}; \; a_2 = \frac{r - qu_m}{2B}.$$

Let us denote by $F^0(x)$ and $F^1(x)$ the profit function $F(x)$ in the intervals $R_0 = [0, x_*]$ and $R_1 = (x_*, \infty)$, respectively. As follows from (8) and (10), these functions satisfy the following linear equations:

$$Bx^2 F^0_{xx} + x\left(r + B - \frac{r}{K}x\right) F^0_x$$
$$- \delta F^0 = 0, 0 \leq x < x_* \qquad (12)$$
$$Bx^2 F^1_{xx} + x\left(r + B - qu_m - \frac{r}{K}x\right) F^1_x$$
$$- \delta F^1 = u_m(c - pqx), x > x_*. \qquad (13)$$

By virtue of the above mentioned smoothness properties of the profit function F both functions F^0 and F^1 must satisfy the condition (9) at the switch point x_* (this is smooth sewing condition). Besides the condition

$$F^0(0) = 0 \qquad (14)$$

follows directly from (5) and (6). This makes it possible to obtain the solutions of equations (12) and (13) in analytical form and express these solutions by generalized power series.

Let us seek the solution of equation (12) in the form of generalized power series

$$F^0(x) = x^\sigma(a_0 + a_1 x + a_2 x^2 + \cdots). \qquad (15)$$

Substituting the series (15) in (12) and equating to zero the coefficients by $x^\sigma, x^{\sigma+1}, \ldots$ the system of equations defining the characteristic exponent σ and the coefficients a_i, $i = 0, 1, 2, \ldots$ is obtained:

$$[B\sigma(\sigma - 1) + (r + B)\sigma - \delta]\, a_0 = 0,$$
$$[B\sigma(\sigma - 1) + (r + B)\sigma - \delta + 2B\sigma + r + B]\, a_1$$
$$- \frac{r\sigma}{K} a_0 = 0,$$
$$\cdots\cdots\cdots$$
$$[B\sigma(\sigma - 1) + (r + B)\sigma - \delta + n(2B\sigma$$
$$+ r + nB)]a_n - \frac{r}{K}(\sigma + n - 1)a_{n-1} = 0, \qquad (16)$$
$$\cdots\cdots\cdots$$

Assuming that $a_0 \neq 0$ the characteristic equation is obtained from the first equality in (16):

$$B\sigma^2 + r\sigma - \delta = 0$$

which roots

$$\sigma^0_1 = \left(-r + \sqrt{r^2 + 4\delta B}\right)/2B,$$
$$\sigma^0_2 = \left(-r - \sqrt{r^2 + 4\delta B}\right)/2B,$$

define two possible values for the characteristic exponent σ in (15). By virtue of negativity of σ^0_2 and the boundary condition (14) only the value σ^0_1 has to be used in (15) and therefore the solution of equation (12) may be represented in the form

$$F^0(x) = a_0 \Psi(x) \qquad (17)$$

where

$$\Psi(x) = x^{\sigma^0_1}\left(1 + \Sigma^\infty_{n=1} \bar{a}_n x^n\right),$$
$$\bar{a}_n = \frac{1}{n!}\left(\frac{r}{KB}\right)^n \qquad (18)$$
$$\times \frac{\sigma^0_1(\sigma^0_1 + 1)\cdots(\sigma^0_1 + n - 1)}{(2\sigma^0_1 + \frac{r}{B} + 1)\cdots(2\sigma^0_1 + \frac{r}{B} + n)}.$$

The following estimate

$$\bar{a}_n \leq \frac{1}{n!}\left(\frac{r}{2KB}\right)^n$$

is valid for the coefficients of the series (18). This implies that the series (18) converges by any finite $x \geq 0$, this series may be termwise differentiated and integrated, its sum $\Psi(x)$ is entire analytical function which satisfies the following upper and lower estimates

$$x^{\sigma^0_1} \leq \Psi(x) \leq x^{\sigma^0_1} \exp\left(\frac{rx}{2KB}\right).$$

The constant a_0 in (17) is found from the boundary condition (9) for the function F^0 at the switch point x_*. Hence the following final expression for the solution of equation (12) is:

$$F^0(x) = \left(p - \frac{c}{qx_*}\right) \frac{\Psi(x)}{\Psi_x(x_*)}. \qquad (19)$$

The solution of nonhomogeneous equation (13) is obtained similarly using the standard method of variation of parameters. Omitting the corresponding simple, but bulky, calculations the final formulas defining the function $F^1(x)$ are given below. This function satisfys the equation (13) and boundary condition (9) at the switch point x_*:

$$F^1(x) = \left[p - \frac{c}{qx_*} - \Phi_x(x_*)\right] \frac{\Psi_2(x)}{\Psi_{2x}(x_*)} + \Phi(x), \qquad (20)$$

where

$$\Phi(x) = \frac{u_m}{B}\left[\Psi_1(x) \int \frac{(c - pqx)\Psi_2(x)dx}{x^2\left(\Psi_2\Psi_{1x} - \Psi_1\Psi_{2x}\right)}\right.$$
$$\left. - \Psi_2(x) \int \frac{(c - pqx)\Psi_1(x)dx}{x^2\left(\Psi_2\Psi_{1x} - \Psi_1\Psi_{2x}\right)}\right],$$
$$\Psi_1(x) = x^{\sigma^1_1}\left[1 + \Sigma^\infty_{n=1} \frac{a^1_n}{n!}\left(\frac{rx}{KB}\right)^n\right],$$
$$a^1_n = \frac{\sigma^1_1(\sigma^1_1 + 1)\cdots(\sigma^1_1 + n - 1)}{(\alpha + 1)(\alpha + 2)\cdots(\alpha + n)},$$
$$\Psi_2(x) = x^{\sigma^1_2}\left[1 + \Sigma^\infty_{n=1} \frac{a^2_n}{n!}\left(\frac{rx}{KB}\right)^n\right],$$

$$a_n^2 = \frac{\sigma_2^1(\sigma_2^1+1)\cdots(\sigma_2^1+n-1)}{(-\alpha+1)(-\alpha+2)\cdots(-\alpha+n)},$$

and numbers σ_1^1, σ_2^1 and α are defined by the expressions:

$$\sigma_{1,2}^1 = \frac{1}{2B}\left(qu_m - r \pm \alpha B\right),$$

$$\alpha = \frac{1}{B}\sqrt{(qu_m - r)^2 + 4\delta B}.$$

Two functions $F^0(x)$ (19) and $F^1(x)$ (20) define the profit function $F(x)$ satisfying the Bellman equation (8) for all $x \in R^+ = [0, \infty)$. These functions contain the parameter x_* not yet defined. To determine the value of x_*, the property of the continuity of the profit function $F(x)$ is used.
Both functions $F^0(x)$ and $F^1(x)$ themselves are continuous. Consequently, to provide the continuity of the profit function F defined by the functions F^0 and F^1, it is sufficient to require the fulfilment of the condition

$$F^0(x_*) = F^1(x_*) \tag{21}$$

at the switch point x_* (condition of smooth sewing). By virtue of (9), (12) and (13) the condition (21) amounts to the continuity of the second derivatives

$$F^0_{xx}(x_*) = F^1_{xx}(x_*) \tag{22}$$

at the switch point x_*.
Obviously that the conditions (21) or (22) do not make it possible to obtain the value of the switch point x_* in the explicit analytical form (like (11)). Nevertheless the conditions (21) and (22) together with formulas (18), (19) and (20) give a way for simple numerical calculation of the value x_*. Moreover due to the above noted convergence of the power series defining the functions F^0 and F^1 this numerical calculation permits to obtain the value x_* with arbitrary preassigned accuracy.

Besides in the special case considered below the equations (21) and (22) may be solved approximately in analytical form.

3. APPROXIMATE FORMULA FOR THE SWITCH POINT x_*

As it follows from (18) and (20) in the case $K \to \infty$ the functions $\Psi(x)$, $\Psi_1(x)$ and $\Psi_2(x)$ are defined by the finite formulas:

$$\Psi(x) = x^{\sigma_1^0},\ \Psi_1(x) = x^{\sigma_1^1},\ \Psi_2(x) = x^{\sigma_2^1}.$$

Respectively for the functions $F^0(x)$ and $F^1(x)$ instead of the series (17) and (20) the finite expressions are obtained:

$$F^0(x) = \frac{1}{\sigma_1^0}\left(px_* - \frac{c}{q}\right)\left(\frac{x}{x_*}\right)^{\sigma_1^0}, \tag{23}$$

$$F^1(x) = u_m\left(\frac{pqx}{\delta - r - B + qu_m} - \frac{c}{\delta}\right)$$
$$+ \frac{1}{\sigma_2^1}\left[\frac{(\delta - r - B)px_*}{\delta - r - B + qu_m} - \frac{c}{q}\right]\left(\frac{x}{x_*}\right)^{\sigma_2^1}. \tag{24}$$

Substituting (23) and (24) in (22) $x_* = x^0$ is obtained, where x^0 is defined by the expression (11) obtained by Kolosov (1996). If capacity of the medium K is a finite number then, as it was noted above, the value x_* can not be expressed in the finite form. However in the case of large values of the parameter K the value x_* will be close to the value x^0 which may be considered as the zero approximation of the root of the equations (21) and (22). Besides the refining corrections to this zero approximation may be calculated in accordance with the following scheme.

If K is large then the number $\varepsilon = \frac{r}{KB}$ may be considered as a small parameter and, as it follows from (18), (19) and (20), the functions $F^0(x)$ and $F^1(x)$ may be represented in the form of the series of powers of ε:

$$F^0(x) = F_0^0(x) + \varepsilon F_1^0(x) + \varepsilon^2\cdots, \tag{25}$$

$$F^1(x) = F_0^1(x) + \varepsilon F_1^1(x) + \varepsilon^2\cdots. \tag{26}$$

The root of the equations (21) and (22), i.e. the co-ordinate of optimal switch point x_*, is also sought in the form of asymptotic series

$$x_* = x^0 + \varepsilon\Delta_1 + \varepsilon^2\Delta_2 + \varepsilon^3\cdots, \tag{27}$$

where the numbers $x^0, \Delta_1, \ldots$ are to be defined. Substituting the expansions (25), (26) and (27) in the equation (21) (or (22)) and sequentially equating the expressions by the different powers of ε in left- and right-hand sides of (21) (or (22)) the chain of equations permitting to calculate sequentially the numbers $x^0, \Delta_1, \ldots$ is obtained in the expansion (27).

The first term x^0 in (27) coincides obviously with the expression (11). When calculating the first correction Δ_1 the expressions of zero and first order in the expansions (25), (26) and (27) are held. As a result from the equation (22) the following expression for the first correction Δ_1 was obtained

$$\Delta_1 = \frac{q(x^0)^3}{c(\sigma_1^0 - \sigma_2^1)}\left\{A_2(\sigma_2^1 - 2)\right.$$
$$+\left(p - \frac{c}{qx^0}\right)\left(\frac{\sigma_2^1+1}{1-\alpha} - \frac{\sigma_1^0+1}{1+\bar{\alpha}}\right)$$
$$\left.-\frac{(p - A1)(\sigma_2^1+1)}{1-\alpha}\right\}, \tag{28}$$

where the numbers $x^0, \sigma_1^0, \sigma_2^1$ and α were defined above, and the constants A_1, A_2 and $\overline{\alpha}$ are defined by the formulas:

$$A_1 = \frac{(\delta - r - B)p}{\delta - r - B + qu_m},$$

$$A_2 = \frac{2qu_m B A_1}{(r+B-\delta)(4B-2qu_m+2r-\delta)},$$

$$\overline{\alpha} = \sqrt{r^2+4\delta B}/B.$$

Thus the expression

$$x_1 = x^0 + \frac{r}{KB}\Delta_1 \tag{29}$$

with x^0 and Δ_1 defined by (11) and (28) may be called as the switch point of the first approximation in the case of large value of capacity of the medium K. The efficiency of using the switch point (29) in control algorithm (10) instead of the optimal value x_* is confirmed by numerous numerical experiments.

Let us denote by $u_i(x)$ the suboptimal control algorithms:

$$u_i(x) = \begin{cases} 0 & , \text{ if } 0 \le x < x_i \\ u_m & , \text{ if } x \ge x_i, \quad i=0,1. \end{cases} \tag{30}$$

Obviously, using these algorithms instead of the optimal control algorithm leads to decrease of the value of the functional (6) with respect to the optimal value $F(x)$. However, this decrease is expected to be small when parameter K is large. This expectation was confirmed by numerical experiments described below. Denote by $G_i(x)$ the value of the functional (6) when control $u_i(x)$ is used and the initial population size $x(0)=x$. Then $G_i(x)$ is continuously differentiated functions of the initial state x and satisfy the linear equations:

$$Bx^2G_i'' + x\left(r+B-\frac{r}{K}x\right)G_i'$$
$$+\left(pqx - c - qxG_i'\right)u_i(x) - \delta G_i = 0, \tag{31}$$
$$G_i(0) = 0.$$

Denoting by $G_{i0}(x)$ and $G_{i1}(x)$ the values of function $G_i(x)$ in the left- and right-hand sides of the switch point x_i the following equations for $G_{i0}(x)$ and $G_{i1}(x)$ are obtained:

$$Bx^2G_{i0}'' + x\left(r+B-\frac{r}{K}x\right)G_{i0}'$$
$$-\delta G_{i0} = 0, \quad 0 \le x \le x_i \tag{32}$$
$$Bx^2G_{i1}'' + x\left(r+B-qu_m-\frac{r}{K}x\right)G_{i1}'$$
$$-\delta G_{i1} = u_m(c-pqx), \quad x \ge x_i. \tag{33}$$

These equations are absolutely analogous to equations (12),(13). Consequently, the solutions of these equations may be found absolutely analogously to the above calculations. And these functions have the form:

$$\begin{aligned} G_{i0}(x) &= C_1\Psi(x), \\ G_{i1}(x) &= C_2\Psi_2(x) + \Phi(x), \end{aligned} \tag{34}$$

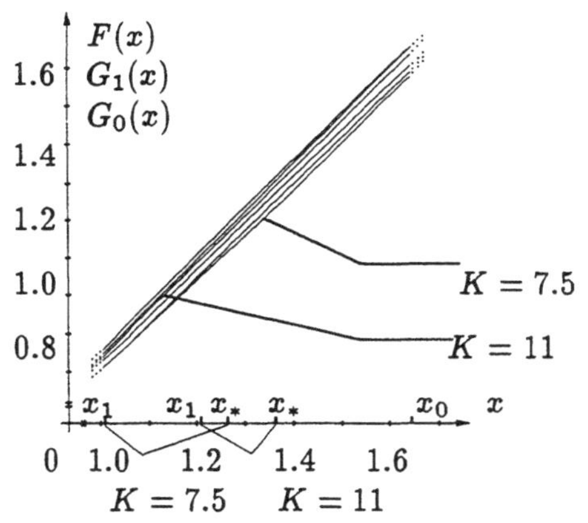

Fig. 1. $r=1,\ \delta=3,\ B=1,\ q=3,\ U_m=1.5,$ $c=3,\ p=2,\quad$ a)$K=7.5$ or b)$K=11$.

where functions Ψ, Ψ_2, Φ are defined by formulas (18),(20). Functions (34) differ from corresponding functions (19) and (20) by the ways of finding C_1 and C_2 in (34).

These constants are defined from the conditions of continuety of the functions G_i and their first derivatives at the switch point x_i:

$$\begin{aligned} G_{i0}(x_i) &= G_{i1}(x_i), \\ G_{i0}'(x_i) &= G_{i1}'(x_i). \end{aligned} \tag{35}$$

It follows from (34) and (35):

$$\begin{aligned} C_1 &= \frac{\Phi(x_i)\psi_2'(x_i) - \Phi'(x_i)\psi_2(x_i)}{\psi(x_i)\psi_2'(x_i) - \psi'(x_i)\psi_2(x_i)}, \\ C_2 &= \frac{\Phi(x_i)\psi'(x_i) - \Phi'(x_i)\psi(x_i)}{\psi(x_i)\psi_2'(x_i) - \psi'(x_i)\psi_2(x_i)}. \end{aligned} \tag{36}$$

Giving the concrete values of the parameters $r, K, q, \ldots$ for the problem (5),(6) the coefficients (36) and functions $G_i(x)$, $i=0,1$ can be calculated with arbitrary preassigned accuracy with help of computer. Also these formulas (35),(36) may be used for numerical construction of $F(x)$ being the solution of Bellman equation (5) if the value x_i is changed by x_*. In this case functions $G_{i0}(x)$ and $G_{i1}(x)$ defined by (35) coinside with $F_0(x)$ and $F_1(x)$ defined by (19) and (20), respectively, i.e. $G_i(x) \equiv F(x)$. Described procedure of numerical construction of the $G_0(x), G_1(x), F(x)$ has been realized as a program for numerical calculations. Diagrams of $G_0(x), G_1(x), F(x)$ shown in figures 1,2 were constructed for the following parameters:

1. $r=1, \delta=3, B=1, q=3, U_m=1.5, c=3,$ $p=2$
   (a) $K=7.5$ or
   (b) $K=11$.
2. $r=1, \delta=20, B=1, q=3, U_m=100, c=3,$ $p=2$

(a) $K = 0.17$ or

(b) $K = 0.3$.

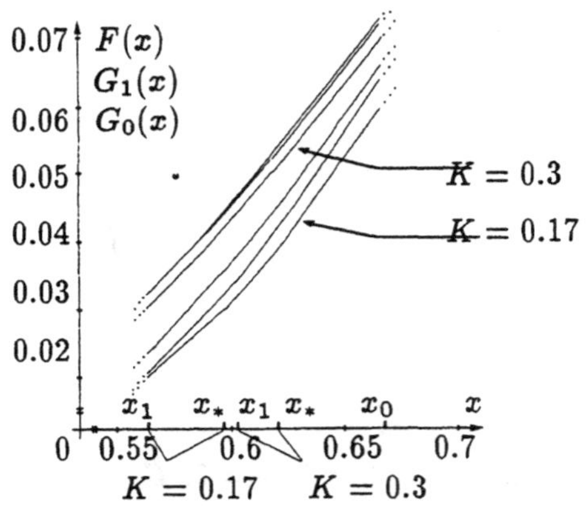

Fig. 2. $r = 1,\ \delta = 20,\ B = 1,\ q = 3,\ U_m = 100,$
$c = 3,\ p = 2,$ a)$K = 0.17$ or b)$K = 0.3$.

These diagrams show the efficiency of suboptimal control algorithms $u_0(x)$ and $u_1(x)$.

REFERENCES

Clark, C. W. (1985). *Bioeconomic Modelling and Fisheries Managements.* J.Wiley.

Goh, B. S. (1980).*Management and Analysis of Biological Populations.* Elsevier Sci.Publ.Company, Amsterdam.

Kolosov, G. E. (1996). Exact Solution of a Stochastic Problem of Optimal Control by Population Size.*Dynamic Systems and Applications*, **5**, 1, pp. 153-161.

Krylov, N. V. (1980).*Controlled Diffusion Processes.* Springer,NY.

Stratonovich, R. L. (1968).*Conditional Markov Processes and their Applications to the Theory of Optimal Control.* Am.Elsevier Co.,Inc.,NY.

RECENT RESULTS OF DYNAMIC CONTRACTION METHOD (SURVEY)

Valery Dm. Yurkevich *,[1]

* *Automation Department, Novosibirsk State Technical University Novosibirsk, 630092, RUSSIA, e-mail: yurkev@nstu.nsk.su*

Abstract. The problem of forming desired output transients for discrete-time system is discussed under assumption of incomplete information about varying parameters of the system and unknown external disturbances. To solve the discussed control problem a design method which was called a dynamic contraction method is used. Such problems as the analysis of fast and slow motions in the discrete-time system with small parameter, choices of control law parameters, influence of varying parameters and control accuracy are considered.

Keywords. Discrete-time system, disturbance rejection, multirate system

1. INTRODUCTION

The problem of forming desired output transients for control plants is discussed under assumption of incomplete information about varying parameters of the system and unknown external disturbances.

The controllers for uncertain nonlinear systems based on Variable Structure Systems (VSS) theory (Utkin, 1978; Utkin, 1994) may be used, where in the sliding phase the motion of the system is insensitive to parameter variations and disturbances in the system. Control laws that are based on the Non-linear Inverse Dynamics (NID) method (Porter, 1970; Boychuk, 1971; Slotine and Li, 1991; Kotta, 1995) may be used if the dynamics of the system are exactly known. An algorithmic approach to solution of the inverse dynamics problem should be used under condition of incomplete information. For example, a solution of the inverse dynamics problem based on the Gradient Descent Method (GDM) usually is used in adaptive control systems (Krutko, 1989; Fradkov, 1990). Another way of the algorithmic solution of the inverse dynamics problem is developed as the Localization Method (LM) (Vostrikov, 1977; Vostrikov *et al.*, 1982; Vostrikov, 1990; Vostrikov and Yurkevich, 1991; Vostrikov and Yurkevich, 1993), which allows to provide the desired transients for nonlinear time-varying systems. The peculiarity of LM is the application of the higher order derivatives (Shchipanov, 1939; Green, 1961) jointly with high gain in the control law (Meerov, 1959). Methods of singularly perturbed equations (Gerashchenko, 1975; Saksena *et al.*, 1984) are used in LM to analyse the closed loop system properties. The digital control algorithms on the basis of the localisation method are discussed in (Mutschkin, 1985; Fehrmann *et al.*, 1989; Vostrikov *et al.*, 1990).

As a result of a development of LM a new view on control problem, called Dynamic Contraction Method (DCM), has been presented (Yurkevich, 1992; Yurkevich, 1993a). At the same time, the DCM can also be seen as a generalization of GDM (Green, 1961; Krutko, 1989; Fradkov, 1990). In particular, the structure of the control law discussed in DCM follows also from a higher order local optimization procedure in continuous case and from a multi-steps local optimization procedure in discrete case (Tsypkin, 1968). Based on DCM, the continuous and discrete control laws have been derived in order to provied output tracing with prescribed dynamics in spite of uncertainty in the system. A distinguishing feature of both LM and DCM is, as opposed to VSS, that the motion of the system is insensitive to parameter variations

[1] Partially supported by the Russian Federation State Committee for Higher Education

and disturbances in a bounded set of the entire state-space. As opposed to LM, DCM allows an integral action to be incorporated in the control loop without increasing the controller's order.

In the paper a short survey of publications which is connected with control system design by DCM for continuous plants is presented and then, as a continuation of (Yurkevich, 1992; Yurkevich, 1994c; Yurkevich, 1995a), the main attention is given to problem of the providing desired output transients for discrete-time system under assumption of incomplete information

2. CONTINUOUS-TIME SYSTEM DESIGN

The procedure of continuous control law design by DCM for continuous nonlinear time-varying SISO systems was given in (Yurkevich, 1993a; Yurkevich, 1995a) and for MIMO systems in (Yurkevich, 1993b; Yurkevich, 1995b). The design of an aircraft motion controllers based on DCM was presented, for example, in (Yurkevich, 1994a; Błachuta *et al.*, 1995a; Błachuta *et al.*, 1995b). Some procedures of digital control law design based on DCM for continuous nonlinear time-varying systems was discussed in (Błachuta *et al.*, 1996; Yurkevich *et al.*, 1997; Błachuta *et al.*, 1997a). The solition of the output tracing problem by DCM under condition of high-frequency sensor noise was presented in (Yurkevich, 1994b; Błachuta *et al.*, 1997b).

3. DISCRETE-TIME SISO SYSTEM

Let us consider a nonlinear time-varying discrete system

$$y_k = f(k, \boldsymbol{y}_k, \boldsymbol{w}_k) + \boldsymbol{b}\tilde{\boldsymbol{u}}_k, \ \boldsymbol{y}_0 = \boldsymbol{y}^0 \quad (1)$$

where y_k is the output available for measurement, u_k is the control, w_k is the external disturbance unavailable for measurement, $f(\cdot)$ is continuous function with respect to y_{k-i}, w_{k-i} for all $i = \overline{1,n}$ and $\boldsymbol{y}_k = \{y_{k-n}, \ldots, y_{k-1}\}'$, $\boldsymbol{w}_k = \{w_{k-n}, \ldots, w_{k-1}\}'$, $\tilde{\boldsymbol{u}}_k = \{u_{k-n}, \ldots, u_{k-1}\}'$, $\boldsymbol{b} = \{b_n, \ldots, b_1\}$. For example, the discrete-time system of (1) may also have the form of the linear difference equation

$$y_k = \sum_{i=1}^{n} a_i y_{k-i} + \sum_{i=1}^{n} b_i u_{k-i} + \sum_{i=1}^{n} g_i w_{k-i} \quad (2)$$

Assumption 1 *Let us assume that the roots of the polynomial* $b_1 z^{n-1} + \cdots + b_{n-1} z + b_n$ *are placed on the outside of some neighbourhood of 1, then* $\sum_{i=1}^{n} b_i \neq 0$.

Assumption 2 *Let us assume that the discrete-time system (1) may be rewritten in the form of the following difference equation*

$$y_k = \boldsymbol{a}_0 \boldsymbol{y}_k + \mu\{\tilde{f}(k, \boldsymbol{y}_k, \boldsymbol{w}_k) + \tilde{\boldsymbol{b}}\tilde{\boldsymbol{u}}_k\}, \boldsymbol{y}_0 = \boldsymbol{y}_\mu^0 (3)$$

where μ is a small parameter, $\boldsymbol{y}_\mu^0 = \boldsymbol{y}^0(\mu)$ is the initial state, $\boldsymbol{a}_0 = \{a_{0,n}, \ldots, a_{0,2}, a_{0,1}\}$, $a_{0,i} = (-1)^{i+1} C_n^{n-i}$, $\boldsymbol{b} = \mu\tilde{\boldsymbol{b}}$, $\tilde{\boldsymbol{b}} = \{\tilde{b}_n, \ldots, \tilde{b}_2, \tilde{b}_1\}$ and

$$\lim_{\mu \to 0} \boldsymbol{y}^0(\mu) = \boldsymbol{y}_0^0, \ \boldsymbol{y}_0^0 = \{y_0^0, y_0^0, \ldots, y_0^0\}' \quad (4)$$

Remark 1 *If $\mu = 0$ then from (3) and (4) it follows that we have the system $y_k = \boldsymbol{a}_0 \boldsymbol{y}_k$, $\boldsymbol{y}_0 = \boldsymbol{y}_0^0$ where its characteristic polynomial has the form $(z-1)^n = 0$ and $y_k = y_0^0 \ \forall \ k = 0, 1, 2, \ldots$*

Remark 2 *If u_k and w_k are bounded in (3) $\forall \ k = 0, 1, \ldots$ then $\lim_{\mu \to 0} y_k(\mu) = y_0^0$ and accordingly $\lim_{\mu \to 0}\{y_k(\mu) - y_{k-i}(\mu)\} = 0$ are fulfilled for the solitions of (3) where $0 \leq i < k < \infty$.*

Remark 3 *The conditions of the assumptions 1 and 2 usually are fulfilled if the discrete-time system (1) follows, for example, from the minimum-phase continuous-time system discretization.*

4. CONTROL PROBLEM

4.1 *Output tracing problem*

The control system is designing to provide the following condition

$$\lim_{k \to \infty} \Delta_k = 0 \quad (5)$$

where Δ_k is the error of the reference input realisation, $\Delta_k = r_k - y_k$; r_k is the reference input. Moreover, the controlled transients $\Delta_k \to 0$ should have a desired performance indices.

4.2 *Insensitivity condition*

The desired behaviour of the output transients can be provided if the output transients satisfy the desired stable difference equation

$$y_k = F(\boldsymbol{y}_k, \boldsymbol{r}_k) \quad (6)$$

where $\boldsymbol{r}_k = \{r_{k-n}, \ldots, r_{k-1}\}'$ and (6) has the form

$$y_k = \sum_{i=1}^{n} a_i^d y_{k-i} + \sum_{i=1}^{n} b_i^d r_{k-i} \quad (7)$$

Here $1 - \sum_{i=1}^{n} a_i^d = \sum_{i=1}^{n} b_i^d$ and $\sum_{i=1}^{n} b_i^d \neq 0$. Parameters of (7) are selected based on the required output transient performance indices.

Assumption 3 *Let us assume that the roots of the characteristic polynomial of (7) are selected into the unit disk and into some small neighbourhood of 1.*

Under condition of the assumption 3, let us rewrite (6) in the form

$$y_k = \boldsymbol{a}_0\,\boldsymbol{y}_k + \mu\tilde{F}(\boldsymbol{y}_k, \boldsymbol{r}_k). \tag{8}$$

Let us denote

$$\Delta_k^F = F_k - y_k \tag{9}$$

where $F_k = F(\boldsymbol{y}_k, \boldsymbol{r}_k)$.

The equation (6) defining the desired behaviour of y_k is fulfilled if and only if the following holds

$$\Delta_k^F(u_{k-n}, \ldots, u_{k-2}, u_{k-1}) = 0 \tag{10}$$

for all $k = 0, 1, \ldots$

If the condition of (10) is held then the output transient performance indices are insensitive to parameter variations and external disturbances in the system (1). So the control problem has been reformulated as a problem of finding the solution to (10) when its varying parameters are unknown.

5. DISCRETE CONTROL LAW STRUCTURE

To fulfil the requirement of (10), let us form the control law in the form of the following difference equation (Yurkevich, 1992)

$$u_k = \boldsymbol{d}\,\boldsymbol{u}_k + \lambda_0\Delta_k^F \tag{11}$$

where $\boldsymbol{d} = \{d_q, \ldots, d_1\}$, $\boldsymbol{u}_k = \{u_{k-q}, \ldots, u_{k-1}\}'$,

$$d_1 + d_2 + \cdots + d_n = 1,\ \lambda_0 \neq 0,\ q \geq n \tag{12}$$

From (12) it follows, that a stable equilibrium of (11) is the solution of equation (10).

Definition 1 *The difference equation (11) is called as a difference contraction equation if its stable equilibrium is the solution of equation (10).*

6. CLOSED-LOOP SYSTEM PROPERTIES

6.1 *Fast motions*

The closed-loop system equations have the form

$$y_k = \boldsymbol{a}_0\,\boldsymbol{y}_k + \mu\{\tilde{f}(k, \boldsymbol{y}_k, \boldsymbol{w}_k) + \tilde{\boldsymbol{b}}\,\tilde{\boldsymbol{u}}_k\},\ \boldsymbol{y}_0 = \boldsymbol{y}_\mu^0$$
$$u_k = \boldsymbol{d}\,\boldsymbol{u}_k + \lambda_0\Delta_k^F,\ \boldsymbol{u}_0 = \boldsymbol{u}^0$$

where we assume that $\lambda_0 = \mu^{-1}\tilde{\lambda}_0$. From (9) it follows that the closed-loop system equations can be rewritten in the form

$$y_k = \boldsymbol{a}_0\boldsymbol{y}_k + \mu\{\tilde{f}(k, \boldsymbol{y}_k, \boldsymbol{w}_k) + \tilde{\boldsymbol{b}}\tilde{\boldsymbol{u}}_k\} \tag{13}$$
$$u_k = \boldsymbol{\beta}\boldsymbol{u}_k + \tilde{\lambda}_0\{\tilde{F}(\boldsymbol{y}_k, \boldsymbol{r}_k) - \tilde{f}(k, \boldsymbol{y}_k, \boldsymbol{w}_k)\} \tag{14}$$

where $\boldsymbol{\beta} = \{\beta_q, \ldots, \beta_1\}$, $\beta_i = d_i - \tilde{\lambda}_0\tilde{b}_i\ \forall\, i = \overline{1,n}$, and $\beta_i = d_i\ \forall\, i = \overline{n+1,q}$.

Let us assume that there is a sufficient time-scale separation, represented by small parameters μ, between the fast and slow modes in the closed loop system. Then fast-motion subsystem (FMS) and slow-motion subsystem (SMS) can be analysed separately (Litkouhi and Khalil, 1985).

If $\mu \rightarrow 0$ then, from (13) and (14), it follows that

$$u_k = \boldsymbol{\beta}\boldsymbol{u}_k + \tilde{\lambda}_0\{\tilde{F}(\boldsymbol{y}_k, \boldsymbol{r}_k) - \tilde{f}(k, \boldsymbol{y}_k, \boldsymbol{w}_k)\} \tag{15}$$

is the FMS equation, where the state vector of the system of (13) is constant during the transients in the system of (15), i.e. $\boldsymbol{y}_k - \boldsymbol{y}_{k-i} \approx 0\ \forall\ i\ \approx \overline{1,q}$.

6.2 *Control law parameters*

The asymptotic stability and desired transients of u_k can be achieved by a proper choice of the control law parameters $\boldsymbol{d}$, $\tilde{\lambda}_0$. For example, the modal approach may be used. Let us form the characteristic polynomial

$$z^q + \beta_1^0 z^{q-1} + \cdots + \beta_{q-1}^0 z + \beta_q^0 \tag{16}$$

where the roots of (16) are selected in a some small neighbourhood of zero in accordance with the requirements to the admissible transients in FMS. Then the characteristic polynomial of FMS (15)

$$z^q + \beta_1 z^{q-1} + \cdots + \beta_{q-1} z + \beta_q$$

has the desired form (16) if

$$\tilde{\lambda}_0 = \{1 - \beta_s^0\}\{\tilde{b}_s\}^{-1} \tag{17}$$
$$d_i = \beta_i^0 + \tilde{b}_i\tilde{\lambda}_0, \quad \forall i = \overline{1,n} \tag{18}$$
$$d_i = \beta_i^0 \quad \forall\ i = \overline{n+1,q} \tag{19}$$

where $\tilde{b}_s = \tilde{b}_1 + \tilde{b}_2 + \cdots + \tilde{b}_n$, $\beta_s^0 = \beta_1^0 + \beta_2^0 + \cdots + \beta_q^0$.

If eqn.(16) has the form $z^q = 0$ then from (17)-(19) follow that $\lambda_0 = 1/(b_1 + b_2 + \cdots + b_n)$, $d_i = b_i\lambda_0$, $\forall i = \overline{1,n}$, and $d_i = 0\ \forall\ i = \overline{n+1,q}$ in eqn.(11) (Yurkevich, 1992; Yurkevich, 1995a).

6.3 *Slow motions*

If we have a steady state (more precisely, a quasi steady state) for the FMS (15), then

$$u_{k-q} = \cdots = u_{k-1} = u_k = u_k^a \tag{20}$$

where $u_k^a = \{\tilde{b}_s^0\}^{-1}\{\tilde{F}(\boldsymbol{y}_k, \boldsymbol{r}_k) - \tilde{f}(k, \boldsymbol{y}_k, \boldsymbol{w}_k)\}$.

If the FMS (15) is in the steady state then eqns.(13), (14) and (20) imply that the SMS equation is equal to (8). So, if a sufficient time-scale separation is provided between the fast and slow modes in the closed loop system then the output transient performance indices are insensitive to parameter variations and external disturbances in (1).

7. INFLUENCE OF VARYING PARAMETERS

7.1 *Fast and slow motion subsystems*

Let us consider the influence of varying parameters on the properties of the linear closed loop system (2),(11), where $q = n$ and

$$\lambda_0 = \lambda_0^0 + \Delta\lambda_0$$
$$d_i = d_i^0 + \Delta d_i,\ b_i = b_i^0 + \Delta b_i \quad \forall\ i = \overline{1,n}$$
$$\lambda_0^0 = 1/(b_1^0 + b_2^0 + \cdots + b_n^0),\ d_i^0 = \lambda_0^0 b_i^0$$

Here Δb_i is the variation of the parameter b_i and Δd_i, $\Delta\lambda_0$ are the errors of the practical realization of the control law parameters d_i^0, λ_0^0. In this case we have that

$$u_k = \sum_{i=1}^{n} \tilde{\beta}_i u_{k-i} + \{\lambda_0^0 + \Delta\lambda_0\} \times$$
$$\times \sum_{i=1}^{n} \{[a_i^d - a_i]y_{k-i} + b_i^d r_{k-i} - g_i w_{k-i}\} \quad (21)$$

is the FMS equation (Yurkevich, 1994c), where

$$\tilde{\beta}_i = \Delta d_i - \Delta\lambda_0\{b_i^0 + \Delta b_i\} - \Delta b_i \lambda_0^0 \quad \forall\ i = \overline{1,n}.$$

If there is a steady state (20) for the FMS (21) then

$$u_k^a = \left\{1 - \Delta d_s + \lambda_0^0 \sum_{i=1}^{n} \Delta b_i\right\}^{-1} \times$$
$$\times \lambda_0^0 \sum_{i=1}^{n} \{[a_i^d - a_i]y_{k-i} + b_i^d r_{k-i} - g_i w_{k-i}\},$$

where $\Delta d_s = \Delta d_1 + \Delta d_2 + \cdots + \Delta d_n$. As a result the closed loop system eqns.(2), (11) imply that

$$y_k = \sum_{i=1}^{n} a_i^d y_{k-i} + \sum_{i=1}^{n} b_i^d r_{k-i} +$$
$$+\Delta d_s \left\{[1 + \Delta\lambda_0/\lambda_0^0][1 + \lambda_0^0 \sum_{i=1}^{n} \Delta b_i] - \Delta d_s\right\}^{-1} \times$$
$$\times \sum_{i=1}^{n} \{[a_i^d - a_i]y_{k-i} + b_i^d r_{k-i} - g_i w_{k-i}\}$$

is the SMS equation (Yurkevich, 1994c).

7.2 *Steady-state error*

Let us assume that $r_k = r = const$, $y_k = y^s = const$, $w_k = w^s = const\ \forall\ k = 0, 1, \ldots$ Denote the steady-state error $\Delta^s = r - y^s$, where

$$\Delta^s = \Delta_r^s + \Delta_w^s.$$

Here Δ_r^s is the steady-state error with respect to r_k, and Δ_w^s is the steady-state error with respect to w_k. Then, from (2) and (11), it follows that

$$\Delta_r^s = -\Delta d_s \left[1 - \sum_{i=1}^{n} a_i\right] \eta^{-1} r^s$$
$$\Delta_w^s = \Delta d_s \left[\sum_{i=1}^{n} g_i\right] \eta^{-1} w^s$$

where

$$\eta = \left[1 - \sum_{i=1}^{n} a_i^d\right]\left[1 - \sum_{i=1}^{n} \tilde{\beta}_i\right] - \Delta d_s \sum_{i=1}^{n} [a_i^d - a_i].$$

7.3 *Velocity error*

Let us assume that $\Delta d_s = 0$, $r_k = r = const$, $y_k = y^v = const$, $w_k = \alpha hk\ \forall\ k = 0, 1, \ldots$ where h is the sampling period. Denote the velocity error with respect to w_k as follows Δ_w^v, where $\Delta_w^v = r - y^v$. Then, from eqns.(2) and (11), it follows that

$$\Delta_w^v = -h \left[\sum_{i=1}^{n} g_i\right]\left[1 - \sum_{i=1}^{n} a_i^d\right]^{-1} \zeta\alpha$$

where

$$\zeta = \left[\sum_{i=1}^{n} i\tilde{\beta}_i\right]\left[1 - \sum_{i=1}^{n} \tilde{\beta}_i\right]^{-1} + \left[\sum_{i=1}^{n} ib_i\right]\left[\sum_{i=1}^{n} b_i\right]^{-1}.$$

Denote the velocity error with respect to r_k as follows Δ_r^v. We can find expression of Δ_r^v as the limit

$$\lim_{k\to\infty} \{r_k - y_k\} = \Delta_r^v$$

where $\Delta d_s = 0$, $w_k = 0$, $r_k = \alpha hk$. As a result we have

$$\Delta_r^v = h \left[1 - \sum_{i=1}^{n} a_i^d\right]^{-1} \times$$
$$\times \left\{\left[\sum_{i=1}^{n} i[a_i^d + b_i^d]\right] + \left[1 - \sum_{i=1}^{n} a_i\right]\zeta\right\}\alpha.$$

8. DISCRETE-TIME MIMO SYSTEM DESIGN

Let us consider discrete-time system

$$y_k = f(k, \bar{y}_k, \bar{w}_k) + \sum_{i=1}^{n} B_i u_{k-i} \quad (22)$$

where $y_k \in R^p$, $u_k \in R^p$, $\bar{y}_k = \{y'_{k-n}, \ldots, y'_{k-1}\}'$, $\bar{w}_k = \{w'_{k-n}, \ldots, w'_{k-1}\}'$. We assume also that the conditions, which are similar to assumptions 1 and 2 are also fulfilled in this case.

Let us form the desired stable difference equation system $y_k = F(\bar{y}_k, \bar{r}_k)$ in the form

$$y_k = \sum_{i=1}^{n} A_i^d y_{k-i} + \sum_{i=1}^{n} B_i^d r_{k-i} \qquad (23)$$

where $\bar{r}_k = \{r'_{k-n}, \ldots, r'_{k-1}\}'$ and

$$I_p - \sum_{i=1}^{n} A_i^d = \sum_{i=1}^{n} B_i^d, \quad \det\left\{\sum_{i=1}^{n} B_i^d\right\} \neq 0.$$

Parameters of (23) are selected in according with the assumption 3. In order to provide the decoupling of controlled channels let us assume also that A_i^d and B_i^d are diagonal matrices $\forall\ i = \overline{1,n}$.

As a result we have that the output transient performance indices are insensitive to parameter variations and external disturbances in the system (22) and the controlled channels are decoupled if and only if the following holds

$$\mathbf{\Delta}^F(u_{k-n}, \ldots, u_{k-2}, u_{k-1}) = 0 \quad \forall\ k \quad (24)$$

where $\mathbf{\Delta}_k^F = F_k - y_k$, $F_k = F(\bar{y}_k, \bar{r}_k)$.

To fulfil the requirement of (24), let us form the following control law (Yurkevich, 1993c)

$$u_k = \sum_{i=1}^{q\geq n} D_i u_{k-i} + \Lambda_0 \mathbf{\Delta}_k^F$$

where $\sum_{i=1}^{q} D_i = I_p$ and $\det \Lambda_0 \neq 0$.

The first, based on the above procedure from the closed loop system equations it follows that

$$u_k = \sum_{i=n+1}^{q>n} D_i u_{k-i} + \sum_{i=1}^{n}[D_i - \Lambda_0 B_i]u_{k-i} +$$
$$+\Lambda_0\{F(\bar{y}_k, \bar{r}_k) - f(k, \bar{y}_k, \bar{w}_k)\} \qquad (25)$$

is the FMS equation, where $\bar{y}_k - \bar{y}_{k-i} \approx 0\ \ \forall\ i \approx \overline{1,q}$, i.e. during the transients in the system (25).

For example. the characteristic polynomial of FMS (25) is equal to $I_p z^q = 0$ if

$$\Lambda_0 = \{B_1 + B_2 + \cdots + B_n\}^{-1}$$

$$D_i = \Lambda_0 B_i\ \forall\ i = \overline{1,n} \text{ and } D_i = 0\ \forall\ i = \overline{n+1,q}$$

At second, if there is a steady state (or more exactly a quasi steady state) for the FMS (25) then we have the SMS in the form of (23) under condition of unknown function $f(k, \bar{y}_k, \bar{w}_k)$.

9. CONCLUSIONS

In the paper, it has been shown that the discussed method may be used for the design of digital controllers which solve the tracing problem for output variables where the desired decoupled output transients are attained under assumption of incomplete information about varying parameters of the system and external disturbances. Obviously, that if the discrete-time model follows from the continuous-time system discretization then it is easy to provide the restrictions of the discussed method by decreasing of the sampling period.

10. REFERENCES

Błachuta, M.J., V.D. Yurkevich and K. Wojciechowski (1995a). Aircraft motion control by dynamic contraction method. In: *Preprints of the Int. Conf. on Recent Advances in Mechatronics - "ICRAM'95"*. Istanbul, Turkey, pp.404–411.

Błachuta, M.J., V.D. Yurkevich and K. Wojciechowski (1995b). Design of aircraft 3D motion control using dynamic contraction method. In: *Proc. of the IFAC-Workshop "Motion Control"*, Munich, German, pp. 323–330.

Błachuta, M.J., V.D. Yurkevich and K. Wojciechowski (1996). Some Aspects of designing a digital realization of aircraft control algorithm by dynamic contraction method. In: *Proc. of the 3-rd Int. Conf. on Mechatronics and Machine Vision in Practice*. Guimarães, Portugal, Vol. 1, pp. 167–172.

Błachuta, M.J., V.D. Yurkevich and K. Wojciechowski (1997a). Design of analog and digital aircraft flight controllers based on dynamic contraction method. In: *Proc. of the AIAA Guidance, Navigation, and Control Conference*. New Orleans. Louisiana.

Błachuta, M.J., V.D. Yurkevich and K. Wojciechowski (1997b). Aircraft control in the presence of sensor noise designed by dynamic contraction method. In: *Proc of IFAC Conf. on Control of Industrial Systems (CIS'97)*. Belfort, France.

Boychuk, L. M. (1971). *A method of the structure synthesis of nonlinear automatic control systems*. Energy, Moscow. (in Russian)

Fehrmann, R., W. Mutschkin and R. Neumann (1989). Entwurf und praktische erprobung eines digitalen *PL*-reglers. *Messen Steuern Regeln*. VEB Verlag Technik, Berlin. Vol. 32, No. 2, pp. 70–72.

Fradkov, A.L. (1990). *Adaptive control in lagrescale systems*. Nauka, Moscow. (in Russian)

Gerashchenko, E. I. and S.M.Gerashchenko (1975). *A method of motion separation and optimization of non-linear systems*. Nauka, Moscow. (in Russian)

Green, W.G. (1961). Logarithmic navigation for precise guidance of space vehicle. *IRE Trans. of Aerospace and Navigational Electronics*. No.2, pp.54,

Kotta, Ü. (1995). Inversion method in the discrete-time nonlinear control systems synthesis problem. *Lecture Notes in Control and Information Sciences.*- New-York: Springer-Verlag. Vol. 205.

Krutko, P.D. (1989). *Inverse problems of control system dynamics: Nonlinear models.* Nauka, Phys. & Math. Publ., Moscow. (in Russian)

Litkouhi, B. and H. Khalil (1985). Multirate and composite control of two-time-scale discrete-time systems. *IEEE Trans. Automat. Contr.*, Vol. AC-30, No. 7, pp. 645–651.

Meerov, M.V. (1959). *Synthesis of structures of high-precision automatic control systems.* Nauka, Moscow. (in Russian)

Mutschkin, W.S. (1988). Direkte digitale regelung nach dem prinzip der lokalisation. *Messen Steuern Regeln.* VEB Verlag Technik, Berlin. Vol. 31, No. 7, pp. 315–317.

Porter, W.A. (1970). Diagonalization and inverses for nonlinear Systems. *Int. J. of Control.* Vol. 11, No. 1, pp. 67–76.

Saksena, V.R., J. O'Reilly and P.V. Kokotovic (1984). Singular perturbations and time-scale methods in control theory: Survey 1976-1983, *Automatica.* Vol. 20, No. 3, pp. 273–293.

Shchipanov, G. V. (1939). The theory and methods of the automatic controller design. *Avtomatica i Telemechanica.* Nauka, Moscow, No. 1, pp. 49–66. (in Russian)

Slotine, J.-J. and W.Li (1991). *Applied nonlinear control.* Prentice Hall.

Tsypkin, Ya.Z. (1968). *Adaptation and learning in automatic control systems.* Nauka, Moscow. (in Russian)

Utkin, V.I. (1978). *Sliding modes and their application in variable structure systems.* Mir Publishers, Moskow.

Utkin, V.I. (1994). Sliding mode control in discrete-time and difference systems. In: *Variable Structure and Lyapunov Control* (A.S.I Zinober, Ed.) Springer-Verlag.

Vostrikov, A.S. (1977). On the synthesis of control units of dynamic systems. *Systems Science.* Techn. Univ., Wrocław, Vol.3, No.2, pp.195–205.

Vostrikov, A.S., V.I. Utkin, and G.A. Frantsuzova (1982). Systems having the derivative of the state vector in control. *Avtomatica i Telemechanica.* No. 3, pp. 22–25. (in Russian)

Vostrikov, A.S. (1990). *Synthesis of nonlinear systems by means of localization method.* Novosibirsk State University, Novosibirsk. (in Russian)

Vostrikov, A.S., A.A. Voevoda, W.S. Mutschkin and V.N. Klevakin (1990). *Discrete systems of automatic control on the basis of localization method,* Novosib. State Techn. Univ., Novosibirsk. (in Russian)

Vostrikov, A.S. and V.D. Yurkevich (1991) Decoupling of multi-channel non-linear time-varying systems by derivative feedback. *Systems Science.* Vol. 17, No. 4, pp. 21–33.

Vostrikov, A.S. and V.D. Yurkevich (1993) Design of control systems by means of localisation method. In: *Preprints of 12-th IFAC World Congress.* Sydney. Vol. 8, pp. 47-50.

Yurkevich, V.D. (1992). Dynamic contraction method and discrete control system design. In: *Proc. Siberian Conf. on Checing and control microprocessor systems* . Novosibirsk, pp. 137–140. (in Russian)

Yurkevich, V.D. (1993a). Control of uncertain systems: dynamic compaction method. In: *Proc. of the 9-th Int. Conf. on Systems Engineering.* University of Nevada, Las Vegas, pp. 636–640.

Yurkevich, V. D. (1993b). On the synthesis of multi-channel systems by means of dynamic contraction method. In: *Automatic control for plants with varying characteristics.* Novosib. State Techn. Univ., Novosibirsk, part 2, pp. 100–109. (in Russian)

Yurkevich, V.D. (1993c). On the design of multichannel discrete systems by means of dynamic contraction method. In: *Abstracts of Papers of the Scientific-Technical Conf.on Problems of Electronic Engineering.*Novosibirsk, pp. 53-57. (in Russian)

Yurkevich, V.D. (1994a). Design of aircraft longitudinal motion control using dynamic compaction method. In: *Proc. of 3-rd Int. Workshop on Advanced Motion Control.* University at Berkeley, pp. 1029–1038.

Yurkevich, V.D. (1994b). On the design of continuous control systems by means of dynamic contraction method: noise influence analysis. In: *Proc. of the 2-nd Int. Conf. on Urgent Problems of Electronic Instrument-making,* Novosibirsk. Vol. 4, pp. 75–79. (in Russian)

Yurkevich, V.D. (1994c). Design of discrete control systems by means of dynamic contraction method. *Izv.Akad. Nauk. Tekhn. Kibernet,* Moscow. No. 6, pp. 223–233. (in Russian).

Yurkevich, V.D. (1995a). A new approach to design of control systems under uncertainty: dynamic contraction method. In:*Preprints of the 3-rd IFAC Symposium on Nonlinear Control Systems Design.* Tahoe City, California. Vol. 2, pp. 443–448.

Yurkevich, V.D. (1995b). Decoupling of uncertain continuous systems: Dynamic contraction method. In:*Proc. of the 34-th IEEE Conference on Decision and Control.* New Orleans. Louisiana, pp. 196–201.

Yurkevich, V.D., M.J.Błachuta and K.Wojciechowski (1997). Design of digital controllers for MIMO non-linear time-varying systems based on dynamic contraction method. In: *Proc. of the European Control Conference (ECC'97).* Brussels.

AN APPROACH TO SOLVING THE MEDICAL-ECOLOGICAL-ECONOMIC STABILITY PROBLEM

Vladimir Baturin, Elena Danilina

Irkutsk Computing Centre of Siberian
Branch of the Russian Academy of Sciences
134, Lermontov Str., Irkutsk, 664033, Russia
e-mail : rozen@icc.ru, tel: (395-2) 31-15-07

Abstract. This article suggests research on a model describing the interaction of three systems: economy, ecology, and population health. The investigation is based on special optimal control methods, enabling the main solution to be obtained.

Keywords. Optimization problem, model, economics, ecology, optimal control.

1. INTRODUCTION

The problem of assessing the anthropogenic influence upon human health, particularly in urban centers with developed industries is gaining in importance. In this case, account must be taken not only of the influence of harmful discharges into the water and air environment upon all urban residents as a whole but also of additional risk factors for those working directly in unhealthy enterprises. Thus, it is necessary to consider the economic, ecological and medical blocks in combination. In the proposed model, a description of enterprise activities contains the distribution of products between the manufacture and development, the final non-productive consumption, and cleaning of resulting pollutants, with various allocations and taxes being also considered here. A set of representative medical characteristics is chosen based on statistical information which is collected according to health conditions of the population. A description of their dynamics takes into account the ability to "self-restoration", the influence of both "peak" and accumulated contaminations, the indirect influence of the unhealthy enterprise upon part of the population, and the effectiveness of treatment in public health institutions.

The intent of this paper is to do a qualitative research into a medical-ecological-economic model for obtaining the solution that is optimal in the functional for a reasonably long time interval. A substantive interpretation of the model research results is presented.

2. MODEL DESCRIPTION

The subjects of study are urban enterprises turning out the bulk of products and being the main pollution sources. The health conditions of the urban residents working in these enterprises is considered separately, which is dictated by the need to take into account harmful factors directly during the production process (humidity, noise, industrial injures, increased or decreased temperatures, monotonicity, etc.). Health of the remaining urban residents, including children, is described by separate indices.

Let $v_i(t)$ be the gross output of the i-th enterprise at time t; $\Phi_i(t)$, basic production assets of the i-th enterprise; $p_i(t)$, profits of the i-th enterprise; $u_i(t)$, rate of investment in the main production at the i-th enterprise; $z_i(t)$, cleaning intensity at the i-th enterprise; $w_i(t)$, rate of investment in the waste treatment facilities at the i-th enterprise; $\Phi_i^{(z)}$, basic assets of waste treatment facilities at the i-th enterprise; $\Phi^{(x)}$, basic assets of public health; x_i, allocations of the i-th enterprise to the public health system; $y(t)$, population health restoration rate in public health institutions; r_i, intensity of pollutant discharges at the i-th enterprise (vector).

A model of a separate (i-th) enterprise consists of the dynamic balance equation which, in addition to productive and development expenditures, allows for expenditures on cleaning and for payment for pollutant discharges, expenditures on medical ensurance and tax on public health, and deductions to the local, regional and federal budgets.

$$v_i = A_i v_i + u_i + A_i^{(z)} z_i + w_i + \rho(r_i) + x_i + \\ + \lambda_i A^{(y)} y + p_i + \beta(v_i - A_i v_i), \qquad (1)$$

$$i = 1, 2, \ldots, M,$$

$$r_i(t) = G_i v_i(t) - z_i(t). \qquad (2)$$

These relationships are complemented with equations for basic assets of the chosen enterprises and for basic assets of cleaning facilities of these enterprises:

$$\dot{\Phi}_i = u_i - \Delta_i \Phi_i, \qquad (3)$$

$$\dot{\Phi}_i^{(z)} = w_i - \Delta_i^{(z)} \Phi_i^{(z)}. \qquad (4)$$

The basic assets of public health are described in a similar manner:

$$\dot{\Phi}^{(x)} = X + \sum_{i=1}^{M} x_i - \Delta^{(x)} \Phi^{(x)}. \qquad (5)$$

For describing the medical block, a vector of indices of population health conditions $-h(t)$ is introduced in terms of deviations from a "natural" level $h^*(t)$ which is set up depending on climatic, weather and social conditions, national peculiarities, etc. Thus, the model estimates the sickness rate associated with the influence of production. The worsening of the health indices is associated with the output of products (a direct influence upon those working in a given enterprise only) and pollutions. The influence of "peak" (single) and cumulative (accumulated) contaminations is considered separately. An improvement of health occurs as a result of a treatment in medical institutions and "self-restoration".

$$\dot{h} = Qh + C_i v_i + F_i r_i + D_1 R + \\ + D_2 R^{int} + D_{3i} r_i^{int} - y, \qquad (6)$$

$$r_i^{int} = \int_{t_0}^{t} r_i(\tau) d\tau, \qquad R^{int} = \int_{t_0}^{t} R(\tau) d\tau.$$

In addition, there exists additional constraints

$$0 \le v_i \le V_i(t, \Phi_i), \qquad 0 \le z_i \le Z_i(t, \Phi_i^{(z)}),$$

$$w_i, u_i, p_i \ge 0, \quad y \le Y(\Phi^{(x)}), \quad \sum_{i=1}^{M} \lambda_i = 1 \qquad (7)$$

and initial conditions

$$X_i(t_0) = X_{i0}, \; X_i^{(z)}(t_0) = X_{i0}^{(z)},$$

$$X^{(x)}(t_0) = X_0^{(x)}, \; h(t_0) = h_0. \qquad (8)$$

In the system (1)-(8), the following designations are additionanlly used:

$\Delta_i, \Delta_i^{(z)}, \Delta^{(x)}$ - depreciation coefficients;

$V_i(t, \Phi_i)$ - function, characterizing a maximum possible output;

$Z_i(t, \Phi_i^{(z)})$ - function, characterizing a maximum possible degree of cleaning;

$Y(\Phi^{(x)})$ - function, characterizing a maximum possible popultion health restoration rate in public health institutions;

R - background pollution intensity;

r_i^{int}, R^{int} - integral indices of discharges;

A_i - coefficient of direct expenditures on the manufacture of a unit product of the i-th enterprise;

$A_i^{(z)}$ - vector of expenditures of products with a single rehabilitation intensity of resources in the i-th enterprise;

$\rho(r_i)$ - function, characterizing payment for discharges;

$A^{(y)}$ - vector of direct expenditures on health restoration;

λ_i - share coefficient of these expenditures for the i-th enterprise;

β - taxation coefficient of the profits of enterprises;

G_i - vector, characterizing pollutant discharges per unit product of the i-th enterprise;

Q - matrix, representing the process of self-restoration and mutual influence of health characteristics;

C_i - vector, characterizing a change of health indices depending on the output of products in the i-th enterprise;

F_i, D_1, D_2, D_{3i} - matrices, representing the influence of pollutants upon population helth indices;

X - federal budget allocations for public health development;

M - number of the city's enterprises involved in the modelling.

As is evident, our description is taken to be linear in order to avoid additional difficulties with identification; however, the research method outlined in the next Section permits also more sophisticated versions of the model to be considered.

A model for the influence of environmental pollutants upon urbal population health that does not contain a direct description of the economic block, was studied in (Baturin et al., 1994). When considering, using the model (1) - (8), control problems, a maximum of the functional was taken as the optimality criterion

$$I = \int_{t_0}^{T} [l^{(p)} \sum_{i=1}^{M} l_i p_i - l^{(h)} h' H h] dt, \qquad (9)$$

which has the meaning of a "reasonable" combination of the total profits (income) of the city's enterprises and deductions for the deviation of population health indices from ideal (desirable) values.

The problem of finding such a solution of the dynamic system of equations and inequalities (1) - (8) with a given control quality criterion (9), which makes it possible to maintain during a long time interval the specified ecological and medical constraints and ensures, at the same time, a reasonably high level of output, will be referred to as the problem of medical-ecological-economic stability.

3. MODEL RESEARCH

We now apply to the problem (1) - (9) the multiple maxima method (Gurman, 1977; Gurman et al., 1981) which, while being a method of specifying the function $\phi(t,x)$ in sufficient Krotov's optimality conditions, warrants the non-uniqueness of a maximum of the function $K = \phi_x' f(t,x,u) - f^0(t,x,u) + \phi_t$ for control. Here $f(.)$ stands for the right-hand sides of a system of differential equations, and $f^0(.)$ is the integrand function in the functional. This makes it possible to find an optimal solution in the form of a function, discontinuous at the points t_0 and T , which is subsequently approximated by a sequence of a set of admissible ones. Such solutions, when considered on a sufficiently long time interval, are customarily referred to as the main solutions. Let us additionally introduce the following assumptions:

- For definiteness $t_0 = 0$;
- In the general case, one may specify the final conditions $\Phi_i(T) = \Phi_{iT}$, $\Phi_i^{(z)}(T) = \Phi_{iT}^{(z)}$, $\Phi^{(x)} = \Phi_T^{(x)}$, $h(T) = h_T$;
- The functions $r_i^{int}(t), R^{int}(t)$ and $R(t)$ - are specified. We designate $\Omega_i = D_1 R + D_2 R^{int} + D_{3i} r_i^{int}$;
- The function X - is specified;
- $V_i(t,\Phi_i), Z_i(t,\Phi_i^{(z)})$ and $Y(\Phi^{(x)})$ - are production functions, representable as $V = \gamma \Phi_i^{\alpha}$, $Z = \gamma {\Phi_i^{(z)}}^{\alpha}$ and $Y = \gamma {\Phi^{(x)}}^{\alpha}$, where $0 < \alpha \le 1, \gamma > 0$.
- $\rho(r_i)$ - is a linear function, $\rho(r_i) = \rho r_i$. The variant $\rho(r_i) = \rho r_i^2$ will also be considered later;
- There exist the constraints $\Phi_{min}^{(x)} \le \Phi^{(x)} \le \Phi_{max}^{(x)}$.

We shall investigate the problem using the multiple maxima method. To begin with, we construct the lower Φ_{ia} and the upper Φ_{ib} bounds Φ_i by solving the equation $\dot{\Phi}_i = -\Delta_i \Phi_i$ from the points $(0, \Phi_{i0})$ and (T, Φ_{iT}) with due regard for the constraint $\Phi_i \ge 0$. Similarly, we find the bounds $(\Phi_{ia}^{(z)}, \Phi_{ib}^{(z)})$, $(\Phi_a^{(x)}, \Phi_b^{(x)})$.

Let us write out the function K:

$$K = \phi_{\Phi_i}(u_i - \Delta_i \Phi_i) + \phi_{\Phi_i^{(z)}}(w_i - \Delta_i^{(z)} \Phi_i^{(z)}) +$$
$$+ \phi_{\Phi^{(x)}}(X + \sum_{i=1}^{M} x_i - \Delta^{(x)} \Phi^{(x)}) +$$
$$+ \phi_h' [Qh + C_i v_i + F_i r_i + \Omega_i - y] +$$
$$+ l^{(p)} \sum_{i=1}^{M} l_i [(1 - A_i)(1 - \beta) v_i - u_i - A_i^{(z)} z_i -$$
$$- w_i - \rho(r_i) - x_i - \lambda_i A^{(y)} y] - l^{(h)} h' H h + \phi_t. \quad (10)$$

According to the multiple maxima method, it will be assumed that the function K is independent of u_i, w_i, x_i and y. From this we obtain

$$\hat{\phi}_{\Phi_i} = l^{(p)}, \ \hat{\phi}_{\Phi_i^{(z)}} = l^{(p)},$$

$$\hat{\phi}_h' = -l^p A^{(y)}, \ \hat{\phi}_{\Phi^{(x)}} = l^{(p)} l_i.$$

In order for the last relationship to hold for any i, the condition $l_i = l_j$ must be satisfied for all i, j. There fore, in view of the requirement $\sum_{i=1}^{M} l_i = 1$, we put $l_i = \frac{1}{M}$, and we ultimately get $\hat{\phi}_{\Phi^{(x)}} = \frac{l^{(p)}}{M}$. It will be further assumed that $\hat{\phi}_t = 0$. We substitute the resulting derivatives of the function ϕ into the expression (10) and investigate the function obtained for a maximum in $(v_i, \Phi_i, z_i, \Phi_i^{(z)}, h, \Phi^{(x)})$ with allowance for the specified constraints. Then

$$\tilde{v_i} = \begin{cases} V_i(t,\Phi_i), & \kappa_i > 0 \\ 0, & \kappa_i \le 0, \end{cases}$$

where $\kappa_i = -A^{(y)}(C_i + F_i G_i) + \frac{(1-A_i)(1-\beta)}{M} - \frac{\rho G_i}{M}$.

In the case $\kappa_i > 0$ we have

$$K_1 = \max_{v_i} K = -l^{(p)} \Delta_i \Phi_i - l^{(p)} \Delta_i^{(z)} \Phi_i^{(z)} +$$
$$+ \frac{l^{(p)}(X - \Delta^{(x)} \Phi^{(x)})}{M} - l^{(p)} A(y)[Qh - F_i z_i +$$
$$+ \Lambda_i] + \frac{l^{(p)}(\rho z_i - A_i^{(z)} z_i)}{M} - l^{(h)} h' H h +$$
$$+ l^{(p)} \kappa_i \gamma \Phi_i^{\alpha}.$$

From this we find $\tilde{\Phi}_i = \text{argmax}_{\Phi_i} K_1$:

$$\tilde{\Phi}_i = \begin{cases} \min \left[\Phi_{ib}, (\frac{\gamma \alpha \kappa_i}{\Delta_i})^{\frac{1}{1-\alpha}} \right], & \kappa_i > 0 \\ \Phi_{ia}, & \kappa_i \le 0. \end{cases}$$

By taking the obvious value of $\Phi_{ia} = 0$ as the lower bound, we obtain for Φ_i a typical main solution.

We carry out a similar investigation for z_i and $\Phi_i^{(z)}$ to give

$$\tilde{z}_i = \begin{cases} Z_i(t, \Phi_i^{(z)}), & \kappa_i^{(z)} > 0 \\ 0, & \kappa_i^{(z)} \leq 0, \end{cases}$$

where $\kappa_i^{(z)} = A^{(y)} F_i - \frac{\rho - A_i^{(z)}}{M}$. In the case $\kappa_i^{(z)} > 0$

$$\tilde{\Phi}_i^{(z)} = \begin{cases} \min \left[\Phi_{ib}^{(z)}, \left(\frac{\gamma_1 \alpha_1 \kappa_i^{(z)}}{\Delta_i^{(z)}} \right)^{\frac{1}{1-\alpha_1}} \right], & \kappa_i^{(z)} > 0 \\ \Phi_{ia}^{(z)}, & \kappa_i^{(z)} \leq 0. \end{cases}$$

Next, we find $\tilde{h}$ and $\tilde{\Phi}^{(x)}$:

$$\tilde{h} = -l^{(p)} A^{(y)} Q (2 l^{(h)} H)^{-1},$$

$$\tilde{\Phi}^{(x)} = \begin{cases} \Phi_{max}^{(x)}, & \kappa^{(x)} > 0 \\ \Phi_{min}^{(x)}, & \kappa^{(x)} \leq 0, \end{cases}$$

where $\kappa^{(x)} = -\frac{l^{(p)} \Delta^{(x)}}{M}$.

With a purely economic criterion, when $l^{(h)} = 0$, it is necessary to introduce an additional constraint of the form $0 \leq h \leq h_{max}$, then

$$\tilde{h} = \begin{cases} h_{max}, & \kappa^{(h)} > 0 \\ 0, & \kappa^{(h)} \leq 0, \end{cases}$$

where $\kappa^{(h)} = -l^{(p)} A^{(y)} Q$.

Besides, on the main solution we have $\tilde{u}_i = 0$, $\tilde{w}_i = 0$ and $\tilde{x}_i = 0$. The value of $\tilde{y}$ is determined from the condition $\dot{\tilde{h}} = 0$. From this we get $\tilde{y} = Q\tilde{h} + C_i \tilde{v}_i + F_i (G_i \tilde{v}_i - \tilde{z}_i) + \Omega_i$.

Let us make a qualitative investigation of the solution obtained by assuming that T is sufficiently large and that transitional segments may be neglected.

The profitability condition of the economy $\kappa_i > 0$ in the absence of direct expenditures on health restoration ($A^{(y)} = 0$) and penal sanctions for environmental pollution ($\rho = 0$), is simplified to a clasical variant $A_i < 1$. The necessity of investing a certain portion of the profits in maintaining population health and preservation of the environment obviously reduces the profitability threshold of enterprises $A_i < 1 - \frac{MA^{(y)}(C_i + F_i G_i) + \rho G_i}{1-\beta}$.

Of course, the nonprofitable economy ($\kappa_i \leq 0$) must not develop ($\tilde{v}_i = 0, \tilde{\Phi}_i = \Phi_{ia}$).

The analysis of the function $\kappa_i^{(z)}$ shows that cleaning operations are profitable if only direct expenditures on them are less than penalties for pollution: when $\kappa_i^{(z)} > 0$ we have $A_i^{(z)} < \rho$, (for the sake of simplicity, it is assumed here that $A^{(y)} = 0$). Otherwise, it is more profitable to pay penalty: $\tilde{\Phi}_i^{(z)} = \Phi_{ia}^{(z)} = 0, \quad \tilde{z}_i = 0$.

The investigation was carried out under the assumption of the negative meaning of health indices; therefore, here $Q < 0, C_i > 0, F_i > 0$ and $h \geq 0$. If $l^{(p)} = 0$ is put in the functional, i.e. if the criterion that assesses only the health conditions is considered, then $\tilde{h} = 0$. When considering a purely economic criterion ($l^{(h)} = 0$), we have $\kappa^{(h)} > 0$ and $\tilde{h} = h_{max}$. As far as the basic assets of public health are concerned, then $\tilde{\Phi}^{(x)} = \Phi_{min}^{(x)}$, because $\kappa^{(x)} < 0$.

Consider several other variants.

(1) The cleaning intensity is proportional to the output of products: $z_i = \delta_i v_i$.
In this case

$$\kappa_i = -MA^{(y)}[C_i + F_i(G_i - \delta_i)] +$$
$$+(1 - A_i)(1 - \beta) - A_i^{(z)} \delta_i - \rho(G_i - \delta_i)$$

and the profitability condition has the form

$$A_i < 1 - -\frac{A_i^{(z)} \delta_i + \rho(G_i - \delta_i)}{1 - \beta} +$$
$$+ \frac{MA^{(y)}(C_i + F_i(G_i - \delta_i))}{1 - \beta}$$

under the natural assumption $G_i - \delta_i > 0$.
Besides, $\tilde{z}_i = \delta_i \tilde{v}_i$, $\tilde{\Phi}_i^{(z)} = \Phi_{ia}^{(z)}$, because $\kappa_i^{(z)} = -l^{(p)} \Delta_i < 0$.

(2) Let $z_i = \delta_i v_i$, $\quad \rho(r_i) = \rho r_i^2 = \rho(G_i - \delta_i)^2 v_i^2$.
In this case

$$\tilde{v}_i = \frac{(1 - A_i)(1 - \beta) - A_i^{(z)} \delta_i - MA^{(y)} C_i}{2(G_i - \delta_i)^2 (MA^{(y)} F_i + \rho)}$$

and the profitability condition has the form

$$A_i < 1 - \frac{A_i^{(z)} \delta_i + MA^{(y)} C_i}{1 - \beta}.$$

Besides, $\tilde{\Phi}_i = \Phi_{ia}$, $\tilde{\Phi}_i^{(z)} = \Phi_{ia}^{(z)}$, $\tilde{\Phi}^{(x)} = \Phi_{min}^{(x)}$, because $\kappa_i^{(\Phi)} = -l^{(p)} \Delta_i < 0$, $\kappa_i^{(z)} = -l^{(p)} \Delta_i^{(z)} < 0$, $\kappa^{(h)} = -\frac{l^{(p)} \Delta^{(x)}}{M} < 0$.

(3) Let $z_i = \delta_i v_i$, $x_i = \eta_i v_i$, $\rho(r_i) = \rho r_i = \rho(G_i - \delta_i) v_i$, equation (5) is not considered.
In this case

$$\kappa_i = -MA^{(y)}[C_i + F_i(G_i - \delta_i)] +$$
$$+[(1 - A_i)(1 - \beta) - \rho(G_i - \delta_i) - A_i^{(z)} \delta_i - \eta_i]$$

and the profitability condition

$$A_i < 1 - \frac{\rho(G_i - \delta_i) + \eta_i + A_i^{(z)} \delta_i}{1 - \beta} +$$
$$+ \frac{MA^{(y)}(C_i + F_i(G_i - \delta_i))}{1 - \beta}.$$

(4) Unlike variant 3, $\rho(r_i) = \rho r_i^2$.
Then

$$\tilde{v}_i = \frac{(1 - A_i)(1 - \beta) - MA^{(y)} C_i - A_i^{(z)} \delta_i - \eta_i}{2(G_i - \delta_i)^2 (MA^{(y)} F_i + \rho)}.$$

In order for the condition $\tilde{v}_i > 0$ to be satisfied

$$A_i < 1 - \frac{MA^{(y)}C_i + A_i^{(z)}\delta_i + \eta_i}{1-\beta}.$$

In all the cases considered above it is evident that additional expenditures of enterprises on cleaning of pollutants and on health restoration reduce the profitability threshold; however, this makes it possible to maintain during long time intervals the necessary ecological and medical constraints.

4. APPROXIMATION OF THE DISCONTINUOUS SOLUTION

The basic main solution, as additional solutions, constructed above do not satisfy in the general case the specified initial and final conditions; therefore, on the segments $[0, t^*)$ and $(t^{**}, T]$, where t^* is the point of reaching the main and t^{**} is the point of leaving it, it is necessary to carry out an approximation by, for example, a sequence of linear functions. To show that such an approximation is, in principle, possible, we now consider the simplest scalar variant of the model where the city's economy, pollution and population health are represented by generalized indices. For $t \in [0, t^*]$ we obtain

$$\Phi_s(t) = \Phi_0 + st, \qquad s = \frac{\tilde{\Phi} - \Phi_0}{t^*},$$

$$\Phi_s^{(x)}(t) = \Phi_0^{(x)} + k(s)t, \qquad k(s) = \frac{\tilde{\Phi}^{(x)} - \Phi_0^{(x)}}{t^*},$$

$$h_s(t) = h_0 + n(s)t, \qquad n(s) = \frac{\tilde{h} - h_0}{t^*},$$

$$\Phi_s^{(z)}(t) = \Phi_0^{(z)} + m(s)t, \qquad m(s) = \frac{\tilde{\Phi}^{(z)} - \Phi_0^{(z)}}{t^*},$$

$$u_s(t) = \dot{\Phi}_s + \Delta\Phi_s =$$
$$= s + \Delta(\Phi_0 + st) = \Delta\Phi_0 + s(1 + \Delta t),$$
$$w_s(t) = \dot{\Phi}_s^{(z)} + \Delta^{(z)}\Phi_s^{(z)} =$$
$$= m(s) + \Delta^{(z)}(\Phi_0^{(z)} + m(s)t) =$$
$$= \Delta^{(z)}\Phi_0^{(z)} + m(s)(1 + \Delta^{(z)}t),$$
$$v_s(t) = V(\Phi_s) = \gamma\Phi_s^{\alpha},$$
$$z_s(t) = V(\Phi_s^{(z)}) = \gamma_1 \Phi_s^{(z)^{\alpha_1}},$$
$$x_s(t) = \dot{\Phi}_s^{(x)} + \Delta^{(x)}\Phi_s^{(x)} - X =$$
$$= \Delta^{(x)}\Phi_0^{(x)} - X + k(s)(1 + \Delta^{(x)}t),$$
$$y_s(t) = Q(h_s - h^*) + Cv_s + F(Gv_s - z_s) + \Omega - n(s),$$
$$p_s(t) = (1-A)(1-\beta)v_s - u_s - A^{(z)}z_s - w_s -$$
$$-\rho(Gv_s - z_s) - x_i - A^{(y)}y_s.$$

The transition regime on the segment $[t^{**}, T]$ can be constructed in a similar manner. The smaller the transitional segments, i.e. the closer the approximating sequence approaches the theoretical discontinuous solution, the larger in absolute value should be the controls u, w and x which make for reaching the main and leaving it to the initial point. Constructively, this may be interpreted as the assumption about the unlimited character of capital investments on a sufficiently short time interval. Negative controls are understood to be the reduction in production.

ACKNOWLEDGEMENTS

The parts of the research are supported by the Russian Fund of Fundamental Researches, Grants Nos. 97-01-960002, 97-01-960005 and 97-01-00047.

REFERENCES

Baturin, V.A., Y.A. Leshchenko and D.M. Rozenraukh (1994). A model for environmental pollutant influence upon health of urban residents. In: *Proceedings of the All-Russia School "Computer Logic, Algebra, and Intelligent Control. Problems of Analyzing Sustainable Development and Strategic Stabilization"*, **2**, 174-185.(in Russian)

Gurman, V.I. (1977) *Singular Problems of Optimal Control.* Nauka, Moscow.

Gurman, V.I., V.A. Baturin, V.A. Dykhta (1981) *Management Models for Natural Resources.* Nauka, Moscow.

ON CONNECTION BETWEEN THE FINDING OF THE GLOBAL FUNCTION'S MINIMUM AND THE OPTIMAL SIMULATED ANNEALLING SCHEDULE

Nataliya P. Belyaeva *

* *Programm Systems Instiute, Russian Academy of Science*

Abstract: The finding of the global function's minimum is a difficult optimization problem. Often the solving of this task is founded on stochactic algorithms. The optimal simulated annealling schedule has been used for the finding of the global function's minimum. It has been shown that the possibility of the getting over the local minimum is due to the system's temperature. For two simple models it has been shown that the choosing of the system's temperature $T(t)$ is a problem of optimal control. Some methods have been represented for finding optimal and near-optimal solutions. Some means of problem's generalizations have been discussed. Some examples have been shown.

Keywords: annealling, optimal control, extremal problems, optimality conditions

Simulated annealling is a global optimization procedure (Andresen, Hoffmann (1988), Hoffmann, Salamon (1990)) for the finding near-optimal solutions to the problem by simulating the cooling of the corresponding physical systems. The problem is to move the physical system into its state of equilibrium with the ambient temperature T. Thus for $T \to 0$ the system moves into its ground state. The most of used methods are based on Monte-Carlo modeling. The disadvantage of these methods is a greate volume of calculations. Let us reduce the problem of finding the global minimum to the optimal control task. For the simplicity let us consider physical system with three different states with energies E_1, E_2, E_3.

Let us assume that $E_3 < E_1 < E_2$.

Denote:
$\Delta E_{ij} = \Delta E_{ji} = E_i - E_j$,
$T(t)$ - a temperature T dependind on time t,
P_i - a probability of finding in a state i,
P_{ij} - a probability of transaction from a state i to a state j.

At each step of the algorithm a new state is selected at ramdom to be a candidate for becoming the next state. It astually becomes the next state only with probability

$$P_{acceptance} = \begin{cases} 1, & if \Delta T \le 0 \\ e^{-\frac{\Delta E}{T}}, & otherwise. \end{cases}$$

Kolmogorov's equations for this system are represented as:

$$\begin{cases} \dot{P}_1 = P_2P_{21}(T) - P_1P_{12}(T) \\ \dot{P}_2 = P_1P_{12}(T) + P_3P_{32}(T) - P_2(P_{21}(T) \\ \quad +P_{23}(T)) \\ \dot{P}_3 = P_2P_{23}(T) - P_3P_{32}(T) \end{cases} \quad (1)$$

Let us take initial conditions:

$$\begin{cases} P_1(0) = \frac{1}{3} \\ P_2(0) = \frac{1}{3} \\ P_3(0) = \frac{1}{3} \end{cases} \quad (2)$$

That is the system may be in any state with equal possibility.

Our assumptions are:

(1) $P_{12} = P_{23} = \frac{1}{2}$ (that is the transaction from state 2 to state 1 ot to state 3 may be done with an equal possibility),

(2)

$$\begin{cases} P_{12}(T) = e^{-\frac{\Delta E_{12}}{T(t)}} = A(t) \\ P_{32}(T) = e^{-\frac{\Delta E_{32}}{T(t)}} = B(t) \end{cases}$$

Let us denote $\Delta E_{12} = a$, $\Delta E_{32} = b$, then $b > a$.

We have:
$B(T) = e^{-\frac{\Delta E_{12}}{\Delta E_{12}}\frac{\Delta E_{12}}{T(t)}} = A(t)^{\frac{b}{a}} = A(T)^{\gamma}$, where $\gamma > 1$.

In new variables our system (1) is writen as

$$\begin{cases} \dot{P}_1 = \frac{1}{2}P_2 - P_1 A(T) \\ \dot{P}_2 = P_1 A(T) + P_3 A(T)^{\gamma} - P_2 \\ \dot{P}_3 = \frac{1}{2}P_2 - P_3 A(T)^{\gamma} \end{cases} \qquad (3)$$

We have the system (3) with initial conditions (2). The system (3) may be considered as a system of differential equations with control $A(T)$.

Different criteria may be choosed, for example:

(1) Maximum of possibility $P_3(\tau)$ at the end of finite time interval $[0;\tau]$;
(2) Maximum of system energy $E(\tau)$ at the end of finite time interval $[0;\tau]$;
(3) Maximum of average energy an so on.

Let us consider the next problem:
To find such control $A(T)$, when

$$P_3(\tau) = \int_0^{\tau} \dot{P}_3 dt \rightarrow max \qquad (4)$$

or

$$P_3(\tau) = \int_0^{\tau} (\frac{1}{2}P_2 - P_3 A(T)^{\gamma})dt \rightarrow max$$

for the case $A(T) > 0$ under the conditions (1), (2).

Let us introduce new notation:
$x_i = P_i; u(t) = A(t)$.

In new terms our system will be represanted as

$$\begin{cases} \dot{x}_1 = -ux_1 + \frac{1}{2}x_2 \\ \dot{x}_2 = ux_1 - x_2 + u^{\gamma}x_3 \\ \dot{x}_3 = \frac{1}{2}x_2 - u^{\gamma}x_3 \end{cases}$$

3 possibilities are connected with norm condition $\sum x_i = 1$, so 1 variable may be eliminated. Our problem will be writen as

$$\begin{cases} \dot{x}_1 = -ux_1 + \frac{1}{2}(1 - x_1 - x_3) \\ \quad = -(u + \frac{1}{2})x_1 - \frac{1}{2}x_3 + \frac{1}{2} \\ \dot{x}_3 = \frac{1}{2}(1 - x_1 - x_3) - u^{\gamma}x_3 \\ \quad = \frac{1}{2}x_1 - (u^{\gamma} + \frac{1}{2})x_3 + \frac{1}{2} \end{cases} \qquad (5)$$

$$\begin{cases} x_1(0) = \frac{1}{3} \\ x_3(0) = \frac{1}{3} \end{cases} \qquad (6)$$

$$x_3(\tau) \rightarrow max_u \qquad (7)$$

So we have a system (5), initial conditions (6) and criteria of optimality (7). A Hamiltonian function will be writen as

$$H = [-(u + \frac{1}{2})x_1 - \frac{1}{2}x_3 + \frac{1}{2}]\psi_1 \qquad (8)$$

$$+[\frac{1}{2}x_1 - (u^{\gamma} + \frac{1}{2})x_3 + \frac{1}{2}]\psi_2 + x_3 \qquad (9)$$

Optimality condition will be

$$\frac{\partial H}{\partial u} = -x_1\psi_1 - \gamma u^{\gamma-1}x_3\psi_2 = 0, \qquad (10)$$

from

$$u^{\gamma-1} = -\frac{x_1\psi_1}{\gamma x_3 \psi_2} \qquad (11)$$

Assosiated system will be

$$\begin{cases} \dot{\psi}_1 = -\frac{\partial H}{\partial x_1} = (u + \frac{1}{2})\psi_1 + \frac{1}{2}\psi_2 \\ \dot{\psi}_3 = -\frac{\partial H}{\partial x_3} = \frac{1}{2})\psi_1 + (u^{\gamma} + \frac{1}{2})\psi_2 - 1 \end{cases} \qquad (12)$$

$$\begin{cases} \psi_1(\tau) = 0 \\ \psi_3(\tau) \rightarrow max \end{cases} \qquad (13)$$

Substituted control (10) into system (11) we can integrate these equation numerically in any symbolic computer system. This aproach can be used for any number of possible states.

REFERENCES

Andresen B., Hoffmann K. H. (1988) On lumped models for thermodinamic properties of simulated annealing problems, *J. Phys. (France)*, N49

Hoffmann K. H., Salamon P.(1990) The optimal simulated annealing schedule for a simple model, *J.Phys A: Math. Gen.* N23 , p. 3511-3523.

ON THE MEASUREMENT ALLOCATION PROBLEM FOR DISTRIBUTED SYSTEM

M.I.Gusev [*,1] **S.A.Romanov** [**]

[*] *Institute of Mathematics and Mechanics, 620219, Ekaterinburg, Russia (gmi@oou.imm.intec.ru)*
[**] *Moscow State University, 119899, Moscow, Russia*

Abstract: A linear distributed-parameter system, described by diffusion equation in R^2 with uncertainty in the right-hand side is considered. It is assumed that all accessible information on the solution of the system is given by the measurements perfomed by the sensors at the finite number of points. The considered model of uncertainty and measurements errors is non-stochastic with set-membership description of unknowns. The problem of allocation of the sensors inside given domain in order to ensure the best possible estimate of the linear functional of solution is considered. The "duality" results, which states that considered problem is equivalent to some impulsive control problem are given. Assuming that unknown data do not depend on time the theorem on sufficient number of sensors is proved and the reduction of measurements allocation problem to the nonlinear programming problem is described.

Keywords: distributed-parameter systems, state estimation, uncertain linear systems, sensors, impulses

1. INTRODUCTION

The sensors placement problems for estimation of distributed system with uncertain parameters were studied by (Chen and Seifeld, 1975; Nakamori *et al.*, 1980; Rafajlowich, 1981; Usinski, 1992; Usinskii *et al.*,1992). These works employ a conventional approach to the study of uncertain systems, related to the assumption that uncertainty may be described as a random process with known characteristics. In many applied problems, however, there may be limited number of observations, incomplete knowledge of the data and no available statistics whatever. An alternative approach to the uncertainty treatment, known as guaranteed (Kurzhanski, 1977; Kurzhanski and Valyi, 1997), is based on set-membership, or unknown but bounded, error description. In considered here problem the set-membership description of uncertainty is employed.

In many applied problems which uncertainty is inherent in it is possible to affect the observation process. As for dynamical systems, the choice of control plant inputs for the parameters identification gives a typical example of observations control. Another one concerns with the measurement allocation problem, arising ,for example, in environmental monitoring. Controlling the observations one can reduce the estimation error. The experiment design problems for dynamic systems with set-membership description of uncertainty were considered in (Gusev, 1981; Gusev,1988). Both the problem of choice of optimal inputs and the measurement allocation problem were considered as the special cases of some abstract experiments design problem under uncertainty. The problems of optimization of the trajectory of the movable sensor for distributed parameter system with integral restriction on uncertain

[1] The research was supported by Russian Fund for Basic Research, grant 97-01-01003

parameters were considered in (Kurzhanski and Hapalov, 1985, Kostousova, 1990). Here the problem of stationary sensors placement is considered for both, geometrical and integral constraints on unknowns.

2. PROBLEM STATEMENT

We assume that distributed field is described as the solution to the following diffusion equation in R^2

$$\frac{\partial \varphi}{\partial t} + div\mathbf{w}\varphi + \sigma\varphi - \mu\Delta\varphi =$$

$$\sum_{i=1}^{k} Q_i(t)\delta(x - a_i), \quad 0 \leq t \leq T, \tag{1}$$

with initial and boundary conditions

$$\varphi(0, x) = \varphi_0(x), \quad \varphi(t, x) = 0$$

under $\quad ||x|| \to +\infty, \quad div\mathbf{w} = 0.$

The system is disturbed at k given points on a plane $a_1, \ldots, a_k$. The perturbations values, described by function $Q_i(t)$, and initial state are assumed to be unknown in advance. Suppose that state φ is observed according to the following measurement equation

$$y_j(t) = \varphi(t, b_j) + \xi_j(t), \qquad j = \overline{1, s}.$$

Here $\xi(t)$ is the measurement error of the j-th sensor, and points $b_1, \ldots, b_s$, describing the "sensors placement", have to be chosen inside given domain Ω^*. Thus, the perturbations in the right hand side of the equation, the measurement errors and initial system state φ_0 are assumed to be unknown. All apriori information on unknowns is given by the inclusion

$$\{Q_i(t),\ i = \overline{1, k},\ \xi_j(t),\ j = \overline{1, s},\ \ \varphi_0(x)\} \in U.$$

This inclusion can be represented either by geometrical constraints

$$0 \leq Q_i(t) \leq \overline{Q}_i, i = \overline{1, k},$$
$$|\xi_j(t)| \leq e_j, j = \overline{1, s}, 0 \leq \varphi_0(x) \leq \bar{\varphi}_0(x),$$

or quadratic constraints

$$\sum_{i=1}^{k} \int_0^T N(t)(Q_i(t) - \overline{Q}_i(t))^2 dt +$$
$$\sum_{j=1}^{s} \int_0^T M(t)(\xi_j(t) - \bar{\xi}_j(t))^2 dt +$$
$$\int_{R^2} K(x)(\varphi_0(x) - \bar{\varphi}_0(x))^2 dx \leq \theta,$$

where $N(t), M(t), K(x) \geq \alpha > 0$.

It is necessary to estimate the value of following integral

$$I = \int_0^T \int_\Omega \varphi(t, x) dt dx$$

on the basis of measurement $y(t), 0 \leq t \leq T$. It can be shown, that the best possible result of estimation may be achieved in the class of continuous affine estimates $h(y(\cdot))$

$$h(y(\cdot)) = \int_0^T \langle z(t), y(t) \rangle dt + z_0, \tag{2}$$
$$z(\cdot) \in C[0, T], \qquad z_0 \in R.$$

We consider the following problem: how to allocate sensors inside domain Ω^* in order to reduce the guaranteed error of estimation? More precisely the problem of sensors placement is as follows

$$F(b_1, \ldots, b_s) \longrightarrow \min_{b_j \in \Omega^*}, \tag{3}$$

$$F(b_1, \ldots, b_s) = \min_{z(\cdot), z_0} \Phi(z(\cdot), z_0),$$

$$\Phi(z(\cdot), z_0) =$$
$$\max \left| \int_0^T \langle z(t), y(t) \rangle dt + z_0 - \int_0^T \int_\Omega \varphi(t, x) dt dx \right|.$$

where the last maximum is taken in all $\{Q(\cdot), \xi(\cdot) \bar{\varphi}(\cdot)\} \in U$.

The considered statement is motivated by the environmental monitoring problems (Marchuk, 1982). The diffusion equation (1) represents two-dimensional model for spreading of air pollution with known air flow velocity $\mathbf{w}$. The sources of polution - a_i are assumed to be known, but their outputs $Q_i(t)$ do not. So the problem consists of allocating sensors inside the region Ω^* in order to get more precisely the value of pollution of the region Ω.

3. SENSORS PLACEMENT FOR NONSTATIONARY SYSTEM: DUALITY THEOREMS

The difficulty of above problem consists in implicit dependence of functional F of it's arguments: to get the value of $F(\mathbf{b}), \mathbf{b} = (b_1, ..., b_s) \in \Omega^*$, one should solve the minmax optimization problem.

However the considered problem may be put in more convenient form if proceed to a adjoint problem. The following theorem shows that considered problem is reduced to some optimal control problem for adjoint distributed system.

Theorem 1. A sensors placement problem (3) for distributed system (1) is equivalent to the following optimal control problem

$$J \longrightarrow \min_{z(\cdot),\mathbf{b}},$$
$$z(\cdot) \in C([0,T]), \quad \mathbf{b} \in R^s,$$

$$-\frac{\partial \psi}{\partial t} - div\mathbf{w}\psi + \sigma\psi - \mu\Delta\psi = \sum_{j=1}^{s} z_j(t)\delta(x-b_j) - \chi_\Omega(x),$$

$$\psi|_{t=T} = 0,\ \psi \to 0,\ ||x|| \to \infty,\ div\mathbf{w} = 0.$$

Here

$$\chi_\Omega(x) = \begin{cases} 1 \text{ under } x \in \Omega^* \\ 0 \text{ under } x \notin \Omega^* \end{cases}.$$

For geometrical constraints $J = J_1$, where

$$J_1 = \sum_{i=1}^{k} \int_0^T \frac{1}{2}\overline{Q}_i|\psi(t,a_i)|dt + \sum_{j=1}^{s} e_j \int_0^T |z_j(t)|dt + \frac{1}{2}\int_{R^2} \bar{\varphi}(x)|\psi(0,x)|dx$$

and for quadratic constraints $J = J_2$,

$$J_2 = \int_0^T \sum_{i=1}^{k} N_i^{-1}(t)\psi^2(t,a_i) + \sum_{j=1}^{s}\int_0^T M_j^{-1}(t)z_j^2(t)dt + \int_{R^2} K(x)^{-1}\psi^2(0,x)dx.$$

Here the control is a sum of s impulses whose placement and values have to be chosen to minimize functional J. The proof is based on the known scheme of the transition to the adjoint equation. Multiplying both side of (1) by function ψ and integrating one gets

$$\int_0^T\int_{R^2} \psi\frac{\partial\varphi}{\partial t}\,dtdx + \int_0^T\int_{R^2} \psi\, div\mathbf{w}\varphi\,dtdx + \int_0^T\int_{R^2} \sigma\psi\varphi dtdx - \int_0^T\int_{R^2} \mu\psi\Delta\varphi\,dtdx = \int_0^T\int_{R^2} \psi\sum_{i=1}^{k} Q_i(t)\delta(x-a_i)\,dtdx. \quad (4)$$

By integrating by parts, applying Ostrogradskii-Gauss and Green's formulas and taking into account that $div\,\mathbf{w} = 0$ and $\psi \to 0$ under $x \to \infty$ the equation (4) may be represented in the following form

$$\int_0^T\int_{R^2} \varphi\Big(-\frac{\partial\psi}{\partial t} - div\mathbf{w}\psi + \sigma\psi - \mu\Delta\psi\Big)\,dtdx = \int_0^T\int_{R^2} \psi\sum_{i=1}^{k} Q_i(t)\delta(x-a_i)\,dtdx + \int_{R^2} \varphi(0,x)\psi(0,x) - \varphi(T,x)\psi(T,x)\,dx.$$

Assume that ψ is the solution of the equation

$$-\frac{\partial\psi}{\partial t} - div\psi\mathbf{w} + \sigma\psi - \mu\Delta\psi = \sum_{j=1}^{s} z_j(t)\delta(x-b_j) - \chi_\Omega(x),$$

with initial and boundary conditions

$$\psi(T,x) = 0, \psi \to 0 \text{ under } x \to \infty. \quad (5)$$

From (5) it follows that

$$\int_0^T\int_{R^2} \varphi\left(\sum_{j=1}^{s} z_j(t)\delta(x-b_j) - \chi_\Omega(x)\right) dtdx = \int_0^T\int_{R^2} \psi\sum_{i=1}^{k} Q_i(t)\delta(x-a_i)dtdx + \int_{R^2} \varphi(0,x)\psi(0,x)dx.$$

Thus

$$\Phi(z(t),z_0) = \max_{\{Q(\cdot),\xi(\cdot)\,\varphi(\cdot)\}\in U} \Big|\sum_{i=1}^{k}\int_0^T Q_i(t)\int_{R^2} \psi\delta(x-a_i)dtdx + \int_{R^2} \varphi_0\psi(0,x)dx + \sum_{j=1}^{s}\int_0^T z_j(t)\xi_j(t)dt + z_0\Big|.$$

Denote $P_i(t) = Q_i(t) - \frac{\overline{Q}_i}{2}$, $\quad p_0 = \varphi_0(x) - \frac{\bar{\varphi}_0(x)}{2}$, then

$$\Phi(z(t),z_0) = \sup_{|P_i(t)|\le\frac{\overline{Q}_i}{2}} \sum_{i=1}^{k}\int_0^T P_i(t)\psi(t,a_i)dt + \sup_{|\xi_j(t)|\le e_j} \sum_{j=1}^{s}\int_0^T z_j(t)\xi_j(t)dt + |C| + \sup_{|p_0|\le\bar{\varphi}_0(x)/2} \int_{R^2} p_0(x)\psi(0,x)dx =$$
$$\sum_{i=1}^{k}\frac{1}{2}\overline{Q}_i\int_0^T |\psi(t,a_i)|dt + \sum_{j=1}^{s}\int_0^T e_j\,|z_j(t)|dt + \int_{R^2}\frac{1}{2}\bar{\varphi}_0(x)|\psi(0,x)|dx + |C|, \quad (6)$$

where

$$C = z_0 + \sum_{i=1}^{k} \frac{1}{2}\overline{Q}_i \int_0^T \psi(t, a_i)dt+$$
$$\frac{1}{2}\int_{R^2} \bar{\varphi}_0(x)\psi(0, x)dx.$$

To complete the proof let us note that after minimization in z_0 the last term in (6) vanishes. The proof for the case of quadratic constrains on uncertainty is analogous.

The differentiability and strong convexity of functional J_2 allow to simplify the considered optimal control problem.

Theorem 2. A sensors placement problem (3) for the case of quadratic constraints is equivalent to the following optimal control problem for distributed system (1)

$$J(b_1, \ldots, b_s) \longrightarrow \min_{\substack{b_1, \ldots, b_s \\ b_j \in \Omega^*}}, \tag{7}$$

$$J(b_1, \ldots, b_s) = \sum_{i=1}^{k} \int_0^T N_i^{-1}(t)\psi^2(t, a_i)dt+$$
$$\sum_{j=1}^{s} \int_0^T M_j(t)p^2(t, b_j)dt+$$
$$\int_{R^2} K^{-1}(x)\psi^2(0, x)dx,$$

$$\frac{\partial p}{\partial t} + div\, p\mathbf{w} + \sigma p - \mu \Delta p =$$
$$\sum_{i=1}^{k} N_i^{-1}(t)\psi(t, a_i)\delta(x - a_i) + K^{-1}(x)\psi(0, x)\delta(t),$$

$$-\frac{\partial \psi}{\partial t} - div \mathbf{w}\psi + \sigma\psi - \mu \Delta\psi =$$
$$\sum_{j=1}^{s} M_j(t)p(t, b_j)\delta(x - b_j) - \chi_\Omega(x),$$

$$\psi(T, x) = 0, \quad p(0, x) = 0,$$
$$\psi, p \to 0 \quad \text{under} \quad ||x|| \to +\infty.$$

4. SENSOR PLACEMENT: THE STATIONARY CASE

Consider the case when outputs Q_i do not depend on time and $\mathbf{w}$ =const. Here the original problem is transformed into the static one in the sense that all considered values don't depend on the time:

$$w_1\frac{\partial \varphi}{\partial x_1} + w_2\frac{\partial \varphi}{\partial x_2} + \sigma\varphi - \mu \Delta\varphi =$$
$$\sum_{i=1}^{k} Q_i\delta(x - a_i). \tag{8}$$

$$\varphi(x) = 0 \quad \text{under} \quad ||x|| \to \infty.$$

Solution of the equation (8) may be represented in the form

$$\varphi(x) = \sum_{i=1}^{k} Q_i\varphi_i(x), \tag{9}$$

where $\varphi_i(x)$ is the solution of (8) with right-hand side being equal to $\delta(x - a_i)$.

4.1 *Reduction to nonlinear programming problem*

In view of the stationarity of yhe system the estimation problem takes the following form. On the base of measurements

$$y_j = \varphi(b_j) + \xi_j \qquad , j = \overline{1, s}, \tag{10}$$

and apriori information on unknowns

$$0 \le Q_i \le \overline{Q}_i, \quad i = \overline{1, k}, \qquad |\xi_j| \le e_j, \quad j = \overline{1, s},$$

denoted briefly as $\{\mathbf{Q}, \xi\} \in W$. it is necessary to estimate the value

$$I = \int_\Omega \varphi(x)dx_1dx_2 = \langle \mathbf{Q}, \mathbf{f}\rangle,$$

where $\mathbf{Q} = (Q_1, \ldots, Q_k)^T$, $\mathbf{f} = (f_1, \ldots, f_k)^T$, $f_i = \int_\Omega \varphi_i(x)dx$.

Hence the problem of sensors placement is as follows

$$\min_{\substack{b_1, \ldots, b_s \\ b_j \in \Omega^*}} F(b_1, \ldots, b_s),$$

$$F(b_1, \ldots, b_s) = \min_{\mathbf{z} \in R^k, z_0 \in R} \Phi(\mathbf{z}, z_0),$$

$$\Phi(z, z_0) = \sup_{\{\mathbf{Q}, \xi\} \in W} |\langle \mathbf{z}, \mathbf{y}\rangle + z_0 + \langle \mathbf{Q}, \mathbf{f}\rangle|. \tag{11}$$

The direct calculations show that

$$\min_{z \in R^k, z_0 \in R} \Phi(z, z_0) =$$
$$\min_{z \in R^k}\Big\{\frac{1}{2}\sum_{i=1}^{k} \overline{Q}_i|\sum_{j=1}^{s} \varphi_i(b_j)z_j - f_i| + \sum_{j=1}^{s} e_j\,|z_j|\Big\}.$$

The further reasoning is based on the following known result.

Lemma 1. The problem

$$\min_{x \in \mathbf{R}^n} \sum_{i=1}^{k} |f_i(x)|$$

is equivalent to the following:

$$\min_{x \in \mathbf{R}^n, r_i, s_i} \sum_{i=1}^{k} (r_i + s_i),$$

under constraints

$$r_i - s_i = f(x), \qquad r_i \ge 0, \quad s_i \ge 0, \quad i = 1, \ldots, k.$$

In view of this lemma the considered problem may be rewritten as follows

$$\min_{\substack{b_j \in \Omega^* \\ \mathbf{r},\mathbf{s},\mathbf{t},\mathbf{p},\mathbf{z}}} \left\{ \frac{1}{2} < \overline{\mathbf{Q}}, \mathbf{r}+\mathbf{s} > + \sum_{j=1}^{s} e_j(p_j+t_j) \right\}$$

under constraints

$$\sum_{j=1}^{s} \varphi_i(b_j) z_j - f_i = r_i - s_i, \quad i = \overline{1,k},$$

$$p_j - t_j = z_j, \quad j = \overline{1,s},$$

$$\mathbf{r},\mathbf{s} \in \mathbf{R}^{\mathbf{k}}_{+};\ \mathbf{p},\mathbf{t} \in \mathbf{R}^{\mathbf{s}}_{+};\ \mathbf{z} \in \mathbf{R}^{\mathbf{s}},$$

or, after exclusion of $\{z_1, \ldots, z_s\}$,

$$\min_{\substack{b_1,\ldots,b_s \\ b_j \in \Omega^*}} \min_{\mathbf{r},\mathbf{s},\mathbf{t},\mathbf{p}} \left\{ \frac{1}{2} < \overline{Q}, r+s > + \sum_{j=1}^{s} e_j(p_j+t_j) \right\} \quad (12)$$

under constraints

$$\sum_{j=1}^{s} \varphi_i(b_j)(p_j - t_j) - f_i = r_i - s_i, \quad i = \overline{1,k},$$

$$\mathbf{r},\mathbf{s} \in \mathbf{R}^{\mathbf{k}}_{+};\ \mathbf{p},\mathbf{t} \in \mathbf{R}^{\mathbf{s}}_{+}.$$

The minimization in $\{\mathbf{r},\mathbf{s},\mathbf{t},\mathbf{p}\}$ in (12) constitutes a linear programming problem. In view of duality theory in linear programming (12) is equivalent to the following problem

$$\min_{\substack{b_1,\ldots,b_s \\ b_j \in \Omega^*}} \max_{\mathbf{y} \in R^k} < \mathbf{f}, \mathbf{y} >$$

under
$$|\sum_{i=1}^{k} y_i \varphi_i(b_j)| = e_j, \quad j = \overline{1,s},$$

$$|y_i| \le \frac{\overline{Q}_i}{2}, \quad i = \overline{1,k}.$$

4.2 *A sufficient number of sensors*

In this case we consider the modified problem statement by assuming information can be received on system state function in entire spatial domain Ω^*.

$$\inf_{\mu,\mu^0} \sup_{\xi(\cdot),Q} \left| \sum_{i=1}^{k} Q_i \int_{\Omega} \varphi_i(x)dx - h(y(x)) \right| \quad (13)$$

$$h(y(\cdot)) = \int_{\Omega^*} y(x)\mu(dx) + \mu^0, \qquad \mu^0 \in R.$$

$$y(x) = \varphi(x) + \xi(x) \qquad x \in \Omega^*, \quad (14)$$

$$0 \le Q_i \le \overline{Q}_i \quad i = \overline{1,k}$$

$$|\xi(x)| \le e(x) \quad , e(x) \in C(\Omega^*),\ e(x) > 0.$$

Here μ denotes a regular measure on Ω^*. The direct calculations give the next equality

$$\inf_{\mu_0} \sup_{\xi(\cdot),Q} \left| \sum_{i=1}^{k} Q_i \int_{\Omega} \varphi_i(x)dx - h(y(x)) \right| =$$

$$= \sum_{i=1}^{k} \frac{\overline{Q}_i}{2} \left| \int_{\Omega^*} \varphi_i(x)\mu(dx) - f_i \right| + \int_{\Omega^*} \sigma(x)|\mu(dx)|.$$

Further, from lemma 1 it follows, that (13) is equivalent to the following problem

$$\inf_{t_i \ge 0, p_i \ge 0, \mu} \left\{ \sum_{i=1}^{k} \frac{\overline{Q}_i}{2}(t_i + p_i) + \int_{\Omega^*} \sigma(x)|\mu(dx)| \right\} \quad (15)$$

under restrictions

$$\int_{\Omega^*} \varphi_i(x)\mu(dx) - f_i = p_i - t_i,\ i = \overline{1,k}. \quad (16)$$

Consider Lagrange function for convex programming problem (15),(16)

$$L(t,p,\mu;\lambda) = \sum_{i=1}^{k} \frac{\overline{Q}_i}{2}(t_i + p_i) + \int_{\Omega^*} \sigma(x)|\mu(dx)| + \sum_{i=1}^{k} \lambda_i (f_i + p_i - t_i - \int_{\Omega^*} \varphi_i(x)\mu(dx)),\ \lambda \in R^k.$$

The minmax theorem being applied to L allows to assert that value (15) coincides with a maximum ν^* for a dual problem

$$\langle \mathbf{f}, \lambda \rangle \longrightarrow \max, \quad (17)$$

under restrictions

$$|\sum_{i=1}^{k} \lambda_i^* \varphi_i(x)| \le \sigma(x), \quad \forall x \in \Omega^*,$$

$$|\lambda_i| \le \frac{\overline{Q}_i}{2} \quad i = 1, \ldots, k.$$

Denote

$$\Gamma = \{x \in \Omega^* \subseteq R^2 : \quad |\sum_{i=1}^{k} \lambda_i^* \varphi_i(x)|/\sigma(x) = \alpha^*\},$$

$$\alpha^* = \max \left\{ \max |\sum_{i=1}^{k} \lambda_i^* \varphi_i(x)|/\sigma(x), \max |\frac{2\lambda_i^*}{\overline{Q}_i}| \right\}.$$

where λ^* is the solution of the dual problem.

Theorem 3. Let μ^* be the optimal measure for the problem (13). Then $\mu^*(\Omega^* \backslash \Gamma) = 0$, that is measure μ^* is concentrated on the set Γ.

The following theorem defines more exactly the structure of the optimal measure.

Theorem 4. There exists the solution μ of optimization problem (13) such that

$$\int_{\Omega^*} y(x)\mu(dx) = \sum_{i=1}^{m} \mu_i y(x_i), m \leq k.$$

This means that optimal measure is concentrated in finite number of points and an information from the rest of points is surplus. It is important that number of optimal measure impulses is no more than k (this is an analog of known theorem of L.Neustadt (Neustadt, 1964)). Thus the sufficient number of sensors don't surpass the number of perturbation sources in the right hand side of the diffusion equation.

5. CONCLUDING REMARKS

Thus, for nonstationary system the sensors placement problem is reduced to the "impulse" optimal control problem for adjoint system. The stationary case is considered in detail for geometrical constraints on unknowns. There are two different situations here.

For the case when there are k or more sensors, by virtue of the "duality" theory a way of determining of optimal sensors coordinates is following. It consists in two stage. The first stage is to solve the following semi-infinite linear programming problem: to maximize the linear function $\langle f, \lambda \rangle$ subject to the constraints

$$|\sum_{i=1}^{k} \lambda_i^* \varphi_i(x)| \leq \sigma(x), \quad \text{for} \quad \forall x \in \Omega^*,$$

$$|\lambda_i| \leq \frac{\overline{Q}_i}{2} \quad i = 1, \ldots, k.$$

A number of computational algorithms there exist for this problems. At the second step it is necessary to find k (or less) points of maximum of the function $|(\lambda^*, \varphi(x))|$ of two variables, where λ^* is the solution of the first task.

If number of sensors is less than k, the original problem is represented in the form

$$F(\mathbf{b}) \longrightarrow \min_{\substack{b_1, \ldots, b_s \\ b_j \in \Omega^*}},$$

where

$$F(\mathbf{b}) = \max_{\mathbf{y} \in R^k} < \mathbf{f}, \mathbf{y} >$$

under

$$|\sum_{i=1}^{k} y_i \varphi_i(b_j)| = e_j, \ j = \overline{1,s}, \ |y_i| \leq \frac{\overline{Q}_i}{2}, \ i = \overline{1,k}.$$

Thus in this case it is necessary to minimize function $F(\mathbf{b})$ of $2k$ variables, each value of this function is obtained as a result of the solution of linear programming problem. The numerical simulation shows that function $F(\mathbf{b})$ is multiextremal in general.

REFERENCES

Chen,W.H. and J.H.Seinfeld. (1975).Optimal allocation of process measurements. *Int. J. Control*, **21**, 6, 1003–1014.

Gusev,M.I.(1981). Multicriteria problems of measurement optimization for dynamical systems under uncertainty conditions. In *Proceedings of 4 congress on theoretical and applied mechanics*, Sofia.

Gusev,M.I.(1988). Measurement allocation problem in estimation of dynamical systems under geometrical constrains on uncertainty, *Diff. uravn.*, **24**, 11, 1862–1870. (in Russian)

Kostousova,E.K.(1990). On approximation of measurement allocation problem for parabolic system. *Journal vichisl. mat. i mat. fiz.*, **30**, 9, 1994–1306. (in Russian)

Kurzhanski,A.B. (1977). *Control and Observation under Uncertainty Conditions*, Nauka, Moscow. (in Russian)

Kurzhanski,A.B. and A.Yu.Hapalov (1985). On estimation of distributed fields through available measurements. In: *Proc. Int. Sobolev Conf.* Novosibirsk.

Kurzhanski,A.B. and I.Valyi (1997). *Ellipsoidal Calculus for Estimation and Control.* Birkhäuser, Boston.

Marchuk,G.I.(1982). *Mathematical Modelling and Environmental Problems.* Nauka, Moscow. (in Russian)

Nakamori,J., S.Miyamoto, S.Ikeda, J.Savaragi (1980). Measurement optimization with sensitivity criteria for distributed parameter systems. *IEEE Trans. Automat.Control*, **AC-25**, 5, 889–900.

Neustadt, L.W. (1964).Optimization, a moment problem, and nonlinear programming. *SIAM J. Control*, Ser. A, **2**, 1.

Rafajlowicz,E.(1981). Design of experiments for eigenvalue in identification distributed parameter system. *Int. J. Control*, **34**, 1079.

Ucinski, D. Optimal sensors location for parameter identification of distributed systems. *Appl.Math.and Comp.Sci.*, **2**, 1.

Ucinski,D.,J.Korbicz and M.Zaremba.(1992) On optimization of sensors motions in parameter identification of two-dimensional distributed systems. In *Proc. 2nd European Control Conference*, v.3, pp. 1359-1364. Groningen, The Netherlands.

THE USAGE OF NEURAL NETWORKS FOR NAVIGATION TASKS SOLVING

Zoya V. Ilyichenkova

Moscow State Institute of Electronics and Mathematics
(Technical University)
109028 Moscow, Russia
E-mail: zv@home.miemstu.msk.su

Abstract: This article deals with the problem of choice the optimal path for autonomous wheel transport robot as a part of control system projection for technical objects. This method makes possible to remove the restrictions on the movement and to choice the optimal path taking into account real dimensions of robot. It makes possible to find a decision in true time.

Keywords: Adaptive control, Convergence analysis, Mobile robots, Neural-network models, Optimal trajectory, Parallel Computers, Path planning.

INTRODUCTION

Nowadays the problem of control system designing for different types of autonomous robots is important. Autonomous robots are devices for different actions. Control systems of such robots are built-in them. Then autonomous robots are controlled without human. Now control systems for such robots were developed by using neural networks (NN).

NN are special parallel structure which is formal human's brain. NN consist of formal neurons. Each neuron has special several input signals. Each of them can have own weight. Output of neuron is an non-linear modification of the sum of input signals. NN can have feed-back control line. Such structure makes possible to solve tasks which are difficult for formalisation, having apriory indefinites, tasks with large dimension, and other problems such as navigation tasks, autonomous control, classification, etc.

In the article the problem of control system building is discussed, using NN for one type of autonomous robots. They are autonomous transport wheel robots (ATWR). These should been move from one point of surface to another point.

Then we have one of the problem of control system designing which contains the finding an optimal safety trajectory for robot movement.

1. CHARACTERISTICS OF TRAJECTORY

The problem of optimal path choice consists of a safety trajectory choice and the minimisation of the functional Φ(P) which characterises energy experience on robot movement along path P and robot turns.

$$\Phi(M) = \sum_j (k_1 d_{i_j i_{j+1}} + k_2 |f_{i_{j-1} i_j i_{j+1}}|),$$

where, $d_{i_j i_{j+1}}$ is the length of *j*-movement part from the point i_j to the point i_{j+1}; $f_{i_{j-1} i_j i_{j+1}}$ is the turn angle in a point i_j when the robot passes from the (*j-1*)-th trajectory section to the *j-th* one; $k_1 \geq 0$,

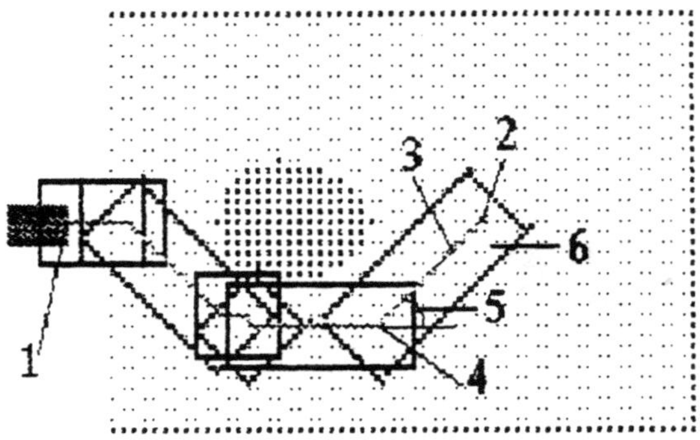

Fig. 1. The view of the robot optimal path.
Light grey points - trafficable points;
Dark grey points - untrafficable points;
1 - the start point; 2 - a finish point;
3 - a movement part;
4 - a point of robot turn;
5 - a angle of robot turn;
6 - a safety passage-way.

$k_2 \geq 0$ are the constants, characterising energy expense for a robot movement and its turn.

We suggest choicing the trajectory as a polyline from the start point - point of the robot start position - to the concrete finish point of the surface (fig. 1). Such view of trajectory makes possible to remove the restriction on the movement directions and to bend obstacles with any configuration.

We have a problem connected with a large dimension of data. Data include the information about the trafficability of each point of surface in the Cartesian coordinate system.

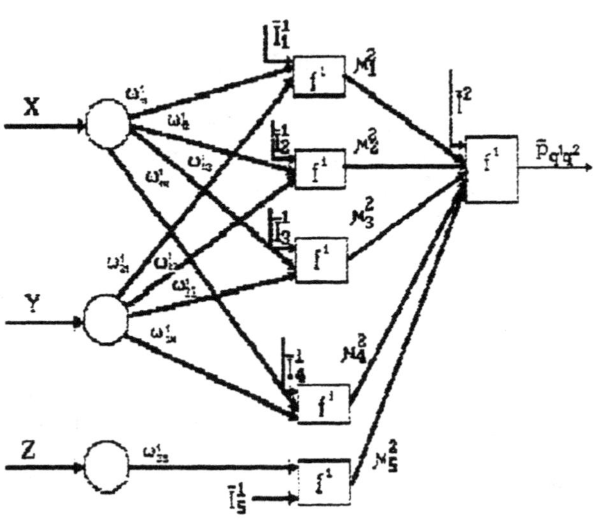

Fig. 2. The neural network for verification of a passage-way trafficability.

2. TO VERIFY OF TRAFFICABILITY OF A TRAJECTORY

2.1 The trafficability of passage-ways.

For solving the task of movement safety we build the safety passage-ways along all movement sections and control the absence of obstacles in them. The safety passage-way is impassable if there exists an obstacle point belonging to a passage-way. For this control, the method of surface classification is used which is a control of getting a point to a field. For the given fields, the two-layers NN.

This net is completed by another neuron controlling the passage of a point. The whole NN for the control of the section trafficability is shown in the figure 2. The output is equal 0 if the passage way doesn't contain obstacles.

2.2 The trafficability of robot position.

The control of the trajectory safety includes the absence of robot's bottom touch of the surface as well as wheels lifting off. For this purpose we must determine the robot position when it moves alone the trajectory.

We build the robot model using the scientific program package "Universal Mechanism". To verify the trafficability of robot position we obtain it dynamic model with help "Universal Mechanism". This model is described by the following ordinary differential equations (DE):

$$\mathrm{M}(x)\ddot{x} + \mathrm{f}(x, \dot{x}) = 0$$

which can be solved by NN too.

Here, M is the symmetric mass matrix; f is the column of inertia forces and generalised forces. The number of equations depends on the robots' model type and corresponding number of freedom degrees. They are different in the wheel numbers, directions of ATWR's movement, etc. Concrete coefficients are determined by the concrete model.

We suggest to solve these differential equations using neural networks. They have an adaptability property which can help to find the correct decision in true time.

Let us consider a 6-wheel robot which has 18 degrees of freedom (n=18). Then we can rewrite DE relatively the coordinates, velocities and accelerations:

$$\sum_{i=1}^{n} m_{ii}^{3}\ddot{x}_i(t)+\sum_{i,j=1}^{n} m_{iij}^{2}\ddot{x}_i(t)x_j(t)$$
$$+\sum_{i,j=1}^{n} m_{iij}^{1}\ddot{x}_i(t)x_j^{2}(t)+\sum_{i,j,k=1}^{n} f_{iijk}^{1}\dot{x}_i(t)\dot{x}_j(t)\dot{x}_k(t)$$
$$+\sum_{i,j,k=1}^{n} f_{iijk}^{2}x_i(t)\dot{x}_j(t)x_k(t)+\sum_{i,j=1}^{n} f_{iij}^{3}x_i(t)\dot{x}_j(t)$$
$$+\sum_{i=1}^{n} f_{ii}^{4}x_i(t)+f_i^{5}=0, \qquad i=\overline{1,n}$$

Consider the approximation of the trajectory **X**(t) by polynomial

$$\mathbf{X}_i(t)=\sum_{m=0}^{\infty} c_{im}(t-t_0)^m=\mathbf{X}_i^0+\sum_{m=1}^{\infty} c_{im}(t-t_0)^m$$

and truncate the series

$$\mathbf{X}_i(t)=\mathbf{X}_i^0+\sum_{m=1}^{M}\mathbf{C}_{im}(t-t_0)^m .$$

Then

$$\dot{X}_i(t)=\sum_{m=1}^{M} m\,\mathbf{C}_{im}(t-t_0)^{m-1}$$
$$\ddot{X}_i(t)=\sum_{m=2}^{M} m\,(m-1)\,\mathbf{C}_{im}(t-t_0)^{m-2} .$$

Then we must find a matrix **C** determining $X(t)$ at each moment t. For this purpose we substitute the expressions $X(t)$, $\dot{X}(t)$, $\ddot{X}(t)$ in the DE and rewrite they relatively $\mathbf{C}_{im}$:

$$a_q+\sum_{im} b_{q,im}\mathbf{C}_{im}+\sum_{im,jp} d_{q,im,jp}\mathbf{C}_{im}\mathbf{C}_{jp}+$$
$$\sum_{im,jp,kr} d_{q,im,jp,kr}\mathbf{C}_{im}\mathbf{C}_{jp}\mathbf{C}_{kr}=0, \qquad q=\overline{1,n}$$

For a concrete t the coefficients in this equations are constant. Then we must solve this system relatively **C**. This equation will have a non-zero error **Y** because we have approximated the trajectory **X**(t) with a definite accuracy:

$$\mathbf{Y}_q=a_q+\sum_{im} b_{q,im}\mathbf{C}_{im}+\sum_{im,jp} d_{q,im,jp}\mathbf{C}_{im}\mathbf{C}_{jp}$$
$$+\sum_{im,jp,kr} d_{q,im,jp,kr}\mathbf{C}_{im}\mathbf{C}_{jp}\mathbf{C}_{kr}, \qquad q=\overline{1,n}$$

Our goal is the minimisation of the error **Y**. For that we consider a functional of quality F:

$$\mathrm{F}(\mathbf{Y})=\mathbf{Y}^T\cdot\mathbf{Y}$$
$$e=\mathrm{F}(\mathbf{Y}),$$

where, e is the mean square error which should be minimised.

We suggest to use the gradient method for solving this problem by neural networks (Wosserman, 1992). When the error **Y** is minimised we obtain our matrix **C** and trajectory X(t) in the whole. If the positions of the robot's body and its wheels are determined, we can do some conclusions about its safety along all movement parts at every instant. If the trajectory is not trafficable, the output of NN u_r is equal to 1. If it is trafficable, then the value is 0.

$$u_r=\mathrm{f}_r(u_2+u_3),$$
$$\mathrm{f}_r(x)=\begin{cases}1, x>0\\ 0, x\le 0\end{cases},$$

where, u_2 is equal to 1 if wheels lifting off; u_3 is equal 1 if robot's bottom touch of the surface.

3. THE CHOICE OF OPTIMAL TRAJECTORY

The optimal trajectory is chosen by another NN - Hopfield net (Hopfield, 1992) - taking into account the control of its safety. Hopfield net is one of the classical type of NN with error back-propagation line. Its feature is the energy function (EF) which has a special view. Value of the EF doesn't increase after each iteration. Then we must write this function for our problem. This EF includes information about some matrix OUT characterised a NN state.

For the task of optimal path choice we suggest comparing this matrix to the trajectory for robot movement. For that we number all points of surface and all time moments when robot turns from one passage-way to another one. The element of matrix $\mathbf{OUT}_{rk}$ is equal to 1 if the robot turns in the point

Table 1 The matrix for path from point 1 to point 6 through points 3 and 2

points of time → points of surface ↓	1	2	3	4	5	6
1	1	0	0	0	0	0
2	0	0	1	0	0	0
3	0	1	0	0	0	0
4	0	0	0	0	0	0
5	0	0	0	0	0	0
6	0	0	0	1	1	1

r in k-time moment. The other elements are equal 0. The example for the path from the start point 1 of the surface to the finish point 6 through points 3 and 2 is shown in table 1.

Then the EF can be written in follow form:

$$E = \mathrm{A}\sum_i \sum_m \sum_{p\neq m} \mathbf{OUT}_{mi}\mathbf{OUT}_{mj}$$
$$+\mathrm{B}\left[\left(\sum_m \sum_i \mathbf{OUT}_{mi}\right) - N\right]^2$$
$$+\mathrm{C}\left[(\mathbf{OUT}_{11} - 1)^2 + (\mathbf{OUT}_{NN} - 1)^2\right]$$
$$+\mathrm{G}\left[\sum_p \sum_r \sum_j \left(\mathbf{k}_1\, d_{pr}\mathbf{OUT}_{pj}\right.\right.$$
$$\left.\cdot\left(\mathbf{OUT}_{r,j+1} + \mathbf{OUT}_{r,j-1}\right)\right)$$
$$+\sum_m \sum_p \sum_r \sum_j \left(\mathbf{k}_2 |f_{mpr}| \mathbf{OUT}_{pj}\right.$$
$$\left.\left.\cdot\left(\mathbf{OUT}_{m,j-1}\mathbf{OUT}_{r,j+1} + \mathbf{OUT}_{m,j+1}\mathbf{OUT}_{r,j-1}\right)\right)\right]$$

But if we build classical Hopfield NN according our EF the NN won't be converged. Then we must do some special modification.

We suggest changing only single element in the column. Then we divide all columns for three groups. Each group contains the series of columns: every third columns beginning from second, third or fourth one for the different groups. After each iteration elements from only single group can be changed. Then we'll complicate NN to provide for convergence. New EF includes the information about trafficability of a trajectory and energy experience on robot movement along it:

$$E = E_1 + E_2;$$

$$E_1 = \mathrm{G}\left[\sum_p \sum_r \sum_j \left(\mathbf{k}_1\, d_{pr}\mathbf{OUT}_{pj}\right.\right.$$
$$\left.\cdot\left(\mathbf{OUT}_{r,j+1} + \mathbf{OUT}_{r,j-1}\right)\right)$$
$$+\sum_m \sum_p \sum_r \sum_j \left(\mathbf{k}_2 |f_{mpr}| \mathbf{OUT}_{pj}\right.$$
$$\left.\left.\cdot\left(\mathbf{OUT}_{m,j-1}\mathbf{OUT}_{r,j+1} + \mathbf{OUT}_{m,j+1}\mathbf{OUT}_{r,j-1}\right)\right)\right]$$
$$E_2 = \mathrm{V}\sum_p \sum_r \sum_j \left(\tilde{p}_{pr}\mathbf{OUT}_{pj}\right.$$
$$\left.\cdot\left(\mathbf{OUT}_{r,j+1} + \mathbf{OUT}_{r,j-1}\right)\right)$$

where, E_1 corresponds to the functional Φ; E_2 controls the safety of path.

The input matrix is a result of the wale algorithm (Sushkov, 1995). When the state of NN doesn't change after some iteration the corresponding matrix **OUT** will characterise the optimal trafficable path.

4. THE RESULTS AND CONCLUSIONS

The surface having the overpass is shown on fig. 3(a). The path bending the overpass is shorter than the path through it (fig. 3(b)).

The labyrinth and the several paths for the functional having the different value of coefficients are shown on fig. 4. On the top figure the coefficient, characterising the turn angle is less than that in the bottom one.

On the fig. 5. the surface with spherical obstacles is shown. The resultant path is a snake-like curve.

Thus we can say about the possibility of using the NN for different problems of control theory. The parallel property makes possible to solve problems of any dimension of the surface. The using of neural networks additives a new quality - adaptability - which is important for a lot of technical system.

This method makes possible to choice the optimal

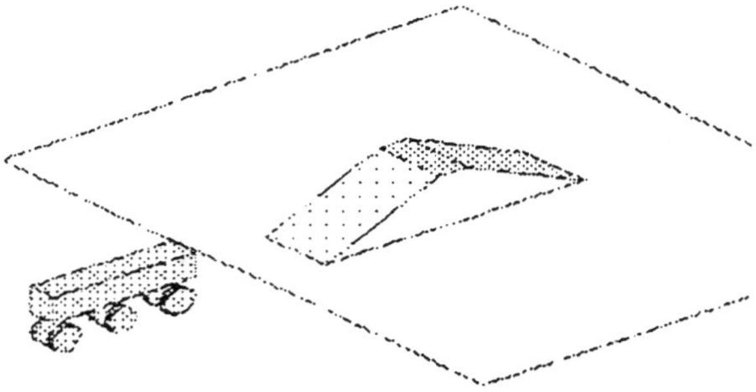

(a)

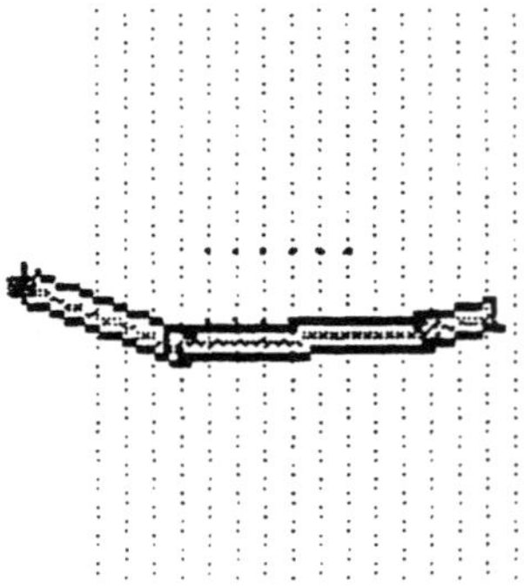

(b)

Fig. 3. The surface, having the overpass (a) and the corresponding optimal path (b).

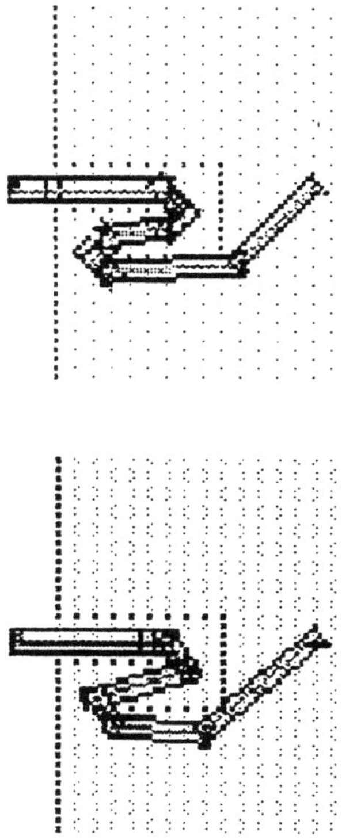

Fig. 4. The optimal paths for the labyrinth.

path taking into account real dimension of robots as well as removing the restriction on the movement. It makes possible to find a decision in true time.

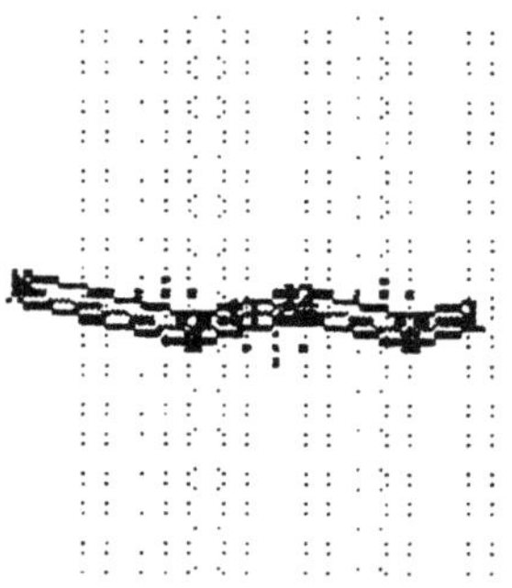

Fig. 5. The optimal path for the surface, having spherical obstacles.

REFERENCES

Hopfield, J.J. (1992). Leaning Algorithm and Probability Distributions in Feed-Forward and Feed-Back Networks. In: *Artificial Neural Networks: Concepts and Theory* (Compl.: P. Mehra, B.W. Wah), pp. 425-429. Los Alamitos (Ca) et al.: IEEE Comp. Soc. Press, 1992 - XI.

Sushkov, B.G. (1995). *Neural Networks for Path Planing on Networks*, 29p. Proceedings of the Russian Academy of Sciences, Moscow. (In Russian).

Wosserman, F. (1992). *Neurocomputer Technics: Theory and Practice (from English)*, 240p. Mir, Moscow. (In Russian).

RECOVERY BY HILBERT TRANSFORM FOR DECENTRALIZED NODES

H. Kang, *K.K.Kang*

470 Carnegie Dr,Milpitas,CA 95035, U.S.A.

Abstract: In decentralized Repeatable stochastic hybrid control system tradeoff exists between complete loss of track control due to continued deferred service or excercising a limited control.To compute location score and to obtain self tuned control of interframe spacing timer to minimize probability of frame collision,technique of probabilistic data association is employed . Recursive stochastic approximation algorithm is obtained to track target in clutter. Proposed design for exception handling is fault tolerant to single outages of control channels.Mathematical model for signature identification of targets aggregate wavelet is obtained. Procedure for obtaining minimum spatial resolution of maneuvering tragets in clutter is given.By application of Hilbert transform mathematical model of clutter of Arteries Vains and Capillaries is obtained. The phase error minimization is employed for signature recovery in capillaries clutter.

Keywords: Singular decomposition,Model management,Fault tolerance,Sequence estimating,Signature analysis.

1. INTRODUCTION

Class of large controlled multiplexing systems (Kang,1995) competing for slot time in receiver frame packet with composite multiple information rates results in severe computational requirements . For the class of stochastic hybrid systems under consideration the continuous time state descriptor (Kang, 1996) is completely decomposable in finite state Markov chain jump diffusion process with active and nonactive (idle) state parameters (figure 1) . Further with increase in number of sources, the state processes converge to limit. In this writeup it is proposed to select performance index to assign heavy weight to loss of track than increased tracking error. Protocol sequencer (figure 2) identifies low priority frames,to assign smaller retries and to defer the low priority frame processing for the period. Intermodal mapping information matrix and activity index is introduced for the purpose. The concept of Repeatability Measure for Repeatable systems was introduced by Kang (1995).A aggregate model is obtained in the following to characterize the global behavior (figure 2) of free terminal time repeatable stochastic hybrid system. Clutter tracking problem is defined as equivalent best fit approximation problem (Kang,1973, Sowder, 1976). Proposed design procedure has considerable computational advantage .The control and management (figure 3) of decentralized control systems consists of fault diagnosis, fault isolation, identification of collision presence, exception handling for fault accommodation, and to provide continued service through reallocation of resources and topological reconfiguration of hierarchical structure of bus controllers / remote terminals (figure 2). Proposed procedure minimizes the probability of frame collision. If the remote terminal is in inactive (idle) state, the performance of the link can be improved by increased interframe spacing and deferred frame processing. It is shown that under incomplete observations the switching type policy with decentralized network controllers is optimal. It is shown that spin echo pulses and refresher pulses initiated and transmitted by offsite supervisor form a markov process. Results presented in the paper are illustrated by the help of decentralized (figure 3) network control problem for a Airborne Refueling System,and Signature Identification in clutter of Arteries,Vains,and Capillaries.

2. MATHEMATICAL FORMULATION

Stochastic hybrid system descriptor (figure 2) is defined by complete probability space (X,x,p^a), $x(t) \in \Re^n$,$\Re^n$ is a borel set of euclidean R^n,$u(t)\in\Re^n$ is the field of piecewise continuous input functons, p^a is the probability measure for joint probability space for aggregated system model.The index r(t) is assumed to be continuous completely decomposable switching process with active and inactive (idle) states taking values in f-

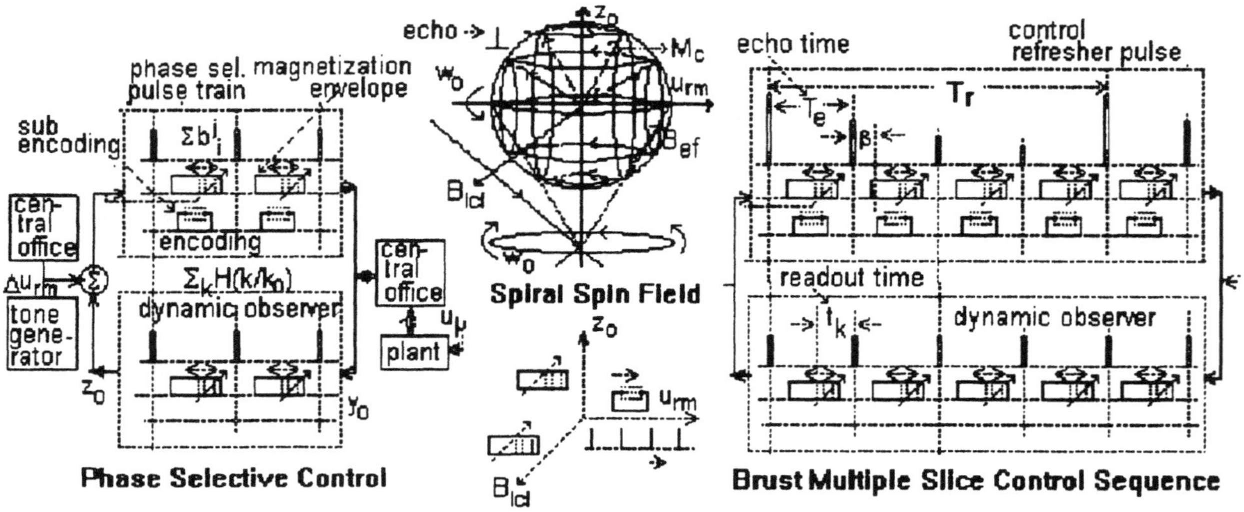

Figure 1. SECURE OFFSITE SINGULAR CONTROL

inite Set S={1,..,N}.The N state Markov process r(t) is decomposable in n groups defined by $\Sigma^{n}_{l=1}(n_l)_l$=N.The process r(t) can be replaced by aggregate process r^a (t)⊂ S^a={1,.,n}.For aggregate model,n groups are characterized by ergodic distribution probability vector (e_r),and the states of groups E_i,i= 1,..,n_i,are defined by unit row vectors $E_i=[e_{il}]$;l=1,.,n_i charactertistic functions

3. SELF TUNED CONGESTION CONTROL

System under consideration (figure 2) consists of cluster of Markov modulated type multicast address receivers with unique signatures.Command block broadcast frames for the integrated network,alternates between receivers with complete observation or low prioriy receivers with partial observation (Kang,1995) sets.

3.1 Tuned Control With Receding Horizone

State of data frames of central bus controller queue buffer is given by the following buffer content equation

$$M_b(t_k)=M_b(t_0)+[\nu\int^{tk}_{t0}B(t)dt- c_b t_k]+a(t_k)-s(t_k-\delta t_k)+ L(t_k)- M_d(t_k)+M_r(t_k)+\varepsilon(t_k)\ ;\ c_b\geq 0 \qquad (1)$$

Here $M_b(t_k)$ are the data frames (waiting processing), ν is the mean rate of requests for processing frames. δt_k is the delay (wait cycles) introduced by the network and the resource management controller.ε(t)→0 the infinitesimal process is introduced by factors as network traffic problems of durations longer than one frame. $a(t_k)$ are the Command frames (waiting processing),$s(t_k)$ are the frames processed,$L(t_k)$ is the empty buffer (correction) factor, $M_d(t_k)$ is the frames alignment/ frame spill/ frame length error correction factor, $M_r(t_k)$ is the frame rerouting differential available from network backoff management and is function of network jammimg/ network collisions. It is assumed the processes $a(t_k)$ and $s(t_k)$ are martingales.Control arbitration signal for active and idle states are distributed exponentially with mean ($1/p_r$) and ($1/p_n$).B(t) is the number of receiving stations being serviced at time (t). To control channel congestion , self tuned control $\int^{tk}u(t-\delta t)dt$ is introduced. The buffer contents can be expressed by

$$dB(t)= -(p_r+p_n)B(t)dt+dw_1(t);\ dM_b(t)=(\nu B(t)-c_b)dt-u(-\delta t)dt+dw_2(t)+dw_3(t)+dL(t)-dM_d(t)+dM_r(t) \quad (2)$$

Here c_b> 0;$w_i(.)$;i=1,2,3, are the independent Wiener processes, with $Ew^2{}_1(t)=2p_rp_nt/(p_r+p_n)$, $Ew^2{}_2(t)=\nu p_nt/(p_r+p_n)$,and $Ew^2{}_3(t)=\varsigma^2\nu p_nt/(p_r+p_n)$,$\varsigma^2$ is the variance of frame processing requests. Policy of scoreboarding (figure 2) on outstanding cacheable frames may be employed to reduce these factors. Concurrent processing,look ahead pipeline controller/scheduler may be employed to decouple and reduce interaction between network delay (δt) and data frames $M_b(t)$ in the pipeline.

3.2 Boundary Variance Control :

To minimize rate of collision and from requirements for frame alignment and frame spill, interframe packet spacing timer (bus throttle timer) index distribution is predicted and continuously tuned to minimize the probability of collision (figure 2).τ_- and τ^+ are characterized by active and idle states of the process. Stochastic approximation algorithm to determine the boundary values of interframe packet spacing timer for ea-

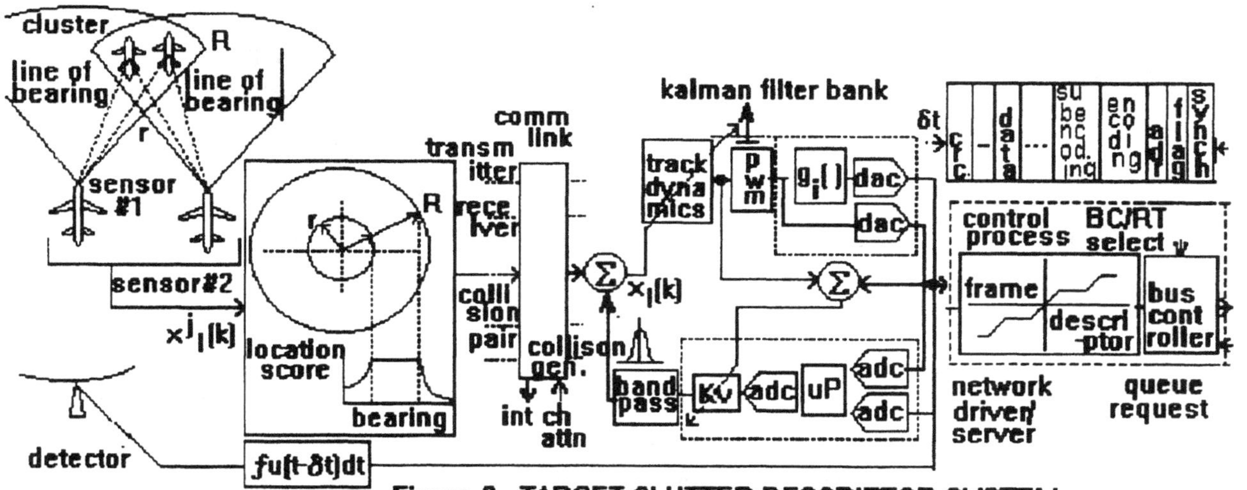

Figure 2. TARGET CLUTTER DESCRIPTOR SYSTEM

ch group field is given by $\tau^{j+1}=\tau^{j}+c_t\delta\tau; E\{\delta\tau\}=\sup_{\tau\in[\tau-,\tau+]}E\{\Psi(\tau^j,\theta^j)\}\leq K_0\leq\infty$; here $\delta\tau$ is mapping transformation $\Psi:(\tau^j,\theta^j)\rightarrow\{\delta\tau\}$ of sequence of random variables, $\tau_-\leq\tau\leq\tau^+$. θ is system descriptor parameter and is element of separable matric space. Parameter $c_t>0$ may be selected from the requirements of asymptotic stability following theorem.

Theorem 3.3 : Given bounded initial conditions x_0, ε_0, the solution of cluster descriptor (equation 4) converges in mean squared sense to the solution of model equation (3).

3.3 Tracking Repeatable Hybrid Systems :

The absorbing group of decomposable switching process r(t) (section 2) is defined by Repeatability cycle (Kang, 1995) and is characterized by its Repeatability Measure. The distributed descriptor system for the repeatable system is given by

$$x_i(k+1/qk_r)= g_i[\theta(t_k),r(t_k)]_{Krc}x_i(k/qk_r)+ b_i[r(t_k)]_{Krc}u_i(k/qk_r)+w_i;\quad r(t_0)=r_0; r(t_f)=r_C; q=0,1..; k=0,..,r \quad (3)$$

Where soft state constraints for maneuvering target are given by $x(qt_r)= x_0+\varepsilon_0$; $x_i\in R^n$; with dynamics of the control process as given in section (3.2). $[g_i]_{Krc}$ and $[b_i]_{Krc}$ are the periodic matrices of integer period K_{rc}, g_i is identically distributed random variable with $N(\theta(t_k),\sigma^2(t_k))$, here θ is the unknown parameter with apriori distribution $N(\theta_0,\sigma^2_0)$, and w_i is the additive process $w_i(0,W_i)$. Using stochastic approximation algorithm given above, decomposable aggregated model of decentralized hybrid system (with $S_j\in S$), and its observation process can be expressed by

$$x^j_i(k+1/qk_r)=[\Sigma^{ni}_{l=1}g^j_{il}(\theta^j(t_k))e_{il}]_{Krc}x^j_i(k/qk_r)+[\Sigma^{ni}_{l=1}b_{il}e_{il}]_{Krc}u^j_i(k/qk_r)+w_i;\; y^j_i(k/qk_r)=[h_i]_{Krc}x^j_i(k/qk_r)+v_i \quad (4)$$

Here v_i is an additive disturbance with covariance V_i (t). Assuming that g_i and b_i are bounded, we state

3.4 Structural Control -with Jump Parameters

In the following mathematical model for repeatable system with uncertain parameters as identified by random variables $N(\theta(t_k),\sigma^2(t_k))$ is obtained. It is assumed that introduced environmental noise (w_i) is infinitesimal. For tracking in cluster it is desired to minimize the performance index $J=E[\{x(k/qk_{rc})-x_d\}^t Q[r(t)]\{x(k/qk_{rc})-x_d\}]$; $Q[r(t)\in S]= Q_i$. During the command control channel congestion period the control on statistics of queue may be excercised by management of decentralized systems hierarchy and reconfiguration of bus controller -remote terminals architecture. The tracking location score (figure 2) is given by exp $(-y^t\Xi y/2)$, where Ξ is the covariance of bearing measurements, and y is the distance matric from the permissible band $r\leq(.)\leq R$. For a perfect observation case it can be shown that the periodic optimal control for repeatable hybrid system is given by

$$u^j_i{}^o(k-1/qk_r)= -[H^j_i{}^\wedge(k-1)]_{Krc}x^j_i(k-1/qk_r);\; H^j_i{}^\wedge(k-1)]_{Krc}=[\Sigma^{ni}_{l=1}g^j_{il}{}^\wedge(\theta^j,k-1)e_{il}]_{Krc}/[\Sigma^{ni}_{l=1}b^j_{il}e_{il}]_{Krc} \quad (5)$$

Minimum value of performance index is given by min $EJ=[\Pi^n_{l=1}\sigma(t_l)^2[x^j_l(k_0)^t x^j_l(k_0)]$. The resulting LQG optimal control law is also periodic.

Modelling Hybrid Cluster Descriptor : Let $\zeta : \zeta_- < \zeta <$

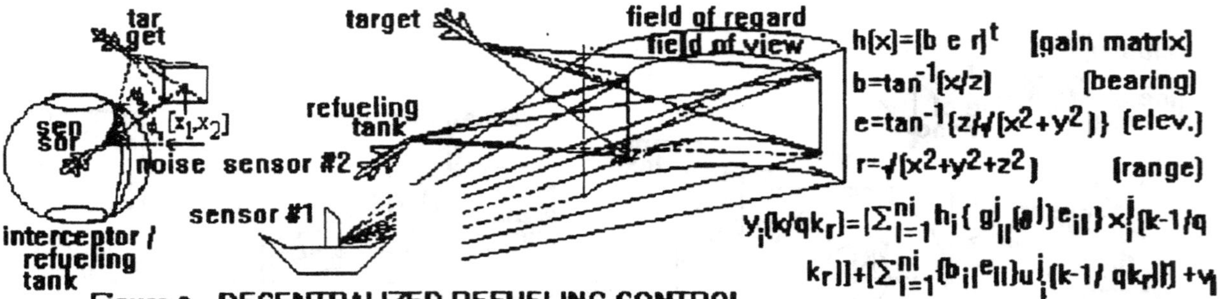

Figure 3. DECENTRALIZED REFUELING CONTROL

ζ^+ be the real valued sequence of step size for the stochastic hybrid system plant parameter θ.Following algorithm may be employed to update plant parameters during each period receiver is served by the server.

$$\theta^j_{l+1}\hat{}(.)=\theta^j_l\hat{}(.)+\zeta^j_l(y^j_l - h_l x^j_l)\ ;\ \zeta^j_{n+1}=\zeta^j_n+\upsilon(\partial\theta^j_n\hat{}(.)/\partial\zeta_n)\ ;\ \theta^j_{n0}\hat{}=\theta^{j-1}_{nf}\hat{};\zeta^j_{n0}=\zeta^{j-1}_{nf};0\leq\upsilon\leq 1 \quad (6)$$

Procedure to compute joint probability density $p^r(\theta^j/g^j_l,..,g^j_0)$ (figure 3)and identification of parameter (θ) is given in the following.Optimum control law (equation 5) for repeatable system can be computed using equation (6).In the following stochastic hybrid formulation is extended to decentralized global space.

Data Compression And System Reliability: For reducing data frames $\{M_b(t_k)\}$ (waiting processing) , traffic jamming and congestion,decentralized control system is configured to process common information transferred from neighborhood receivers.Let $p^r_{il}(x,t)$ be the joint probability density function for neighboring receivers,the aggregate model joint probability density function for the decentralized system in global space is given by $p^a_{il}(x,t)=e_r e_{il} p_{ril}(x,t)$ (section 2). For data compression $p^a_{il}(x,t)$ can be expressed by

$$p^a(x/y_{il}(k)..y_{il}(0))=\ln p^a(y_i(k)/x)+\ln p^a(x/y_i(k-1)..y_i(0))+\ln p^a(y_l(k)/x)+\ln p^a(x/y_l(k-1)..y_l(0))-\ln p^a(x/y_{il}(k-1)..y_{il}(0)) \quad (7)$$

4.0 FAULT DIAGNOSIS IN CLUTTER

For reliable control not compromised by control channel failure, Bus Control arbitrator may reconfigure topology by concurrent processing, reallocating low priority frames and switching remote terminal transfer mode. An important factor in decentralized control systems fault diagnosis is detection of cable failure of control channel,and computing failure distance $t_s(V_s/2F_s)$ from network server,where V_s is the wave propogation speed (m/s), F_s is the propogation carrier frequency (Hz), and t_s is the return carrier sense delay obtained by application of time domain reflectometry.Activity Index $AI_d=[p^{j,j-1}_i{}^r(\theta^j_i/g^j_i,..,g^0_i,t)]/[p^r(\theta^j_i/g^j_i)p^r(g^j_i/g^{j-1}_i,..,g^0_i,t)]$ may be employed to detect possible control channel failures.On fault detection and diagnosis,controller may update network scoreboard and rerouting differential factor $M_r(t_k)$ to process commands.

4.1 Fault Identification And Accomodation :

For repeatable system, to identify fault status hypothesis-conditional probability $p(\theta^j|g^j)$ matches closely the measurements (figure 2) may be verified.The aggregate joint probability density function $p^a_{il}(\theta,t)d\theta = prob\{\theta\leq\theta(t)\leq\theta+d\theta;r(t)=l\in S\}$for process r(t) is given by

$$p^{j,j-1}_i{}^a(\theta^j_i,t)=e_r e_{il} p^{j,j-1}_i{}^r(\theta^j_i/g^j_i,..,g^0_i,t)$$
$$p^{j,j-1}_i{}^r(\theta^j_i/g^j_i,..,g^0_i,t)=C_g\exp\{-(\theta^j_i-\varsigma^j_i)^2/2\rho^j_i{}^2\}$$
$$1/\rho^j_i{}^2=(1/\rho^{j-1}_i{}^2+1/\sigma^2) \quad (8)$$
$$p^r(g^j_i/g^{j-1}_i,..,g^0_i,t)=C_g\exp\{-(g^j_i-\varsigma^{j-1}_i)^2/2(\sigma^2+\rho^{j-1}_i{}^2)\}$$
$$\varsigma^j_i=(\varsigma^{j-1}_i/\rho^{j-1}_i{}^2+g^j_i/\sigma^2)/(1/\rho^{j-1}_i{}^2+1/\sigma^2) \quad (9)$$

Where $p^{j,j-1}_i{}^r(\theta^j_i/g^j_i,..,g^0_i,t)$ is the joint probability density function. The changed process status for conditional probability $p^{j,j-1}_i{}^a(\theta^j_i/g^j_i,..,g^0_i,t)$,and $p^{j,j-1}_i{}^r(\theta^j_i/g^j_i,..,g^0_i,t)$ results in lower value of ρ^j_i during measurements (figure 2). Equations (8-9) and parameter ρ^j_i may be employed to reconfigure bus controller/ remote terminal for fault isolation.

Outage In Control Channel : Structural control design of decomposable aggregated model for outage in control channel (figure 2) is considered in the following.Assuming $[b_i][b_i]^t\leq[b_l][b_l]^t,[h_i][h_i]^t\leq[h_l][h_l]^t$ holds ,here $[b_i]$ is obtained from $[b_l]\in R^{mxm}$ by omitting i th column,and $[h_i]$ is obtained from $[h_l]\in R^{n\,xn}$ by omitting i th row. It can be shown that for a single channel outage hybrid system maintains stability,avoavoiding pipeline stalls by initiation of alternative pipeline for frame (figure 2),minimizing collision,retransmitting frame for collision or miss.Further system performance index has upper bound for H_∞ norm .

4.2 Exception Handling In Clutter :

In the following,procedure to maximize (P^a_d) the probability of detection in clutter is outlined. Let ι_i ;$i\leq m$ be clutter density, let $v_k=\Sigma^m{}_{i=1}\iota_i v_i(k)$ be innovation process for Kalman filter bank (figure 2),and Sv be associated innovation covariance matrix. Let $K^j{}_i$ be the gain matrix, the state estimate $x^j{}_i{}^\wedge(\,)$ and tracking error covariance matrix $P(k+qk_r/k+qk_r)$,are given by

$$x^j{}_i{}^\wedge(k+qk_r/k+qk_r)=x^j{}_i{}^\wedge(k+qk_r/k-1+qk_r)+K^j{}_i[\theta^j(t_k)]v(k) \quad (10)$$

$$P^j{}_i(\theta^j{}_i,k+1+qk_r/k+1+qk_r)=P^j{}_i(\theta^{j-1}{}_i,k+1+qk_r/k+qk_r) -C(P^a{}_d)K^j(\theta^j{}_i,k+1+qk_r)S_v(\theta^j{}_i,k+1+qk_r)K^j(\theta^j{}_i,k+1+qk_r)^t \quad (11)$$

Where $0\leq C(P^a{}_d)\leq 1$.$\theta^j{}_i(t_k)$ is selected to maximize $P^a{}_d$ .From the expression for tracking error covariance it can be concluded,if inactive (idle) receiver frame is serviced with innovation process $v_i(k)=0$,bias error is increased.Available slot time for iterframe spacing timer (throttle timer) is considerably rduced.

5.ARTERY VAIN CAPILLARY-CLUTTER

The proposed boundary variance control for wavelet descriptor system employs variable length encoding .

5.1 Modelling Aggregate Wavelets :

Hilbert transform property is employed for identification of secure offsite controller (figure 1) parameters. Let angular spin $\Theta^s(\omega)$ be the fourier transform of rotational spin frame $\{\theta(k\Delta t)\}$ resulting from admisible magnetization B_{lcl} at plant site and received frame from the offsite controller.$\{\Theta^s(k\Delta t)\}$ is characterized by

$$\theta(k\Delta t)=\Upsilon B_{lc}k\Delta t+\Upsilon\int_{t0}^{k\Delta t}(x_1+x_2t)\Delta u_{rm}\partial t+\int_{t0}^{k\Delta t}dw(t) \quad (12)$$

Here $x_1,x_2=\partial x_1/\partial t$ are the states (spatial and velocity) of maneuvering targets in clutter,$\Delta u_{rm}=[(\partial u_{rm}/\partial x)_e+(\partial u_{rm}/\partial x)_{se}]$;where $(\partial u_{rm}/\partial x)_e$ and $(\partial u_{rm}/\partial x)_{se}$ are the spatial and velocity unipolar gradient (convolued subsegment/manchester encoded/frame signatures) components of transmitted frame by the offsite controller. The separable brownian motion process $w(\,)\sim(0,W)$ is class of phase deviation (β) resulting from network resolution,fluctuations or jitter introduced by the transmission network,and is identified by $<W>=2\Upsilon^2C_n\Sigma_{xyz}[\int_0^{Tr}\{\delta(\Delta u_{rm})\}^2\partial t]$;$C_n$ is network coefficient,Υ is the gyro magnetic ratio,T_r is the signature refresh time (figure 1).It can be shown that resulting rotational spin magnetization M_c and desired output of nonideal Hilbert transformer can be expressed by parabolic equation

$$dM_c/dt=\Upsilon M_c(q_f,h_p,\Upsilon)\{(\Theta^s(k\Delta t)/t)/\Upsilon+\omega_o/\Upsilon\}$$
$$I_m\{\Theta^s(\omega)\}=\Im^{\,j,j-1}{}_i(\omega)R_e\{\Theta^s(\omega)\}= -je^{j\beta}sgn(\omega)R_e\{\Theta^s(\omega)\} \quad (13)$$

Where q_f is the quantum factor,ω_o is the Larmor frequency of magnetization field.From Hilbert transformpair ($\Im$) and equation (13),it can be verified that it is sufficient to transmit $R_e\{\Theta^s(\omega)\}$ the real component to the offsite plant site,reducing considerably $\{M_b(t_k)\}$.

5.2 Signature Identification In Clutter :

From the resolution of angular spin rotation $\{\Theta^s(\omega)\}$, and phase detectors employed by the supervisor to monitor adjacent maneuvering Targets in clutter, resolution for signature identifcation can be determined. It can be shown that the minimum spatial resolutions are given by $(\Delta x^j)_{mn}=\{[\Delta\Theta^s(k\Delta t)]_x\}_{mn}/\Upsilon\int^{\tau j+1}{}_{\tau j}[\Delta u_{rm}]_x\partial t$;$(\Delta y^j)_{mn}=\{[\Delta\Theta^s(k\Delta t)]_y\}_{mn}/\Upsilon\int^{\tau j+1}{}_{\tau j}[\Delta u_{rm}]_y\partial t$;$(\Delta z^j)_{mn}=\{[\Delta\Theta^s(k\Delta t)]_z\}_{mn}/\Upsilon\int^{\tau j+1}{}_{\tau j}[\Delta u_{rm}]_z\partial t$;where τ^j is obtained by application of boundary variance control.The set of admissible signatures Γ_j is given by $E\{\Gamma_j\}=\inf\Gamma_{j\in ds}\sup_{\tau\in[\tau-,\tau+]}E\{J[\Theta^s{}_e],J[E_u]\}$;$J[\Theta^s{}_e]$ is the index for phase error,$J[E_u]$ is the index for signal energy.The signature identification of maneuvering targets in clutter is dependent on spread functions $E\{(\tau^j)(\tau^j)^t\}$,$E\{(\Delta x^l)(\Delta x^l)^t\}$,$E\{(\Delta y^l)(\Delta y^l)^t\}$,$E\{(\Delta z^l)(\Delta z^l)^t\}$.Tracking of targets in clutter is function of carrier frequency (F_s).It can be shown that minimum spatial resolutions can be also expressed as $(\Delta x)_{mn}=2\pi(\Delta F_s)_{mn}/\Upsilon[\Delta u_{rm}]_x$;$(\Delta y)_{mn}=2\pi(\Delta F_s)_{mn}/\Upsilon[\Delta u_{rm}]_y$,$(\Delta z)_{mn}=2\pi(\Delta F_s)_{mn}/\Upsilon[\Delta u_{rm}]_z$.

Phase Velocity And Echo Spin : In the following state reconstruction of maneuvering target in clutter with multiple slice exogenous system (figure 1) is considered. Mathematical model is obtained for spin echo decay . The design of refresher conrol sequence for controller operating in supervisory mode is also given.It can be shown that the angular velocity for multi slice wavelets is given by $\omega=\Upsilon\int_{t0}^{tf}\Delta u_{rm}\partial t$; $\omega=[\omega_1\ \omega_2]$;$\omega_2=\partial\omega_1/\partial t$.Let $y_o(t_0)$ be the echo generated by unipolar gradient $\Delta u_{rm}(t_0)$, C_a be the coefficient of network .Resulting brownian motion output echo sequence $z_{oi}(t):g^{j,j-1}{}_i[u_{rm},\delta(t-k\pi/2)]$;$t_k=(k\,\Delta t)$ is given by

$$z_{oi}(t)=z_{oi}(t_0)\exp\{\Upsilon^2C_a-\int_0^{Tr}[\int_0^{\Delta t}\Delta u_{rm}dt]^2{}_{t-k\pi/2}dt_k\}$$
$$C_a=\ln\{[y_o(t_k)/y_o(t_0)]/\Upsilon^2(\int_0^{Tr}[\int_0^{\Delta t}\Delta u_{rm}dt]^2{}_{t-k\pi/2}dt)\} \quad (14)$$

5.3 Capillaries clutter-Hilbert transform model

In the following wavelet aggregate of offsite clutter of Arteries, Vains, and Capillaries is considered. The problem is formulated as design of time variance controller to regulate kienetics of radioactive process (figure 1) by amplitude and phase of maganetization of excited spin produced by random motion positrons emission. Let $z_o{}^j{}_i$ be the measure of flow in clutter of capillaries resulting from interaction of positrons. The exogenous system consisting of amplitude echo spin is given by

$$z_o{}^j{}_i(k)=\Sigma_i H^{j,j-1}{}_i(k)z_o{}^{j-1}{}_i(k)+C^j{}_i y_o{}^j{}_i(k)+w(k);\quad H^{j,j-1}{}_i(x,k/k_0)=f^{Tr}p_i(x,k/k_0)\{1+\exp(j\alpha)\}\exp\{j\Theta^s(x,k\Delta t)\}dx \tag{15}$$

where Fourier encoding matrix $H^{j,j-1}{}_i(x,k/k_0)$ is nonideal Hilbert Transformer, α is phase perturbation introduced by offsite controller network. The $p_i(x,k)$ probability distribution of velocity $x^j{}_{2i}$ for i th capillary is a measure of correlation of flow through Arteries, Vains, Capillaries and strength of positron emission. w(k) is zero mean gaussian disturbance process. In the following time multiplexed technique with Gradient (amplitude) encoded/subencoded transmittle frame is employed for signature recovery of subset targets in capillaries clutter (figure 1). Refresh pulses are employed for encoding select and handshake by restructuring command, data and status frames in extended data segments. The controller mapping space is extended by joint process$\{(u_{rm},z_o{}^j{}_i),x^j{}_i\}$. The $x^j{}_i$ is not Markov process, however the joint process $\{(u_{rm},z_o{}^j{}_i),x^j{}_i\}$ is a Markov process. Generated spin echo pulses $x^j{}_i(t_k)=\{\Sigma_k g^{j,j-1}{}_i[\vartheta^j{}_i(x,y,z),\delta(tk\pi/2)]\}\exp(-C_{df}t_k)$ are employed as successive select initiation pulses, here echo signal $g^{j,j-1}{}_i[\vartheta^j{}_i(x,y,z),\delta(t-k\pi/2)]$ is function of flow spin density distribution $\vartheta^j{}_i(x,y,z)$, C_{df} is envelope decay factor. As spin echo decays, the offsite controller initiates supervisory refresh pulse sequence for $x^j{}_i\subseteq S_{th}$, here S_{th} is threshold subset in Fourier space (figure 1). Proposed technique may be employed to initiate refresh pulses to service and to release active process to transfer control to supervisor. It may be verified Fourier space with spiral spin rotational frame is dense, Further $\{M_b(k)\}$ is reduced considerably.

Signature Recovery -Capillaries Clutter : From the consideration of network transmission losses and line reflection, refraction controller is operated in Auto negotiating mode. The refresher time T_r for target signature in clutter may be obtained from $T_r=-\tau_d\ln[(u_{rm}(t_k)-y_{th})/(u_{rm}(t_k)-y_o(t_k))]$; here y_{th}: sup $y_o(k)\in S_{th}$ is spin echo threshold, τ_d is signal envelope decay factor, $y(t_k)$ is the spin echo output at time $t(k+)$. It can be shown, the exogenous system with Fourier encoding matrix $\Sigma_i H^{j,j-1}{}_i(k)$ can be implemented on computer controlled system in real time by $z_o(k+1)=\{(2^{15}-32E_{df})z_o(k)+(32E_{df}*y_o(k+1))\}/2^{15}+w(k)$, where E_{df} is the envelope decay factor. Let $\Gamma:\Gamma_i\in d_s:\{r_d\in R^{[0,rmx]}\}$, be the disk Set of admissible signatures, r_{mx} is signature separation band of capillaries in clutter. Recovery process for signature of capillaries in clutter employing Hilbert Transform can be formulated as phase error minimization problem

$$J[\Theta^s{}_e]=\inf_{\Gamma\, i\in ds}E[\{\Theta^s{}_e(k\Delta t)\}^t Q\{\Theta^s{}_e(k\Delta t)\}];\quad \Theta^s{}_e(\omega)=\mathrm{sgn}[I_m\{\Theta^s(\omega)\}]R_e\{\Theta^s(\omega)\}-\mathrm{sgn}[R_e\{\Theta^s(\omega)\}]I_m\{\Theta^s(\omega)\} \tag{16}$$

The phase error minimization problem stated can be reformulated as design of self tuned control (figure 1) such that Set condition $2R_e[b^j{}_i u_{rm}+b^j{}_i\Sigma_i H^{j,j-1}{}_i(k)z_o{}^{j-1}{}_i(k)+C^j{}_i y_o{}^j{}_i(k)]\subset ds$ holds. It can be shown that signal energy level E_u received by target in clutter is given by

$$E_u=[R_e\{b^j{}_i u_{rm}+b^j{}_i(\Sigma_i H^{j,j-1}{}_i(k\Delta t)z_o{}^{j-1}{}_i(k\Delta t)+C^j{}_i y_o{}^j{}_i(k\Delta t))\}]^2+[I_m\{b^j{}_i u_{rm}+b^j{}_i(\Sigma_i H^{j,j-1}{}_i(k\Delta t)z_o{}^{j-1}{}_i(k\Delta t)+C^j{}_i y_o{}^j{}_i(k\Delta t))\}]^2 \tag{17}$$

The optimal control can be obtained from index $J[E_u]_{max}$. Proposed design is independent of initial values.

EXAMPLE (Air Born Refueling) : For Air Borne refueling system (figure 3) in axial coordinates, the rigid body dynamics fusion process observation descriptor for multitarget maneuvering in cluster, is given in figure 3. Transmitted frames are prefetched sequentially. From observation descriptor we can obtain $\Delta Y=\{\partial h(.)/\partial X\}X(k)+v$. Tracking of the maneuvering target is continued in the maneuvering mode (in line acceleration/inplane/out of plane rotation) until crossover of matrix Jacobian $J_x h$ threshold. After target identification, system maneuvers in constant velocity mode.

REFERENCES

Kang.H.S. (1996)"VLSI-embedded neural systems taxonomy of association"IFIP WG7.6 conf,Opt.based comp.aided design,Noisy-legrand,France

Kang.H.S. (1995),"Game of pursuit with zero stop probability ", IFAC -17th IFIP Conf on Syst modelling and opt., Prague, Czech Rep

Kang.H.S.(1997),"Modelling fabrication control of semiconductor interconnect Structure"2,IFAC symp on Robust cont. Budapest, Hungary.

Kang.H.S,(1973)"Sensitivity of the performance of optimal stochastic systems"Proc IEEE,No 2

Sowder.D(1976)"Control of systems subject to sudden changes in character"Proc IEEE,64,No 8,1219-25

THE OPTIMAL ADAPTIVE ALGORITHMS FOR ESTIMATION OF A DYNAMIC OBJECT STATE VECTOR WITH INCOMPLETE A PRIORI INFORMATION

Robert A. Ashiniants, Svetlana M. Ivanova

Moscow State Institute of Electronics and Mathematics (Technical University),
109028 Moscow Russia
E-mail: dmitry@bogolub.msk.ru

Abstract: Adaptive algorithms for optimal identification of a restoration system are described which allow to estimate the coordinates of a linear dynamic system from a restricted set of observable coordinates in the presence of noise. The algorithms are based on the stochastic approximation method. The estimations obtained converge to the optimal ones in the sense of minimum mean square deviation from the exact value.

Keywords: adaptation, control systems, convergence, identification algorithms, optimal estimation, sensitivity functions.

INTRODUCTION

The problem of restoration of a signal estimations in linear dynamic systems can be reduced to the classical filtration problem. Given a structure of a dynamic object, the estimations of the system phase coordinates with restricted number of observations in the presence of noise are determined by the structure of the optimal Kalman filter in the sense of minimum mean square error (Calman and Bucy, 1961).

$$\dot{\hat{x}}(t) = F\hat{x}(t) + K(t)(z(t) - H\hat{x}(t)), \qquad (1)$$

where $\hat{x}$ is the n-dimensional vector of estimations of phase coordinates, F is the $n{\times}n$ matrix of the initial parameters of the dynamic system, z is m-dimensional vector of observations in the presence of the Gaussian noise, H is the matrix of observation $n{\times}m$, K is the matrix of optimization coefficients $m{\times}n$. The K matrix is defined by the solution of the Riccati equation for the estimations variance P. The matrix coefficients are computed as follows:

$$K(t) = P(t)H^{T}R^{-1}, \qquad (2)$$

where R is $m{\times}n$ matrix of the observations noise variances values.

The problem statement described above and its classical solution has several features making the application of the solution algorithm too complicate for systems and observation conditions of the practical interest.

1. The exact knowledge of parameters in the initial dynamic system is often unknown in practice. Only the structure of the initial dynamic system up to the coefficients is known. In this case, the algorithm of optimal filtration does not provide any significant results, and the estimations can diverge.

2. The constraint in the form of the Gaussian distribution for the observation noise is rather severe. An assumption on the presence of the white noise with the normal distribution and a "good" intensity matrix R in the observed values of the available coordinates of the system state vector, is a strong idealization. If the initial problem statement differs from those used in the classical problem, the R matrix may be badly-posed, and then its inversion will cause additional errors in the estimations. In this case, the matrix

inversion belongs to the class of ill-posed problems.

3. The solution of the Riccati equation is a complicated computational problem. The computational resources expenditures and the time required to define the variances and therefore the coefficients of the K filter, make the practical implementation senseless. Taking into account the considerations in items 1, 2, we expect that the solution of the Riccati equation may diverge.

THE PROBLEM STATEMENT

The synthesis of the identification algorithms for the restoration system in the presence of noise is as follows:

- We assume that the structure and the equation describing the object are known:

$$\dot{x} = Fx + u, \tag{3}$$

where x is the n-dimensional vector of phase coordinates, F is the $n \times n$ matrix of object parameters, u is the l-dimensional vector of the input signals.

There is a class of objects whose parameters may vary in a wide range of values, or for which not all the parameters defining the object dynamics are known.

- The observed able coordinates of the object state vector are defined by the following equation:

$$z = Hx + v, \tag{4}$$

where H is the observation matrix $n \times m$, v is the additive m-dimensional stationary observation noise. However, there are no severe constraints for the noise characteristics.

Now we explain the term "restoration" used below. Using (4) we can observe (measure) only restricted number of coordinates. From equation (1) it follows that the system of optimal estimation contains the exact repetition of the object structure. System (1) guarantees obtaining the optimal estimations for the object state vector in the sense of minimum of square criterion in stationary conditions. Thus nothing prevents us to observe all the components of vector $\hat{x}$ which is a prototype of the described above vector x. Therefore the restoration is just obtaining the optimal estimations for non-observable coordinates of the object state vector. Equation (1) describes the closed system for the input signal z and it may be produced like control system with the tuning unit K.

- We note that the restoration system structure (the object model) is known and it is described by equation

$$\dot{\hat{x}} = F\hat{x} + u^{1}, \tag{5}$$

where u stands for the input signal of object in the model.

- Some elements of F-matrix which form the vector $a = (a_1, a_2, \dots a_n)$ are unknown. Obviously the problem of obtaining of the optimal equation a such wide statement has to be accompanied with solving of the identification problem. In so doing we solve the following restoration problem: to obtain the optimal estimation for the state vector and estimations for the object parameters, which is the identification problem.

- As an optimum criterion J we use the mean value for the loss function

$$J = \mathcal{M}\{W(z, \hat{x}(a, t))\}, \tag{6}$$

where W is a convex function differentiable in $\hat{x}$. We have to define such values a^* of parameters a, that provide the validity of the following equality:

$$\min_{a} J = J^* = J(\hat{x}^*(a^*))$$

The problem is reduced to computation of the mean value for the loss function and its subsequent minimization with respect to a. While solving the problem in the given statement, we shall use the square function for the loss function.

Thus, the solution of the synthesis problem is reduced to the parametric optimization, with the optimal tuning of the restoration system being accompanied by the object identification. It is obvious that the identification problem requires involvement of the sensitivity theory methods (Kokotovich, 1961; Rutman, 1968), and development of realizable algorithms for almost optimal in the sense of criterion (6) restoration system identification. In this work we use the stochastic approximation method (Nevelson and Hasmiskiy, 1972).

THE PROBLEM SOLVING

Ya. Z. Tsypkin showed possibility of the use of the sensitivity theory methods in case of synthesis of adaptive discrete systems with noise. The application of the stochastic approximation method for definition of the properties of dynamic objects described by difference equations is considered in (Tsypkin, 1970; Tsypkin, 1995). In

---

[1] The vector equation (5) has the structure of n equations of the first order or of single equation of the n-order with coefficients $a_0, a_1, \dots a_{n-1}$.

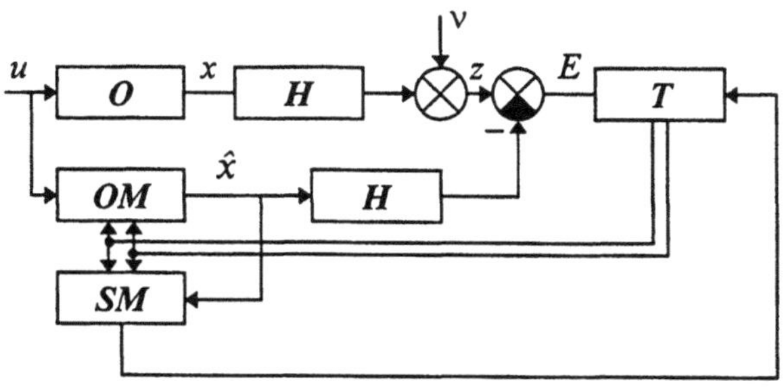

Pic. 1

several works by Ya. Z. Tsypkin's successors, the convergence of discrete adaptive algorithms is proved (Devyaternikov, et al., 1969). The convergence of continuous adaptive algorithms has not been considered however.

The system of restoration-identification is shown at the following block-scheme (pic. 1), where ***O*** stands for the object, ***H*** stands for the observation matrix, ***OM*** stands for the model of objects, ***SM*** stands for the sensitivity model, ***T*** stands for the tuner, *E* stands for the error signal.

If we denote *T* by *K* then the block-scheme shown in pic. 1 can be transformed into another block-scheme nearly repeating the equation (1) structure (pic. 2).

We note that the sensitivity model works during the period of tuning parameters. Hence, we choose the optimal structure for the restoration system described by equation (1). We define an extended vector *b* as, including the tuning parameters for the object model *a* and parameters *K*.

The optimum condition is achieved when the gradient of the functional (6) for the extended vector *b* is equal to zero:

$$\nabla J(b) = \mathcal{M}\{\nabla W(z, \hat{x}(b, t)\} = \mathcal{M}\{\nabla_b W\}=0,$$

or in the extended form:

$$\nabla_b J(t, b)= \mathcal{M}((\partial\hat{x}(t)/\partial b)^{T}\nabla_{\hat{x}} W(x(t), \hat{x}(t, b)))=0.$$

As it often appears in practice when the probability densities of the noise and the observations are not known, the gradient of the

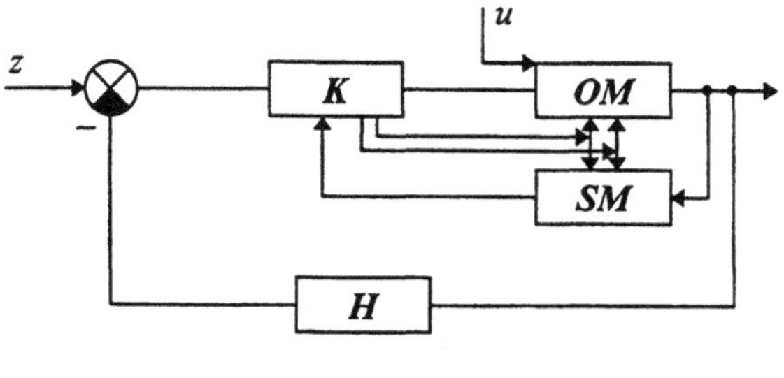

Pic. 2

mean loss $\nabla J(b)$ is not completely defined. In this case, the iterative algorithms cannot be used to get the optimal solution b^*. That is why the adaptive algorithms based on the stochastic approximation method are justified. The peculiarity of such algorithms is that instead of the gradient of the mean loss, the gradient of the loss function $\nabla_b W(z, \hat{x}(b, t))$ is used, which depends on the observation $z(t)$.

We introduce an designation *S* for the sensitivity function matrix $(\partial\hat{x}/\partial b)$ and use the square function for the loss function *W*. In doing so we obtain an algorithm for stochastic tuning of the parameters

$$\dot{b} = \Gamma(t)S^{T}H^{T}(z - H\hat{x}(t)), \tag{7}$$

where this superscript ^{T} means transposition.

Below we describe properties of matrix $\Gamma(t)$. Now we calculate the full estimation derivative

$$\dot{\hat{x}} = \frac{\partial\hat{x}}{\partial t} + \left(\frac{\partial\hat{x}}{\partial b}\right)\dot{b}. \tag{8}$$

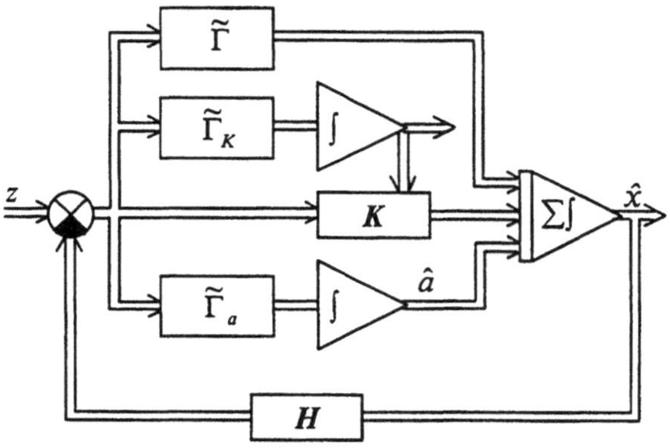

Pic. 3

Using (1) and (7) we obtain

$$\begin{aligned}\dot{\hat{x}} &= F\hat{x}(t) + K(t)(z(t) - H\hat{x}(t)) \\ &+ S(t)\Gamma(t)S^{T}(t)H^{T}(z - H\hat{x}(t)),\end{aligned} \tag{9}$$

We introduce the following notations

$$\tilde{\Gamma}(t) = S(t)\Gamma(t)S^{T}(t),$$

$$\tilde{\Gamma}_a(t) = S_a(t)\Gamma(t)S_a^{T}(t),$$

$$\tilde{\Gamma}_k(t) = S_k(t)\Gamma(t)S_k^{T}(t).$$

Then the tuner structure system will be as follows (pic. 3)

When $t \to \infty$ the third component of the right-hand part of (9) go to zero, $b \to b^*$, $\hat{x}(t) \to \hat{x}^*(t)$.

THE CONVERGENCE OF CONTINUOUS ADAPTIVE ALGORITHMS

We shall use the following notation:

$$\dot{\tilde{x}}_t = (F - K_t)\tilde{x}_t + u_t - Kv_t, \tag{10}$$

$$\dot{S}_t = (F - K_t)s_t + \tilde{x}_t - v_t. \tag{11}$$

Now one can formulate the following theorem for convergence of the algorithms obtained:

Theorem. Let the restricted continuous stochastic processes $\tilde{x}_t$ and s_t be described by equations (10) and (11), for which the following conditions are satisfied in the probability space $\{\Omega, \sigma, P\}$, where P is probability measure on σ-algebra of set Ω:

$$\begin{gathered} P\left\{\omega: \lim_{t\to\infty} \frac{1}{t}\int_0^t \tilde{x}(\omega,\tau)\, s(\omega,\tau)\, d\tau \le D < \infty\right\} = 1, \\ P\left\{\omega: \lim_{t\to\infty} \frac{1}{t}\int_0^t v(\omega,\tau)\, s(\omega,\tau)\, d\tau \le C < \infty\right\} = 1. \end{gathered} \tag{12}$$

And let some determined function $\gamma(t)$ be defined such that:

$$\begin{gathered} \gamma(0) > 0, \quad \int_0^\infty \gamma(t)\, dt = \infty, \\ \int_0^\infty \gamma^2(t)\, dt < \infty, \quad 0 < N_1 < t\cdot\gamma(t) < N_2. \end{gathered} \tag{13}$$

Then the solution of equations (7) and (9) exists with probability 1 and the following equality holds:

$$P\left\{\omega:\ \lim_{t\to\infty} K(\omega, t) = K^*\right\} = 1$$

with

$$P\{\omega: \mathcal{M}[\tilde{x}(\omega, K^*, t)] = 0\} = 1$$

To simplify the proof, we shall consider an one-dimensional object with determined parameters. The complete proof is given in Appendix.

The degree of the stochastic approximation algorithms convergence depends on the Γ matrix. Besides, the form of the Γ matrix defines the specific approximation method to be used. In particular, if the Γ matrix has the form $\Gamma = \gamma(t)I$ in which I is the identity matrix, $\gamma(t) > 0$, we obtain the gradient method.

The Γ matrix which is the reversed Hesse matrix of form

$$\Gamma(t) = [\nabla^2 J(b(t))]^{-1}$$

corresponds to the Newton method.

If the Γ matrix is diagonal of form

$$\Gamma = \gamma(t)B \tag{14}$$

in which B is a positively defined matrix, and $\gamma(t)$ satisfies the Robins-Monrous conditions:

$$\gamma(t) > 0, \quad \int_0^\infty \gamma(t)\, dt = \infty, \quad \int_0^\infty \gamma^2(t)\, dt < \infty.$$

For example, $\gamma(t)$ is of form $1/t$, we have the pseudo-gradient method of stochastic approximation. In this case, the algorithms are referred to as almost optimal ones.

It is obvious that the sensitivity function depends on the noise characteristics. Therefore, the adaptation algorithms with the best choice of the Γ matrix coefficients allow to decrease the impact of the noise on the optimal estimations, and at the same time to influence the degree of convergence for the estimations. Under the proper choice of the Γ matrix, it is possible to obtain the maximum degree of convergence. It is known that the maximum degree can be obtained on the base of the Kramer-Rao inequality (Liptzer and Shiryaev, 1974). However, to make it possible, the noise characteristics must be defined in order to construct the Fisher's information matrix. In (Tsypkin, 1995), the form of B matrix is obtained that corresponds to the maximum degree of convergence.

$$B = \frac{1}{\mathcal{M}\{W''[v(t)]_\varepsilon\}}\left[\mathcal{M}\left\{\left(\frac{\partial \hat{x}}{\partial b}\right)\cdot\left(\frac{\partial \hat{x}}{\partial b}\right)^{\mathrm{T}}\right\}\right]^{-1}, \tag{15}$$

where W is the square loss function and

$$\varepsilon = z - H\hat{x}(t).$$

With $b = b^*$ we have $\varepsilon(z(t), b^*) = v(t)$ and for multi-dimensional noise v $\mathcal{M}\{W''(v(t))\}_\varepsilon$ is the identity matrix.

The implementation of the stochastic approximation algorithms with F matrix of the form (14) is complicated due to the matrix inversion. This problem can be avoided by the use of an additional differential equation with respect to F. For this purpose, we shall introduce an empirical quality functional with the same form of the loss function W:

$$J = \frac{1}{t}\int_0^t W(z(\tau), \hat{x}(b(t)))\, d\tau \tag{16}$$

Having differentiated the gradient of the functional (16) by t and made several transformations we get the following:

$$\Gamma_{opt}\left[\left(\frac{\partial\hat{x}}{\partial b}\right)^{T}\left(\frac{\partial\hat{x}}{\partial b}\right)t\right] = I. \qquad (17)$$

The differential equation for Γ_{opt} takes the form:

$$\dot{\Gamma}_{opt} = -\Gamma_{opt}(\dot{S}^{T}St + S^{T}\dot{S}t + S^{T}S)\Gamma_{opt}. \qquad (18)$$

The solution of this matrix equation depends on the initial condition Γ_0. However, since in the limit $t \to \infty$ the impact of the initial conditions on the convergence of the algorithms practically disappears, the exact definition of the initial conditions becomes unnecessary. Thus, the system of equations (7), (9) presents the optimal adaptive algorithm for obtaining the estimations of the coordinates of the object state vector. In the course of obtaining the estimations, the object identification is carried on.

The restoration system includes a rather complex and resource consuming sensitivity model that contains the object model as its part. To simplify the sensitivity model, we suggest to use the sign functions instead of the exact ones. As the sign of the sensitivity function in algorithms (7), (9) is determined by the direction of vector $\hat{x}$ and coefficients b variation (i.e. the change of the corresponding derivatives signs), it is possible to use the sign sensitivity function.

$$\widetilde{S} = \operatorname{sign}\dot{\hat{x}} / \operatorname{sign}\dot{b}$$

Such sensitivity function can be implemented with simple logic elements. The output signal of such element is +1 if the signs of $\dot{\hat{x}}$ and $\dot{b}$ coincide, and is −1 if the signs differ. The advantage of the sensitivity model obtained is that there are no dynamic elements with unknown parameters that have to be identified in the course of the adaptation. It is obvious that with such choice of sensitivity function, the degree of convergence of algorithms (7), (9) decreases. Thus, with considerable simplification of the restoration system its quality retains (in the sense of the optimum criterion), but the degree of convergence fails. This shortcoming can be reduced by the use of fast adaptation algorithms (Tsypkin, 1995).

APPENDIX

Proof of the theorem of convergence. Let's choose positive numbers ε_1 and ε_2 and we shall designate t_0 such moment, that for all of $t > t_0$ and almost all of $\omega \in \Omega$ (except a set of a measure 0) the inequalities take place

$$J_1(t) = \left|\frac{1}{t}\int_0^t \widetilde{x}(\omega,\tau)\, s(\omega,\tau)\, d\tau - D\right| < \varepsilon_1, \qquad (19)$$

$$J_2(t) = \left|\frac{1}{t}\int_0^t v(\omega,\tau)\, s(\omega,\tau)\, d\tau - C\right| < \varepsilon_2. \qquad (20)$$

Let's integrate (7) in limits from t up to $t + \Delta t$

$$K(t+\Delta t) - K(t) = \int_t^{t+\Delta t} \gamma(\tau)\, s(\omega,\tau)\, [\widetilde{x}(\omega,\tau) + v(\omega,\tau)]\, d\tau. \qquad (21)$$

Let's square both parts

$$\begin{aligned}\sigma^2(\tau,\Delta\tau) &= \left[\int_t^{t+\Delta t}\gamma(\tau)\, s(\omega,\tau)\,\widetilde{x}(\omega,\tau)\, d\tau + \int_t^{t+\Delta t}\gamma(\tau) s(\omega,\tau) v(\omega,\tau)\, d\tau\right]^2 \\ &\le N_2^2\left[\int_t^{t+\Delta t}\frac{1}{\tau} s(\omega,\tau)\,\widetilde{x}(\omega,\tau)\, d\tau + \int_t^{t+\Delta t}\frac{1}{\tau} s(\omega,\tau)\, v(\omega,\tau)\, d\tau\right]^2. \end{aligned} \qquad (22)$$

Integrals in square brackets we shall take piecemeal, taking into account inequalities (19) and (20).

$$\begin{aligned}\sigma^2(t,\Delta t) \le N_2\Bigg[&\frac{1}{t+\Delta t}\int_0^{t+\Delta t}\widetilde{x}(\omega,\tau) s(\omega,\tau)\, d\tau - \frac{1}{t}\int_0^{\tau} s(\omega,\tau)\,\widetilde{x}(\omega,\tau)\, d\tau + \int_t^{t+\Delta t}\frac{1}{\tau^2}\int_0^{\tau}\widetilde{x}(\omega,\lambda)\, s(\omega,\lambda)\, d\lambda d\tau \\ &+ \frac{1}{t+\Delta t}\int_0^{t+\Delta t} s(\omega,\tau)\, v(\omega,\tau)\, d\tau - \frac{1}{t}\int_0^{\tau} s(\omega,\tau)\, v(\omega,\tau)\, d\tau \\ &+ \int_t^{t+\Delta t}\frac{1}{\tau^2}\int_0^{\tau} s(\omega,\lambda)\, v(\omega,\lambda)\, d\lambda d\tau\Bigg]^2 \\ \le N_2^2\Bigg[&2\varepsilon_1 + (D+\varepsilon_1)\ln\frac{t+\Delta t}{t} + 2\varepsilon_2 + (C+\varepsilon_2)\ln\frac{t+\Delta t}{t}\Bigg]^2. \end{aligned}$$

And $\lim_{t\to\infty}\sigma^2(t,\Delta t) = 0$. Hence, with probability 1 there is a limit.

Let's show, that $\lim_{t\to\infty} K(\omega,t) = K^*$ for all $\omega \in \Omega$.

Let's assume an inverse

$$a)\ \lim_{t\to\infty} K(\omega,t) = \theta_1 \le \lambda_1 < K \qquad (23)$$

$$b)\ \lim_{t\to\infty} K(\omega,t) = \theta_2 \le \lambda_2 > K \qquad (24)$$

Let's show, that the suppositions (23) reduce in an inconsistency.

Let for each of t_0 there is such of t', that for all of $t > t'$ an inequality is correct.

$$K(\omega,t) \le \lambda_1 < K^*. \qquad (25)$$

Because of continuities of processes $\tilde{x}_t$ and s_t for a fixed value t' we have

$$K(\omega,t)-K(\omega,t')=\int_{t'}^{t}\gamma(\tau)\,\tilde{x}(\omega,\tau)\,s(\omega,\tau)\,d\tau$$
$$+\int_{t'}^{t}\gamma(\tau)\,\nu(\omega,\tau)\,s(\omega,\tau)\,d\tau$$
$$\geq N_1\int_{t'}^{t}\frac{1}{\tau}\tilde{x}(\omega,\tau)\,s(\omega,\tau)\,d\tau+\int_{t'}^{t}\frac{1}{\tau}\nu(\omega,\tau)\,s(\omega,\tau)\,d\tau$$

Because of lemmas from Driml and Nedoma (1960) we have

$$K(\omega,t)-K(\omega,t')\geq N_1\left[\left(D-\varepsilon_1\right)\ln\frac{t}{t'}\right.$$
$$\left.+\left(C-\varepsilon_2\right)\ln\frac{t}{t'}\right]$$

Thus, that with fixed t' and $t\to\infty$ $K(\omega, t')\to\infty$, but it contradicts a condition (25). The inconsistency of the supposition (24) is similarly proved. Hence, $\lim_{t\to\infty} K(\omega,t)\to K^*$ for all of $\omega\in\Omega$.

It is necessary to show, that $\mathcal{M}\{\tilde{x}(K^*, t)\} = 0$, i.e. $\hat{x}(t, K^*)$ is unbiassed estimation $x(t)$.

Let impulse transition function of the equation (9) is $\psi(t,\tau)$. Because of continuous convergence $K(t)$, continuity and boundedness of process $\tilde{x}_t$, the impulse function is limited and $\lim_{t\to\infty}\psi(t,\tau)=0$. The solution of the equation (9) is

$$\tilde{x}(t)=\psi_{t_0}^{t}\left(\tilde{x}_0+\int_{t_0}^{t}\left(\psi_{t_0}^{\tau}\right)^{-1}\left[u(\tau)-K^*\nu(\tau)\right]d\tau\right)$$

By taking expectation from both parts, we shall receive $\lim_{t\to\infty}\mathcal{M}\left\{\tilde{x}(K^*,t)\right\}\to 0$. The proof is over.

REFERENCES

Calman R. and R. Bucy (1961). New Results in Linear Predication and Filtering Theory. *J. Basic Engr. (Trans. ASME, Ser, D),* **83D.**

Devyaternicov I.P., A.I. Kaplinskiy and Ya.Z. Tsypkin (1969). On the Convergence of the Teaching Algorithms. *Avtomatika i Telemechanika.* **Vol. 10.** In Russian.

Driml M. and J. Nedoma (1960). Stochastic Approximation for Continuous Random Processes. *Tr. Of the 2 Prague conf. Inf. Theory, Stat. Decision Function Random Processes.* Czechoslovak Academy of Science, Prague.

Kokotovich P. (1964). Method of Three Points in the Research and Optimization of Linear Control System. *Avtomatika i Telemechanika.* **Vol. 12.** In Russian.

Liptzer, R.Sh. and A.N. Shiryaev (1974). *Statistics of the probabilistic process.* Nauka. Moscow. In Russian.

Nevelson, Hasminskiy (1972). *Stochastic Approximation and Recurrence Estimation.* Nauka. Moscow. In Russian.

Rutman R.S. (1968). Method of Three Points in the Sensitivity Theory. *Izvestiya AN SSSR. Tekhnicheskaya Kibernetika.* **Vol. 4.** In Russian.

Tsypkin, Ya.Z. (1970). *Base of the Theory of the Teaching Systems.* Nauka. Moscow. In Russian.

Tsypkin, Ya.Z. (1995). *Information Theory of the Identification.* Nauka. Moscow. In Russian.

STRUCTURES AND METHODS OF STOCHASTIC OPTIMAL CONTROL UNDER UNCERTAINTY

V.V. Baranov[*],
V.I. Salyga[**]

[*]*SNDRC of RAS*
109147 Moscow, Russia
[**]*INTERNA UKA Centre*
113833 Moscow, Russia

Abstract: In this paper we expose the problem of stochastic optimal control in complex systems at different variants of uncertainty.

Keywords: Complex systems, decision making, Markov models, stochastic control, uncertainty.

The topic is the task of stochastic optimal control in complex systems at different variants of uncertainty. The system approach to the problem is developing, according to this approach the meanings are introduced: base, environment and solver. The base is the determining part of the structure and permits different variants of its task, which constructions of environment and solver depend of. The meaning of fullness of base is introduced and its general composition, which is concretized then depending of suppositions on conditions of decision making and their results, is determined. That creates a set of variants of bases of different complexity, among them markovian base and the primary markovian base, which have a set of useful attributes, are selected. First of all they are the simplest among the other variants of bases, but their main features diffuse to the other more general bases. That permits to investigate at first the solve-ability of the problem in conditions of markovian bases and to build for them appropriate solvers, and then to add and diffuse results to the other more complicated variants of bases. That opens the possibility to develop general methodology of structurization, formalization and building of optimizational methods for systems of decision making with different bases. This methodology permits to investigate the problem of dynamic decision making not at particular suppositions of concrete tasks, but as a general problem for sufficiently broad class of systems marked by appropriate postulates and bases.

Different variants of bases depending of suppositions on incertainty of conditions of decision making and their outlets were investigated. Their fullness is found out and the appropriate solvers are built.

The developing methodology is based on the following main constructions and results.

It is established a regular bases determined by a set of objects

$$\mathbf{SBDR} = \{\Theta,\ G,\ X,\ Y,\ [Y_x \subset Y, x \in X],\ Z,\ P_{(\theta,g)}(Z|X),\ Q_{(\theta,g)}(X|X\times Z\times Y),\ w_{(\theta,g)}(Y\times Z\times X),\ (\theta,g) \in \Theta\times G\}, \tag{1}$$

where Θ – multitude of strategic alternatives; G – multitude of tactic alternatives; X – multitude of situations; Y – multitude of operational alternatives; $Y_x \subset Y$ – limits for admissibility of operational alternatives depending of situations $x \in X$; Z – multitude of outlets, $P_{(\theta,g)}(Z|X)$ – distribution of connection; $Q_{(\theta,g)}(X|X\times Z\times Y)$ – transitional function from $X\times Z\times Y$ to X; $w_{(\theta,g)}(Y\times Z\times X)$ – function of usefulness that represents preferences to Y in condition $Z{\times}X$.

If in this base we suppose, that $X \equiv Z$, markovian base **SBDRM** takes place, and in this base the multitude of situations coincides to the multitude of outlets. In this case situations have a meaning of states.

If strategic and tactic imperatives exist, that base comes to the primary markovian base

$$\mathbf{SBDM} = \{X, Y, Q(X|X{\times}Y), w(X{\times}Y)\}. \qquad (2)$$

Markovian and the primary markovian bases play important role in the proving of fullness of arbitrary base according to the following result.

Theorem. *Let the base comes by some reorganizations to markovian or to the primary markovian base, or it induces some joining succession of the primary markovian bases. Then initial base is full.*

This theorem opens the way of supplementing incomplete bases. For example, if in a regular base sutuations are not accessible for observation, such base is not full. Its supplementing is made by entering of apriory distribution $\alpha(X)$ to X. That permits to bring initial base to markovian base of aposteriori risk.

The variants of deeper uncertainty appear if some structural objects of initial base apriory are not set. Then supplementing of base is made by entering of multitude of hypothesa on unknown object of the base. And hypothesa are used as a component of operational alternative or as a tactic alternative. For these conditions game ideology and constructive methods of successive identification and adaptive best in the meaning of balance decisions making are developing. The results are distributing to complex systems with the hierarchy structure.

MINIMAX FILTERING IN UNCERTAIN-STOCHASTIC SYSTEMS DESCRIBED BY STOCHASTIC DIFFERENTIAL EQUATIONS WITH A MEASURE

Andrey V. Borisov *and* **Alexei R. Pankov**

Department of Applied Mathematics
Moscow State Aviation Institute
4, Volokolamskoje sh., Moscow 125871, Russia
e-mails: Borisych@k804.mainet.msk.su and Pankov@k804.mainet.msk.su

Abstract: The problem of estimation of the random processes described by stochsatic differential equation with a measure in presence of unknown but partially observable input signals under incomplete *a priori* information about the distribution characteristics of random noises is considered. The solution of the problem is based on the theory of minimax statistical estimation of random elements with values in separable Hilbert spaces.

Keywords: stochastic differential equation, measure, uncertain random process, minimax filtering.

1. PROBLEM STATEMENT

Let a random process under consideration $\theta(t)$ be described by a system of stochastic differential equations with a measure of the following type

$$\theta(t) = \theta_0 + \int_{(0,t]} a(\tau)\theta(\tau-)d\mu(\tau) + \int_{(0,t]} b(\tau)u(\tau-)d\mu(\tau) + \int_{(0,t]} c(\tau)d\xi(\tau) + w(t), \quad (1)$$

where $\mu(t)$ is a nondecreasing function of bounded variation with jumps at the time instants $\{\tau_i\}$; $a(\cdot)$, $b(\cdot)$ and $c(\cdot)$ are known matrix-valued functions with piecewise-continuous components; θ_0 is a random initial condition for $\theta(t)$ with unknown expectation and covariance matrix; $u(\cdot)$ is a square integrable random process with unknown expectation and covariance function (i.e., uncertain process); $w(t)$ is a centered martingale with unknown but bounded from above quadratic characteristic; $\xi(t)$ is an observable random process connected with $\theta(t)$ by means of the following equation:

$$\xi(t) = A_0\theta_0 + W_0 + \int_{(0,t]} A(\tau)\theta(\tau-)d\mu(\tau) + \int_{(0,t]} B(\tau)u(\tau-)d\mu(\tau) + W(t). \quad (2)$$

where $A(\cdot)$,$B(\cdot)$ are known piecewise-continuous matrix-valued functions; W_0 is a centered random vector with unknown but bounded from above covariance matrix $R_0 = cov(W_0, W_0) \leq \overline{R}_0$, $\overline{R}_0$ is a positive semidefinite symmetric matrix; $W(t)$ is a centered martingale with bounded from above quadratic characteristic. We suppose θ_0 and W_0 to be independent of $w(\cdot)$ and $W(\cdot)$.
The considered martingale $\eta(t) = col(w(t), W(t))$ describes a random disturbance with partially unknown characteristics. Always below we assume the $\eta(t)$ quadratic characteristic to satisfy the following matrix inequality

$$R_\eta(t) \leq \overline{R}_\eta(t) = \begin{bmatrix} \overline{R}_w(t) & \overline{R}_{wW}(t) \\ \overline{R}_{Ww}(t) & \overline{R}_W(t) \end{bmatrix} \quad (3)$$

where the known matrix-valued function $\overline{R}_\eta(t)$ is supposed to be absolutely continuous with respect to $\mu(t)$.
Let us introduce a random element γ with components θ_0, W_0, $u(\cdot)$ and $\eta(\cdot)$ which are supposed to be independent, and satisfy all listed above conditions. Denote Γ the set of all possible distribution laws $p_\gamma(\cdot)$ of γ. Hence Γ defines the rate of uncertainty in defining the processes $\theta(\cdot)$, $\xi(\cdot)$ by the equations (1),(2).

Our problem is to construct the linear filter for the process $\theta(t)$ estimation given the observation process $\xi(t)$, $t \in [0,T]$, which provides us the optimal in the minimax sense estimate of $\theta(t)$. The general expression for such an estimate is

$$\hat{\theta}(t) = \phi_0\xi(0) + \int_{(0,t]} \phi_1(t,\tau)d\xi(\tau) + \qquad (4)$$

$$+ \int_{(0,t]} \phi_2(t,\tau)\hat{\theta}(\tau-)d\mu(\tau),$$

where ϕ_0, $\phi_1(\cdot)$ and $\phi_2(\cdot)$ should be chosen to satisfy the optimality condition

$$\begin{aligned} &\sup\nolimits_{p\in\Gamma} E_p\|\theta(t) - \hat{\theta}(t)\|^2 \leq \\ &\leq \sup\nolimits_{p\in\Gamma} E_p\|\theta(t) - \tilde{\theta}(t)\|^2, \quad \forall t \in [0,T], \end{aligned} \qquad (5)$$

where $\tilde{\theta}(t)$ is an arbitrary linear esimate of $\theta(t)$ given $\xi(\cdot)$.

Note that using the ordinary mean-square optimality criterion for filtering of random processes under *a priori* uncertainty is incorrect because of criterion value dependance on the unknown distribution $p_\gamma(\cdot)$ of γ. Obviously, the purely stochastic filtering problem is a particular case of the minimax one. Indeed, if the uncertainty set Γ contains only unique element $p_\gamma = p^*$, the minimax filtering problem coincides with the ordinary one.

It should be mentioned, that the constraint for $R_\eta(t)$ presumes the observation noises could be degenerate. The corresponding filtering problem for purely stochastic linear systems has been solved (Miller and Rubinovich, 1995) for special class of observation systems. In our paper this solution is extended to the case of uncertain-stochastic differential systems described by the linear stochastic differential equations with a measure, and generalizes the corresponding results for minimax estimation in ordinary stochastic differential equations presented in (Pankov and Borisov, 1994).

2. MINIMAX ESTIMATION OF RANDOM ELEMENTS

It can be easily shown that the stated above problem is a particular case of the more general problem of the minimax estimation of uncertain-random elements with values in the separable Hilbert spaces (H-spaces). In this section we formulate and solve the corresponding estimation problem which is an extension of the one considered in (Borisov and Pankov, 1996).

Let $\mathcal{L}_2(H)$ be the H-space of square-integrable random elements $\theta(\omega)$, $\omega \in \Omega$ with values in H-space H, and defined on the complete probability space $(\Omega, \mathbf{F}, \mathbf{P})$, with scalar product $<\theta,\xi>_{\mathcal{L}_2(H)} = E\{<\theta,\xi>_H\}$, and corresponding norm $\|\theta\|_{\mathcal{L}_2(H)}$. Let also $\mathcal{L}(H_i, H_j)$ be a space of linear bounded operators from H_i into H_j, and $\mathcal{L}_1^+(H)$ be a set of all linear positive-semidefinite selfadjoint kernel operators on H.

Consider the random elements $u \in \mathcal{L}_2(H_1)$ and $\eta \in \mathcal{L}_2(H_2)$ which are partially observable according to the following linear scheme

$$\xi = \Phi u + \Lambda\eta, \qquad (6)$$

where $\xi \in \mathcal{L}_2(H_3)$ is an observable random element; $\Phi \in \mathcal{L}(H_1, H_3)$, $\Lambda \in \mathcal{L}(H_2, H_3)$ are known operators. Concerning u and η the following *a priori* informaion is available:

(i) $E\{u\} = m_u \in H_1$, and $cov(u,u) = K_u \in \mathcal{L}_1^+(H_1)$ exist but unknown;

(ii) $E\{\eta\} = 0$, and the covariance operator $K_\eta = cov(\eta,\eta)$ is unknown but bounded from above, i.e. $K_\eta \leq \overline{K}_\eta$, where $\overline{K}_\eta \in \mathcal{L}_1^+(H_2)$ (note, that for $A, B \in \mathcal{L}_1^+(H)$ the condition $A \leq B$ means that $B - A \in \mathcal{L}_1^+(H)$), and $\overline{K}_\eta$ is supposed to be known.

The random element $\theta \in \mathcal{L}_2(H_4)$ to be estimated is connected with u and η by means of the linear infinite-dimensional model

$$\theta = Au + B\eta, \qquad (7)$$

where $A \in \mathcal{L}(H_1, H_4)$, and $B \in \mathcal{L}(H_2, H_4)$ are some known operators.

The model (6), (7) is, in fact, infinite-dimensional linear regression model with uncertain (u) and uncertain-stochastic (η) parameters. We suppose below u and η to be uncorrelated. Let us consider a random element $\gamma = col(u,\eta)$, the exact distribution $p_\gamma(\cdot)$ of which satisfy the stated above conditions. We denote the corresponding set of all admissible distributions as Γ.

Definition 1 The estimationg operator $\psi(\xi)$ for θ given the observations ξ is admissible if there exists a sequence of operators $\{\psi_n\}_{n=1}^\infty$, $\psi_n \in \mathcal{L}(H_3, H_4)$ such that $E\{\|\psi_n(\xi) - \psi(\xi)\|_{H_4}^2\} \to 0$ as $n \to \infty$ for every $p_\gamma(\cdot) \in \Gamma$.

Obviously, the estimate $\hat{\theta}$ is admissible if $\hat{\theta} = \hat{\psi}(\xi)$ where $\hat{\psi} \in \Psi$, and Ψ is a set of all admissible estimating operators for the uncertainty set Γ.

Definition 2 The estimate $\hat{\theta} = \hat{\psi}(\xi)$, $\hat{\psi} \in \Psi$ is called the minimax-optimal one if

$$\sup_{p_\gamma(\cdot)\in\Gamma} E\{\|\theta - \hat{\theta}\|_{H_4}^2\} \leq \sup_{p_\gamma(\cdot)\in\Gamma} E\{\|\theta - \tilde{\theta}\|_{H_4}^2\} \quad (8)$$

for any $\tilde{\theta} = \tilde{\psi}(\xi)$, $\tilde{\psi} \in \Psi$. Note, that if the solution to (8) exists then $\hat{\theta} = \hat{\psi}(\xi)$, where

$$\hat{\psi}(\cdot) = arg \min_{\tilde{\phi}(\cdot)\in\Psi} \sup_{p_\gamma(\cdot)\in\Gamma} E\{\|\theta - \tilde{\phi}(\xi)\|_{H_4}^2\}.$$

Let G be a densely defined linear operator (Beutler and Root, 1976). Denote a continuous closure of G as $cl[G]$. Let also Φ^+, be a pseudoinverse operator of Φ and $l.i.m._{n\to\infty}\theta_n = \theta$ means that θ_n,

$\theta \in \mathcal{L}_2(H)$, and $||\theta_n - \theta||_{\mathcal{L}_2(H)} \to 0$ as $n \to \infty$.

Theorem 1 Let the following conditions hold

$$cl[A\Phi^+]\Phi = A, \quad (9)$$
$$cl[A\Phi^+] \in \mathcal{L}(H_3, H_4) \quad (10)$$

then the minimax-optimal estimate $\hat{\theta}$ is given by

$$\hat{\theta} = cl[A\Phi^+]\xi + \quad (11)$$
$$+ l.i.m._{n\to\infty} U\overline{K}_\eta \Lambda^* P (PDP + \gamma_n I)^{-1} P\xi,$$

where $D = \Lambda \overline{K}_\eta \Lambda^*$, $U = B - cl[A\Phi^+]\Lambda$, $P = cl[I - \Phi\Phi^+]$, and $\gamma_n \downarrow 0$ as $n \to \infty$. The estimate $\hat{\theta}$ is unbiased, and the covariance operator of its error $K_\Delta = cov(\theta - \hat{\theta}, \theta - \hat{\theta})$ satisfies the inequality

$$K_\Delta \leq \overline{K}_\Delta = \quad (12)$$
$$= U\overline{K}_\eta U^* - U\overline{K}_\eta \Lambda^*(PDP)^+ \Lambda \overline{K}_\eta U^*.$$

The worst situation for estimation of θ is $K_\eta = \overline{K}_\eta$. In this case $K_\Delta = \overline{K}_\Delta$.

The result presented in Theorem 1 generalizes the corresponding one which has been obtained for the purely stochastic model without uncertain component (Borisov and Pankov, 1996).

Equations (11), (12) make possible to obtain the estimate $\hat{\theta}$ in a strightforward manner, but often it is much more convenient to check the necessary and sufficient conditions to show the estimate $\hat{\theta}$ is minimax one. The corresponding conditions are presented in the following theorem.

Theorem 2 Let $\{\psi_n : \psi_n \in \mathcal{L}(H_3, H_1)\}$ be an arbitrary sequence of operators such that $\exists\ \nu = l.i.m._{n\to\infty}(\xi - \Phi\psi_n\xi)$ for any $p_\gamma(\cdot) \in \Gamma$ with a property $K_\eta = \overline{K}_\eta$, and $E\{\eta\} = 0$. The estimate $\hat{\theta} = \hat{\psi}(\xi)$, $\hat{\psi} \in \Psi$ is a minimax-optimal one iff

$$E\{\theta - \hat{\theta}\} = 0, \quad (13)$$
$$cov(\theta - \hat{\theta}, \nu) = \overline{0}. \quad (14)$$

The corresponding result for finite-dimensional case has been presented in (Borisov and Pankov, 1996).

Remark. It can be shown, that it is enough to check (13), (14) only for one arbitrary sequence $\{\psi_n\}$ and corresponding limit ν, since the result will be the same for all other sequences of described type.

The conditions (13), (14) provide an opportunity to prove the optimality of the estimate $\hat{\theta}$, which has been obtained from, for instance, empirical considerations by means of checking its unbiasedness (13) and the generalized condition of Wiener-Hopf (14). Exactly this approach has been used to construct the minimax -optimal filtering algorithm for the model (1), (2).

3. THE STRUCTURE OF MINIMAX FILTERING ALGORITHM

Assume the following regularity conditions to be fulfilled:

R1: there exist the finite matrix-valued functions $s(t)$ and $q(t)$ such that $\mu-$ everywhere

$$A(t) = \frac{d\overline{R}_W(t)}{d\mu(t)} s(t), \quad and \quad b(t) = q(t)B(t),$$

R2: $A_0^+ A_0 = I$

then the estimate $\hat{\theta}(t)$ which satisfies (4) and (5), exists and is given by the differential equation

$$\hat{\theta}(t) = \hat{\theta}_0 +$$
$$+ \int_{(0,t]} (c(\tau) + c^1(\tau) + \alpha(\tau) + \beta(\tau)) d\xi(\tau) + \quad (15)$$
$$+ \int_{(0,t]} (a^1(\tau) - (\alpha(\tau) + \beta(\tau))) A(\tau)\hat{\theta}(\tau-) d\mu(\tau),$$

where the functions $a^1(\tau)$, $b^1(\tau)$, $\alpha(\tau)$ and $\beta(\tau)$ are as follows:

$$a^1(\tau) = a(\tau) - c^1(\tau)A(\tau),$$
$$b^1(\tau) = b(\tau) - c^1(\tau)B(\tau)\beta(\tau),$$
$$\beta(\tau) = b^1(\tau)B^+(\tau),$$
$$\alpha(\tau) = \{[I + (a^1(\tau) -$$
$$-\beta(\tau)A(\tau))\Delta\mu(\tau)]k(\tau-)A^*(\tau) - \beta(\tau)\overline{R}_W(\tau)\} \times$$
$$\times \left[(I - B(\tau)B^+(\tau))\, Z(\tau)\, (I - B(\tau)B^+(\tau))\right]^+.$$

Here we denote

$$Z(\tau) = A(\tau)k(\tau-)A^*(\tau)\Delta\mu(\tau) + \overline{R}_W(\tau),$$
$$\Delta\mu(\tau) = \mu(\tau) - \mu(\tau-),$$
$$c^1(t) = \frac{d\overline{R}_{wW}(t)}{d\mu(t)} \left[\frac{d\overline{R}_W(t)}{d\mu(t)}\right]^+.$$

The matrix-valued function $k(t)$ defines the guaranteed accuracy of the estimate $\hat{\theta}(t)$, and satisfies the generalized Riccati equation of the following form

$$k(t) = k_0 + \int_{(0,t]} (G(\tau)k(\tau-) + k(\tau-)G^*(\tau) +$$
$$+\beta(\tau)\overline{R}_w(\tau)\beta^*(\tau) + \overline{R}_W(\tau) - \alpha(\tau)Z(\tau)\alpha^*(\tau)) d\mu(\tau) +$$
$$+ \sum_{\tau_i \leq t} G(\tau_i)k(\tau_i-)G^*(\tau_i)\Delta\mu^2(\tau_i), \quad (16)$$

where $G(\tau) = a^1(\tau) - \beta(\tau)A(\tau)$.

The initial conditions $\hat{\theta}_0$ and k_0 for (15) and (16), respectively, are as follows:

$$\hat{\theta}_0 = A_0^+ D\xi(0), \quad (17)$$
$$k_0 = A_0^+ D\overline{R}_0 (A_0^+)^*, \quad (18)$$

where $D = I - \overline{R}_0 \left((I - A_0A_0^+)\overline{R}_0(I - A_0A_0^+)\right)$.

Now we summarize the main result in the following Theorem.

Theorem 3. Let the conditions $R1$, $R2$ hold, then the minimax-optimal estimate $\hat{\theta}(t)$ for the uncertain-stochastic process $\theta(t)$ exists and is given by the equations (15)-(18), and has the properties: $E\{\theta - (t)\hat{\theta}(t)\} = 0$, $k_\theta(t) = cov(\theta(t) - \hat{\theta}(t), \theta(t) - \hat{\theta}(t)) \leq k(t)$, where $k(t)$ is defined by (16), (18).

Note, the function $k(t)$ is nonrandom since it does not depend upon the observations $\{\xi(\cdot)\}$. Hence, it is possible to estimate from above an accuracy of the estimate $\hat{\theta}(t)$ *a priori*.

The proof of Theorem 3 requires to check the conditions of Theorem 2, where $\theta = \theta(t) \in \mathcal{L}_2(R^n)$, $u = \{u(t)\}$ is the square-integrable random process with unknown moment characteristics, $\xi = \{\xi(\tau)\}_{\tau \in [0,t]} \in \mathcal{L}_2(\mathcal{L}_2^m[0,t])$ is an observable uncertain-stochastic $m-$dimensional process, $\eta = \{\eta(\tau)\}_{\tau \in [0,t]}$, where $\eta(\tau) = col(w(\tau), W(\tau))$, is a square-integrable centered martingale with unknown but bounded quadratic characteristic $R_\eta(\tau) \leq \overline{R}_\eta(\tau)$, and $\overline{R}_\eta(0) = 0$, $tr(\overline{R}_\eta(T)) < \infty$.

To conclude this paper let us note that for purely stochastic system with completely known characteristics, the equations (15)-(18) coincide with the equations of the mean-square optimal filtering presented in (Miller and Rubinovich, 1995). In the case $b(t) \equiv 0$, $B(t) \equiv 0$, $\theta_0 = 0$, and $\overline{R}_W(t) = \overline{R}_W^N(t) + \epsilon^2 I$, where $\overline{R}_W^N(t)$ is some known "nominal" quadratic characteristic, the equations (15)-(18) provide the estimate $\hat{\theta}^\epsilon(t)$ which coincides with the regularized mean-square optimal estimate for the purely stochastic differential system with degenerated observation noises (Liptser and Shiryayev, 1978).

REFERENCES

Beutler, F.J. and W.L. Root (1976). The operator pseudoinverse in control and system identification, In: *General inverses and applications*(Nashed M.Z., (Ed.)), 397-494. Academic Press, New York.

Borisov, A.V. and A.R. Pankov (1996). Problems of minimax estimation of random elements with values in Hilbert spaces, *Automation and remote control*, **6**, 61-75.

Liptser, R.Sh. and A.N. Shiryayev (1978). *Statistics of random processes.* Springer Verlag, New York.

Miller, B.M. and E.Ya. Rubinovich (1995). Regularization of a generalized Kalman filter. *Mathematics and computers in simulation,* **39**, 87-108.

Pankov, A.R. and A.V. Borisov (1994). A solution of the filtering and smoothing problems for uncertain-stochastic linear dynamic systems. *International journal of control,* **60**, 413-423.

This work is supported by RFFR Grants 95-01-00573 and 95-01-00789

OBSERVATION CONTROL PROBLEM FOR DISCRETE-CONTINUOUS SYSTEMS AS A SINGULAR CONTROL PROBLEM

Boris M. Miller, Karen V. Stepanyan

Institute for Information Transmission Problems
GSP-4, B. Karetny Per. 19, 101447 Moscow, Russia
e-mail: bmiller@ippi.ras.ru, kvs@ippi.ras.ru

Abstract: The problem of the estimation and observation control is considered for the stochastic system with a state-estimation dependent noise in observations. The optimal estimation in the class of linear filters was obtained and the separation principle for the linear-quadratic optimization problem was also proved. It gives the opportunity to solve simultaneously the control problem for the process and observations. This new problem belongs to a class of singular control problems and can be solved by standard methods of impulsive control theory.

Keywords: stochastic systems, non-Gaussian processes, nonlinear control systems, filtering problems, optimal control.

1. INTRODUCTION

The problem of the estimation and observation control is considered for a class of stochastic systems with state-estimation dependent noise in observation. These systems arise in image processing and in correlation traking systems (Miller, 1991). Here the problem of optimal estimation was solved in the class of linear filters for systems with affine dependence on the estimation error. This optimal linear estimation has a Kalman filter form, however, the equation for covariance matrix differs from a standard one of a Riccati type. Meanwhile, the closed form of filtering equations gives the opportunity to prove a separation principle for this kind of problems and to expand a well-known separation result (Kuznetsov, et.al., 1980) onto a class of nonlinear stochastic control problems. Moreover, if the solution is looking for in the class of linear filters and linear control laws, the originally stochastic problem of simultaneous process and observation control can be reduced to a deterministic one which can be solved by a standard methods of impulsive control theory (Miller, 1985; 1991; 1995).

The structure of the paper is as follows. In Section 2 we present the system model and statement of the observation control problem. In Section 3 we consider the estimation of a Kalman type and derive the equation for the optimal estimation and covariance matrix. In Section 4 we discuss the properties of the estimation derived in the previous section and prove that this estimation is really optimal in the class of linear filters. In Section 5 we consider the problem of simultaneous process and observation control and prove the separation principle. So we can reduce the originally stochastic problem to a deterministic one which has a form of standard impulsive control problem.

2. SYSTEM MODEL

Consider a dynamic stochastic system described

by a stochastic differential equation

$$dx(t) = A(t)x(t)dt + C(t)u(t)dt + B(t)dW_t^0, \quad x(0) = x_0, \quad t \in [0,T], \tag{1}$$

where $x(t) \in R^n$ be a system state, the initial condition x_0 is Gaussian with parameters

$$Ex_0 = m_0, \quad cov(x_0, x_0^*) = \gamma_0,$$

and $u(t)$ be a process control which depends on the observation process described by equation

$$\begin{aligned} dy(t) &= H(t,\alpha(t))x(t)v(t)dt + \\ &G(t, x(t) - \hat{x}(t))v^{1/2}(t)dW_t^1 \\ y(0) &= 0, \end{aligned} \tag{2}$$

In (1) and (2) $\{W_t^0\}$ and $\{W_t^1\}$ are standard Wiener processes, and $\hat{x}(t)$ is some estimation of $x(t)$ based on the values $\{y(s) : 0 \le s \le t\}$. The initial condition x_0, processes $\{W_t^0\}$ and $\{W_t^1\}$ are independent. Observation control is described by the functions $\alpha(t)$ and $v(t)$, satisfying the constraints:

$$\begin{aligned} &\alpha(t) \in U \quad (U \text{ is a compact set }), \\ &v(t) \ge 0, \quad \textstyle\int_0^T v(t)dt \le M < \infty. \end{aligned} \tag{3}$$

Our aim is twofold: the first one is to obtain the estimation process for $x(t)$ which would be optimal in some sence, and the second one is to consider the problem of observation and process control simultaneously. It should be noted, that model (1), (2) doesn't belong to the class of a conditionally-Gaussian processes (Liptser and Shiriayev, 1977) due to the dependence of the observation noise from the process $x(t)$ and its estimation $\hat{x}(t)$.

3. LINEAR FILTERING PROBLEM

Since the solution of the optimal estimation problem for models like (1), (2) is unknown we consider a problem of optimal linear estimation in special case of affine dependence of $G(\cdot)$ upon the estimation error, i.e.,

$$\begin{aligned} &G(t, x(t) - \hat{x}(t)) = \\ &\quad G_0(t) + < G_1(t), x(t) - \hat{x}(t) >; \\ &\text{where} \quad G_0(t) > 0 \quad \text{and} \\ &< G_1(t), x(t) - \hat{x}(t) >= \\ &\quad \sum_{k=1}^{n} G_1^{ijk}(t)(x_k(t) - \hat{x}_k(t)) \end{aligned} \tag{4}$$

We also will consider the class of linear estimations

$$\hat{x}(t) = F_0(t) + \int_0^t L(t,s)dy(s) \tag{5}$$

with deterministic functions $F_0, L(t,s)$, which, however, could depend on the given observation controls $\alpha(t), v(t)$.

Definition 1. The estimation of a type (5) with appropriate error covariance matrix

$$\gamma^0(t) = \mathbf{E}\{(x(t) - \hat{x}(t))(x(t) - \hat{x}(t))^*\} \tag{6}$$

will be called optimal in the class of linear estimations if

$$z^*\gamma^0(t)z \le z^*\gamma(t)z \quad \forall z \in R^n, \ \forall t \in [0,T] \tag{7}$$

where $\gamma(t)$ is the error covariance matrix for any estimation of a type (5).

To derive the equation for the optimal linear estimation we first find the optimal Kalman-type estimation, which will be described by equation

$$\begin{aligned} d\hat{x}(t) &= A(t)\hat{x}(t)dt + C(t)u(t)dt + \\ &K(t)(dy(t) - H(t,\alpha(t))\hat{x}(t)v(t)dt) \end{aligned} \tag{8}$$

with initial condition $\hat{x}(0) = m_0$. So, to define the optimal estimation according to Def. 1, we have to find the coefficient $K^0(t)$ which guaranties that relation (7) holds for any arbitrary $K(t)$.

Theorem 1. The optimal value of $K^0(t)$ is equal to

$$\begin{aligned} K^0(t) &= \gamma(t)H^*(t,\alpha(t)) \times \\ &[G_0(t)G_0^*(t) + < G_1(t)\gamma(t)G_1^*(t) >]^{-1}. \end{aligned} \tag{9}$$

where $\gamma(t)$ is the solution of equation

$$\begin{aligned} d\gamma(t) &= A(t)\gamma(t)dt + \gamma(t)A^*(t)dt + \\ &B(t)B^*(t)dt - \gamma(t)H^*(t,\alpha(t)) \times \\ &[G_0(t)G_0^*(t) + < G_1(t)\gamma(t)G_1^*(t) >]^{-1} \times \\ &H(t,\alpha(t))\gamma(t)v(t)dt, \end{aligned} \tag{10}$$

with initial condition $\gamma(0) = \gamma_0$ and the optimal estimation is described by equation

$$\begin{aligned} d\hat{x}(t) &= A(t)\hat{x}(t)dt + C(t)u(t)dt + \\ &\gamma(t)H^*(t,\alpha(t)) \times \\ &[G_0(t)G_0^*(t) + < G_1(t)\gamma(t)G_1^*(t) >]^{-1} \times \\ &\times (dy(t) - H(t,\alpha(t))\hat{x}(t)v(t)dt). \end{aligned} \tag{11}$$

with $\hat{x}(0) = m_0$.

Remark 1. Equations (10), (11) look like a standard Kalman filter, however, equation for $\gamma(t)$ differs from a standard Riccatti one. Meanwhile, if $G_1 = 0$ we obtain a classical Kalman filter.

Proof. For given $K(t)$ denote by $Z(t) = x(t) - \hat{x}(t)$ and $A_1(t) = A(t) - K(t)H(t,\alpha(t))v(t)$. Then $Z(t)$ satisfies the equation

$$\begin{aligned} dZ(t) &= A_1(t)Z(t) + B(t)dW_t^0 - \\ &K(t)G_0(t)v^{1/2}(t)dW_t^1 - \\ &K(t) < G_1(t), Z(t) > v^{1/2}(t)dW_t^1, \end{aligned} \tag{12}$$

and by applying the Ito's formula to $Z(t)Z^*(t)$ we obtain the following equation for $\gamma(t) =$

$\mathbf{E}Z(t)Z^*(t)$

$$\begin{aligned}
d\gamma(t) = & A_1(t)\gamma(t)dt + \gamma(t)A_1^*(t)dt+ \\
& B(t)B^*(t)dt+ \\
& K(t)G_0(t)G_0^*(t)K^*(t)v(t)dt+ \\
& K(t) < G_1(t)\gamma(t)G_1^*(t) > K^*(t)v(t)dt
\end{aligned} \quad (13)$$

$$\gamma(0) = \gamma_0.$$

To derive the equation for the optimal value of $K(t)$ we put

$$K^\varepsilon(t) = K^0(t) + \varepsilon\Delta K(t)$$

and define the appropriate value of $\gamma^\varepsilon(t)$ in the form of expansion

$$\gamma^\varepsilon(t) = \gamma^0(t) + \varepsilon\Delta\gamma(t) + O(\varepsilon^2).$$

Applying the standard algebraic techique to equation (13), we obtain the following equation for $\Delta\gamma(t)$,

$$\begin{aligned}
d\Delta\gamma(t) = & (A_1(t)\Delta\gamma(t) + \Delta\gamma(t)A_1^*(t)+ \\
& K^0(t) < G_1(t)\Delta\gamma(t)G_1^*(t) > (K^0(t))^*v(t)+ \\
& \Delta K(t)Q_1(t)v(t) + Q_1^*(t)\Delta K^*(t)v(t))dt
\end{aligned}$$

$$\Delta\gamma(0) = 0,$$

where

$$\begin{aligned}
Q_1(t) = & G_0(t)G_0^*(t)(K^0(t))^*+ \\
& < G_1(t)\gamma(t)G_1^*(t) > (K^0(t))^*- \\
& H(t,\alpha(t))\gamma(t).
\end{aligned}$$

As follows from condition (7) if $K^0(t)$ is optimal, then $\Delta\gamma(t) = 0$ for an arbitrary $\Delta K(t)$. Therefore, $Q_1 \equiv 0$ and we obtain relation (9) for the optimal value of $K^0(t)$. By substitution of (9) into (13) we obtain also equation (10) and the optimal estimation in the form of equation (11).

4. PROPERTIES OF ESTIMATION

In the section above we have found the optimal linear estimation of a Kalman type. Now we are in position to prove some additional properties of this estimation.

Theorem 2. Suppose that control law has a form

$$u(t) = L(t)\hat{x}(t).$$

Then

1. the estimation described by equation (11) is nonbiased, i.e.

$$\mathbf{E}(x(t) - \hat{x}(t)) = 0, \quad (14),$$

2. the covariance of the estimation error, namely, $\gamma(t) = cov((x(t) - \hat{x}(t)), (x(t) - \hat{x}(t))^*)$ satisfies the equation (10),
3. the estimation is ortogonal to the estimation error

$$\mathbf{E}\hat{x}(t)(x(t) - \hat{x}(t))^* = 0. \quad (15)$$

Proof. First statement easily follows from equation (12), the second one is a direct collorary of Theorem 1. To prove the ortogonality property consider a variable $\bar{Z}(t) = \hat{x}(t)(x(t) - \hat{x}(t))^*$. By applying Ito's formula we obtain

$$\begin{aligned}
d\bar{Z}(t) = & A(t)\hat{x}(t)(x(t) - \hat{x}(t))^*dt+ \\
& K(t)H(t,\alpha(t))(x(t) - \hat{x}(t))\times \\
& (x(t) - \hat{x}(t))^*v(t)dt+ \\
& K(t)(G_0(t)+ < G_1(t), x(t) - \hat{x}(t) >)\times \\
& v^{1/2}(t)dW_t^1(x(t) - \hat{x}(t))^*+ \\
& C(t)u(t)(x(t) - \hat{x}(t))^*dt+ \\
& \hat{x}(t)(x(t) - \hat{x}(t))^*A^*(t)dt+ \\
& \hat{x}(t)(B(t)dW_t^0)^*- \\
& \hat{x}(t)(x(t) - \hat{x}(t))^*\times \\
& H^*(t,\alpha(t))K^*(t)v(t)dt- \\
& \hat{x}(t)(K(t)[G_0(t)+ \\
& < G_1(t), x(t) - \hat{x}(t) >]\times \\
& v^{1/2}(t)dW_t^1)^*- \\
& K(t)(G_0(t)+ < G_1(t), x(t) - \hat{x}(t) >)\times \\
& (G_0(t)+ < G_1(t), x(t) - \hat{x}(t) >)^*\times \\
& K^*(t)v(t)dt
\end{aligned}$$

And for $Z(t) = \mathbf{E}\bar{Z}(t)$

$$\begin{aligned}
dZ(t) = & A(t)Z(t)dt + \gamma(t)H^*(t,\alpha(t))\times \\
& [G_0(t)G_0^*(t)+ < G_1(t)\gamma(t)G_1^*(t) >]^{-1}\times \\
& H(t,\alpha(t))\gamma(t)v(t)dt+ \\
& C(t)L(t)Z(t)dt + Z(t)A^*(t)dt+ \\
& Z(t)H^*(t,\alpha(t))K^*(t)v(t)dt- \\
& \gamma(t)H^*(t,\alpha(t))\times \\
& [G_0(t)G_0^*(t)+ < G_1(t)\gamma(t)G_1^*(t) >]^{-1}\times \\
& [G_0(t)G_0^*(t)+ < G_1(t)\gamma(t)G_1^*(t) >]\times \\
& \{\gamma(t)H^*(t,\alpha(t))\times \\
& [G_0(t)G_0^*(t)+ \\
& < G_1(t)\gamma(t)G_1^*(t) >]^{-1}\}^*v(t)dt
\end{aligned}$$

From previous equation after algebraic transformations, we have

$$\begin{aligned}
dZ(t) = & (A(t) + C(t)L(t))Z(t)dt+ \\
& Z(t)(A^*(t) + H^*(t,\alpha(t))K^*(t)v(t))dt
\end{aligned} \quad (16)$$

where $Z(0) = 0$. Thus, $Z(t)$ satisfies the homogeneous linear differential equation (16) with zero initial condition, hence $Z(t) = 0$. This completes the proof.

Remark 2. Since the estimation $\hat{x}(t)$ belongs to the class of linear estimations of a type (5) and satisfies the ortogonality condition (16) we can conclude that the estimation (11) is really optimal in the class of all linear estimations according to Def. 1.

The rigorous proof of this statement follows from ortogonality condition (16) and differential equations (10), (11), which describe the estimation. However, this proof is long enough and immaterial for further consideration. Due to the lack of space this proof is omitted.

5. SIMULTANEOUS PROCESS AND OBSERVATION CONTROL PROBLEM

In this section we consider the control problem for system described by equation (1) with observation process (2). Our aim is to control simultaneously the process (1) and oservation (2) to minimize a performance criterion

$$J = \mathbf{E}\int_0^T (x^*(t)P(t)x(t)+ u^*(t)R(t)u(t))dt \to \min \tag{17}$$

Since the model (1), (2) is nonlinear we have to simplify the problem to make it possible to obtain the solution in a closed form. The idea of such simplification is that control law could be taken in the form

$$u(t) = L(t)\hat{x}(t),$$

where $\hat{x}(t)$ be the best linear estimate of a process (1) by observation (2). So the process $\hat{x}(t)$ satisfies the equation (11). Therefore, the problem of simultaneous process and observation control can be formulated as a problem of search the triple $\{L(t), \alpha(t), v(t)\}$ which satisfies (3) and minimize the criterion (17). In linear case the solution of this problem is known and can be formulated in the form of a separation principle (Kuznetsov, et.al., 1980), which gives the opportunity to solve the process and observation control problem separately. The same result also takes place in our case of nonlinear system (1), (2).

Theorem 3. Assume that $R(t)$ in (17) are positively definite for all $t \in [0,T]$, then the minimum value of performance criterion can be achieved by triple $\{L(t), \alpha(t), v(t)\}$ where:

1. $L(t)$ is given by relation

$$L(t) = -R^{-1}(t)C^*(t)N(t)$$

where

$N(t)$ is the solution of equation

$$\dot{N}(\tau) = -A^*(\tau)N(\tau) - N(\tau)A(\tau)+ P^*(\tau) + N^*(\tau)C(\tau)R^{-1}(\tau)C^*(\tau)N(\tau) \tag{18}$$

with terminal condition $N(T) = 0$;

2. $\{\alpha(\cdot), v(\cdot)\}$ minimizes the performance criterion

$$J = \int_0^T Sp(P(t)\Gamma_0(t)+ L^*(t)R(t)L(t)(\Gamma_0(t) - \gamma(t)))dt \tag{19}$$

where $\Gamma_0(t)$ satisfies the following equation

$$\dot{\Gamma}_0(t) = A(t)\Gamma_0(t) + \Gamma_0(t)A^*(t)+ B(t)B^*(t)+ C(t)L(t)(\Gamma_0(t) - \gamma(t))+ (\Gamma_0(t) - \gamma(t))L^*(t)C^*(t), \tag{20}$$

with initial condition $\Gamma_0 = 0$,

$L(t)$ is given by (18), and $\gamma(t)$ is the solution of equation (10).

Remark 3. This theorem really shows the separation principle for this kind of problems. Indeed, the optimal process control $L(t)$ does not depend on the observation as in classical linear problem. Meanwhile, to find the optimal observation control we have to solve the separate problem of optimal control for system with dynamic, described by equations (10), (20) and performance criterion (19).

Remark 4. The problem of optimal control for system (10), (20) with performance criterion belongs to a class of impulsive control problems, because the constraints (3) for variable $v(t)$ allow the using of impulsive controls. However, this problem can be treated by standard methods of impulse control theory as for observation control problem for discrete-continuos systems in linear case (Miller, 1991).

Proof. First, we notice that by virtue of ortogonality condition (15) we have the relation

$$\gamma(t) = \mathbf{E}\{(x(t) - \hat{x}(t))(x(t) - \hat{x}(t))^*\} = \mathbf{E}x(t)x^*(t) - \mathbf{E}\hat{x}(t)x^*(t)- \mathbf{E}x(t)\hat{x}^*(t) + \mathbf{E}\hat{x}(t)\hat{x}^*(t) = \mathbf{E}(x(t)x^*(t)) - \mathbf{E}(\hat{x}(t)\hat{x}^*(t)), \tag{21}$$

Thus, we have

$$\mathbf{E}\hat{x}(t)\hat{x}^*(t) = \mathbf{E}(x(t)x^*(t)) - \gamma(t). \tag{22}$$

Later we derive the equation for $\Gamma_0(t) = \mathbf{E}(x(t)x^*(t))$. Taking into account the choosen type of control law $u(t) = L(t)\hat{x}$, we obtain:

$$dx(t) = A(t)x(t)dt + C(t)L(t)\hat{x}(t)dt+ B(t)dW_t^0$$

and

$$d(x(t)x^*(t)) = x(t)dx^*(t) + dx(t)x^*(t)+ B(t)B^*(t)dt = (x(t)x^*(t)A^*(t) + x(t)\hat{x}^*(t)L^*(t)C^*(t)+ A(t)x(t)x^*(t) + C(t)L(t)x(t)\hat{x}^*(t)+ B(t)B^*(t))dt + dM_t$$

where M_t is a martingale. After the substitution of relation $\mathbf{E}\hat{x}(t)x^*(t) = \mathbf{E}\hat{x}(t)\hat{x}^*(t)$, which is a consequence of (15), into previous relation, we obtain by Ito's formula:

$$\dot{\Gamma}_0(t) = A(t)\Gamma_0(t) + \Gamma_0(t)A^*(t)+ B(t)B^*(t)+ C(t)L(t)(\Gamma_0(t) - \gamma(t))+ (\Gamma_0(t) - \gamma(t))L^*(t)C^*(t). \tag{23}$$

Now the performance criterion (17) can be rewritten in the form

$$\begin{aligned}J = \int_0^T Sp(P(t)\mathbf{E}x(t)x^*(t)+ \\ L^*(t)R(t)L(t)\mathbf{E}\hat{x}(t)\hat{x}^*(t))dt = \\ \int_0^T Sp(P(t)\Gamma_0(t)+ \\ L^*(t)R(t)L(t)(\Gamma_0(t)-\gamma(t)))dt \to \min\end{aligned} \tag{24}$$

and we can find the optimal $L(t)$ for fixed $\gamma(t)$, as the optimal matrix-valued control for the system (23) with performance criterion (24). Suppose that $L(t)$ is the optimal control and

$$L^\varepsilon(t) = L(t) + \varepsilon \Delta L(t),$$

if $\Gamma_0(t)$ and value of J^0 correspond to the optimal $L(t)$ we can find the expansions for $\Gamma^\varepsilon(t)$ and J^ε which correspond to $L^\varepsilon(t)$ in the forms

$$\Gamma^\varepsilon(t) = \Gamma_0(t) + \varepsilon\Delta\Gamma(t) + O(\varepsilon^2)$$

and

$$J^\varepsilon = J^0 + \varepsilon \Delta J + O(\varepsilon^2).$$

Using the standard perturbation methods we obtain the following equation for $\Delta\Gamma(t)$ and relation for ΔJ :

$$\begin{aligned}\frac{d}{dt}\Delta\Gamma(t) = (A(t)+C(t)L(t))\Delta\Gamma(t)+ \\ \Delta\Gamma(t)(A(t)+C(t)L(t))^*+ \\ C(t)\Delta L(t)(\Gamma_0(t)-\gamma(t))+ \\ (\Gamma_0(t)-\gamma(t))\Delta L^*(t)C^*(t),\end{aligned} \tag{25}$$

and

$$\begin{aligned}\Delta J = \int_0^T Sp[P(t)\Delta\Gamma(t)+ \\ \Delta L^*(t)R(t)L(t)(\Gamma_0(t)-\gamma(t))+ \\ L^*(t)R(t)\Delta L(t))(\Gamma_0(t)-\gamma(t))+ \\ L^*(t)R(t)L(t)\Delta\Gamma(t)]dt\end{aligned} \tag{26}$$

For optimal solution $\Delta J = 0$ for any $\Delta L(t)$. To check this condition we can apply to the equation in variations (25) $\Delta L(t) = \Delta L\delta(t-\tau)$. This type of variation correspond to a Pontriagin variation type in the representation of $L^\varepsilon(t)$ on the interval of the lenght ε in the neiborhood of the point τ.

Then:

$$\begin{aligned}\Delta\Gamma(t) = \Phi(t,\tau)[C(t)\Delta L(\Gamma_0(t)-\gamma(t))+ \\ (\Gamma_0(t)-\gamma(t))\Delta L^* C^*(t)]\Phi^*(t,\tau),\ t \geq \tau\end{aligned}$$

$$\begin{aligned}\Delta J = Sp(\Delta K^* R(\tau)K(\tau)(\Gamma_0(\tau)-\gamma(\tau))+ \\ L^*(\tau)R(\tau)\Delta K(\Gamma_0(\tau)-\gamma(\tau))+ \\ \int_\tau^T Sp((P(t)+L^*(t)R(t)K(t))\times \\ \Phi(t,\tau)[C(t)\Delta L(t)(\Gamma_0(t)-\gamma(t))+ \\ (\Gamma_0(t)-\gamma(t))\Delta K^*(t)C^*(t)]\times \\ \Phi^*(t,\tau))dt \equiv 0\ \forall \Delta L(t), \forall\tau\in[0,T].\end{aligned} \tag{27}$$

where

$$\begin{aligned}\Phi(t,\tau) = (A(t)+C(t)L(t))\Phi(t,\tau) \\ \Phi(\tau,\tau) = I, \quad t \geq \tau.\end{aligned} \tag{28}$$

The derivate of ΔJ over ΔL is equal to zero due to the optimality condition, hence we have a relation

$$\begin{aligned}R(\tau)L(\tau)(\Gamma_0(\tau)-\gamma(\tau))^*+ \\ \int_\tau^T C^*(\tau)\Phi^*(t,\tau)(P(t)+L^*(t)R(t)L(t))^*\times \\ \Phi(t,\tau)(\Gamma_0(\tau)-\gamma(\tau))^*)dt \equiv 0\ \forall\tau\in[0,T].\end{aligned}$$

Since $\gamma(t)$ corresponds to covariation matrix of the optimal estimation, we have $\Gamma_0(\tau)-\gamma(\tau) > 0\ \forall\tau\in[0,T]$ and by virtue of condition $R(\cdot) > 0$, we obtain:

$$\begin{aligned}L(\tau) = -R^{-1}(\tau)C^*(\tau)\times \\ \int_\tau^T \Phi^*(t,\tau)(P(t)+L^*(t)R(t)L(t))^*\Phi(t,\tau)dt\end{aligned} \tag{29}$$

Denote

$$\begin{aligned}N(\tau) = \int_\tau^T \Phi^*(t,\tau)\times \\ (P(t)+L^*(t)R(t)L(t))^*\Phi(t,\tau)dt\end{aligned}$$

and derive the differential equation for $N(\cdot)$.

$$\begin{aligned}\frac{dN(\tau)}{d\tau} = \Phi^*(\tau,\tau)(P(\tau)+ \\ L^*(\tau)R(\tau)L(\tau))^*\Phi(\tau,\tau)+ \\ \int_\tau^T \frac{\partial\Phi^*(t,\tau)}{\partial\tau}(P(t)+ \\ L^*(t)R(t)L(t))^*\Phi(t,\tau)dt+ \\ \int_\tau^T \Phi^*(t,\tau)(P(t)+ \\ L^*(t)R(t)L(t))^*\frac{\partial\Phi(t,\tau)}{\partial\tau}dt\end{aligned} \tag{30}$$

Matrix-valued function $\Phi^*(t,\tau)$ as a fundamental solution of linear differential equation has the following properties:

$$\begin{aligned}\frac{\partial\Phi(t,\tau)}{\partial\tau} = -\Phi(t,\tau)A(\tau) \\ \frac{\partial\Phi^*(t,\tau)}{\partial\tau} = -A^*(\tau)\Phi(t,\tau)\end{aligned} \tag{31}$$

By substitution of (31) into (30) we obtain:

$$\begin{aligned}\frac{dN(\tau)}{d\tau} = (P(\tau)+L^*(\tau)R(\tau)L(\tau))^*- \\ \int_\tau^T A^*(\tau)\Phi(t,\tau)(P(t)+ \\ L^*(t)R(t)L(t))^*\Phi(t,\tau)dt- \\ \int_\tau^T \Phi^*(t,\tau)(P(t)+ \\ L^*(t)R(t)L(t))^*\Phi(t,\tau)A(\tau)dt = \\ P^*(\tau)+L^*(\tau)R^*(\tau)L(\tau)- \\ A^*(\tau)N(\tau)-N(\tau)A(\tau) = \\ P^*(\tau)+N^*(\tau)C(\tau)(R^{-1}(\tau))^*R^*(\tau)\times \\ R^{-1}(\tau)C^*(\tau)N(\tau)- \\ A^*(\tau)N(\tau)-N(\tau)A(\tau).\end{aligned}$$

which completes the proof.

6. CONCLUSION

The new, really nonlinear observation control problem has been considered. By appliing the idea of linear filtering it is possible to reduce this problem to a deterministic control one, for which all well-known methods could be applied.

Acknowledgements This work was supported in part by INTAS Grants 94-697 and 93-2622, and Russian Basic Research Foundation Grant No 95-01-00573.

REFERENCES

Liptser R. Sh. and Shiryaev A. N., (1977) *Statistics of Random Processes I, II* New York: Springer-Verlag.

Kuznetsov N. A., Liptser R. Sh., and Serebrovskii A.P. (1980). Optimal control and data processing in continuous time (linear system and quadratic functional) *Automat. Remote Control*, **41**, No. 10, 1369–1374.

Miller B. M. (1985). Optimal control of observations in the filtering of diffusion processes. I, II *Automat. Remote Control*, **46**, No. 2, 207–214; No. 6, 745–754.

Miller B. M. (1991). Generalized optimization in problems of observation control *Automat. Remote Control*, **52**, No. 10, 83–92.

Miller B. M. (1995). Generalized solutions of nonlinear optimization problems with impulse control I, II *Automat. Remote Control*, **55**, No. 4, 62–76, No. 5, 56–70.

MINIMAX GENERALIZED LINEAR-QUADRATIC STOCHASTIC CONTROL PROBLEM WITH INCOMPLETE INFORMATION

Eugene Ya.Rubinovich *

* *Institute of Control Sciences, 65 Profsoyuznaya Str., Moscow, 117806, Russia, E-mail: rubinvch@ipu.rssi.ru*

Abstract: The classic results on stochastic linear-quadratic problems of control with incomplete information are generalized to the case where observable and unobservable processes are described by stochastic differential equations with measure and the perturbing processes are not necessarily Gaussian. To top at all, the second observer, which creates a complemented noise in observed channel, is introduced.

Keywords: Stochastic control, Optimal control, Incomplete data, Quadratic performance indices, Minimax techniques.

1. INTRODUCTION

The statement and solution of generalized linear-quadratic stochastic control problem with incomplete information was conceded in (E.Ya.Rubinovich, 1977). The term "generalized" means the next generalizations with respect to the well known classical statements.

I. The controlled stochastic system was described by stochastic differential equations with measure, i.e., in a universal manner for continuous, discrete and discrete-continuous time.

II. It was conceded a more extensive class of stochastic processes – the semimartingales that may have non-Gaussian martingale parts.

In given work the results are obtained in (E.Ya.Rubinovich, 1977) are generalized in the following direction.

III. It is considered a minimax problem (stochastic game) in which the second observer (called the opponent) is introduced. This observer creates a complemented noise in observed channel interferes by that with a control process.

2. STATEMENT OF THE PROBLEM

Let on a stochastic basis $(\Omega, \mathcal{F}, (\mathcal{F}_t)_{0\leq t\leq T}, P)$ be given an unobservable vector-process $X = (X_t)_{t\geq 0} \in \mathbf{R}^m$ (that describes the evolution of phase coordinates of dynamic plant) and an observable vector-process $Y = (Y_t)_{t\geq 0} \in \mathbf{R}^k$ with Ito's differentials

$$\mathrm{d}X_t = [a_t X_{t-} + b_t u_t(Y)]\mathrm{d}\mu_t + \mathrm{d}\eta_t, \qquad (1)$$

$$\mathrm{d}Y_t = [A_t X_{t-} + B_t v_t(\xi)]\mathrm{d}\mu_t + \mathrm{d}\zeta_t, \qquad (2)$$

and initial conditions $X_{0-} = X_0$, $Y_{0-} = 0$. Here $X_{t-} = \lim X_s$, $s \uparrow t$, and $\xi = (\xi_t)_{t\geq 0}$, $\eta = (\eta_t)_{t\geq 0}$, $\zeta = (\zeta_t)_{t\geq 0}$, are independent square integrable martingales with quadratic characteristics $\langle\xi\rangle_t$, $\langle\eta\rangle_t$, $\langle\zeta\rangle_t$; a_t, b_t and A_t are dμ-integrated matrix-functions; $u = (u_t(Y))_{t\geq 0} \in \mathbf{R}^r$ and $v = (v_t(\xi))_{t\geq 0} \in \mathbf{R}^{r_1}$ are vector-control processes; X_0 is a random vector (independent on ξ, η and ζ) with given mean value and covariance matrix, $\mathbf{E}|X_0|^2 < \infty$. It is assumed that there exists matrix-fanctions α_t, β_t, κ_t and ρ_t satisfying the following consistency conditions

$$\mathrm{d}\mathbf{E}\langle\eta\rangle_t = \alpha_t \mathrm{d}\mu_t,\ \mathrm{d}\mathbf{E}\langle\zeta\rangle_t = \beta_t \mathrm{d}\mu_t,\ \mathrm{d}\mathbf{E}\langle\xi\rangle_t = \kappa_t \mathrm{d}\mu_t,$$

$$\mathbf{Sp} \int_{[0,T]} A_t^* \beta_t^+ A_t \mathrm{d}\mu_t < \infty, \quad A_t = \beta_t \rho_t.$$

The last equation means that this algebraic system must be solvable relative to the matrix ρ_t almost everywhere (a.e.) with respect to the measure $d\mu^c$. Here and later the superscript "c" denotes a continuous component of the measure $d\mu$, **Sp** is a sign of matrix trace, + is a pseudoinvers symbol, * means transposition, **E** is an expectation symbol. All the functions of t are assumed continuous from the right and having limits from the left. To top at all the matrices a_t, b_t, A_t are featured by standard Lipschitz conditions that provide the existence of strong solution (X_t, Y_t) of (1), (2).

Now let us define the classes $\mathcal{U}$ and $\mathcal{V}$ of admissible controls. For this let consider the family of Y-predictable σ-algebras $\mathcal{P}^Y = (\mathcal{P}_t^Y)_{t\geq 0}$ where

$$\mathcal{P}_t^Y = \sigma\{]s,t] \times M : s < t, M \in \mathcal{F}_s^Y\},$$
$$\mathcal{F}_s^Y = \sigma\{Y_\tau, \tau \leq s\}.$$

The family $\mathcal{P}^\xi$ is defined analogously. Let $\mathcal{L}_2(Y_0^t)$ be a closed linear manifold consists of all the linear combinations $\sum_{i,j} \lambda_{ij} Y_{t_i}(j)$ $(0 \leq t_i \leq t)$, $j = 1, ..., k$; $i = 1, 2, ...$ and their mean square limits (here k = dim Y, λ_{ij} = const). Analogously $\mathcal{L}_2(\xi_0^t)$ is defined.

Definition 1. $\mathcal{P}^Y$-measurable (resp. $\mathcal{P}^\xi$-measurable) process u (resp. v) be called admissible control process if it's components belong to $\mathcal{L}_2(Y_0^t)$ (resp. $\mathcal{L}_2(\xi_0^t)$) ($d\mu$ a.e.). In addition, process v is subject to the constraint ($d\mu$ a.e.)

$$\mathbf{E}[v_t^*(\xi)v_t(\xi)] = ||v_t(\xi)||^2_{\mathcal{L}_2} \leq V_t^2, \tag{3}$$

where V_t^2 – $d\mu$-integrable given function.

The problem is to construct an admissible control process u^o that realize

$$J^o = J_u(u^o(Y)) = \inf J_u(u(Y)) \text{ by } u \in \mathcal{U},$$

where

$$J_u(u(Y)) = \sup J_{uv}(u(Y), v(\xi)) \text{ by } v \in \mathcal{V},$$
$$J_{uv}(u(Y), v(\xi)) =$$
$$= \mathbf{E} \int_{[0,T]} [X_{t-}^* K_t X_{t-} + u_t^*(Y) R_t u_t(Y)] d\mu_t. \tag{4}$$

Here K_t, R_t are symmetric $d\mu$-integrable and non-negatively definite ($d\mu$ a.e.) matrix functions. In the case of degenerate payoff of the control process the following conditions must be fulfilled $\mathbf{Sp} \int_{[0,T]} b_t R_t^+ b_t^* d\mu_t < \infty$.

3. SOLUTION OF THE PROBLEM

Solution of this problem is carried out by the scheme represented in (E.Ya.Rubinovich, 1997) and consists of 5 steps.

1-st step – constructive description of the classes of admissible control processes.

Lemma 1. Any $u \in \mathcal{U}$ and $v \in \mathcal{V}$ admit the representation

$$u_t(Y) = \int_{[0,t[} G(t,s) dY_s, \quad v_t(\xi) = \int_{[0,t[} F(t,s) d\xi_s, \tag{5}$$

where matrices $G(\cdot,\cdot) \in \mathbf{R}^{r\times k}$ and $F(\cdot,\cdot) \in \mathbf{R}^{r_1 \times l}$ (l=dim ξ) are $d\mu$-integrable in the totality of variables and for $s \in [0,t]$ ($d\mu$ – *a.e.*)

$$\mathbf{Sp} \int_{[0,t]} G(s,\tau)\beta_\tau G^*(s,\tau) d\mu_\tau < \infty,$$

$$\mathbf{Sp} \int_{[0,t]} F(s,\tau)\kappa_\tau F^*(s,\tau) d\mu_\tau \leq V_t^2.$$

Here and later a stochastic integral with respect to dY is understood as the sum of integrals

$$\int_{[0,t[} G(t,s) dY_s = \int_{[0,t[} G(t,s) A_s X_{s-} d\mu_s + \int_{[0,t[} G(t,s) d\zeta_s.$$

Proof of Lemma 1 has a technical nature and follows the scheme is described in (M.H.A.Davis, 1977) (see §4.3).

2-nd step – convenient approximation of the admissible controls.

The matrix $G(t,s)$ in (5) is representable as

$$G(t,s) = \lim_n G_n(t,s), \quad \text{with} \tag{6}$$

$$G_n(t,s) = h_t^n g_t^n = \sum_{\alpha=1}^{N_n} h_t^n(\alpha) g_s^n(\alpha). \tag{7}$$

$h_t^n = [h_t^n(1) \vdots \ldots \vdots h_t^n(N_n)]$, $g_t^n = [g_t^n(1) \vdots \ldots \vdots g_t^n(N_n)]^*$ are the block-matrices (row and column), where $h_t^n(\alpha)$ and $g_t^n(\alpha)$ are respectively some $(r \times l_n)$- and $(l_n \times k)$-dimensional matrix-functions $d\mu$-integrable on [0,T]; N_n and l_n are some numbers. A convergence in (6) is regarded in the next sense: for any $t \in [0,T]$

$$\lim_n \int_{[0,t]} [G_n(t,s) - G(t,s)] \beta_s [G_n(t,s) - G(t,s)]^* d\mu_s = 0.$$

Analogously $F(t,s) = \lim_n F_n(t,s)$, where $F_n(t,s) = \sum_{\alpha=1}^{\bar{N}_n} \bar{h}_t^n(\alpha) \bar{g}_s^n(\alpha)$. Denote

$$u_t^n(Y) = \int_{[0,t[} G_n(t,s) dY_s = \sum_{\alpha=1}^{N_n} h_t^n(\alpha) \int_{[0,t[} g_s^n(\alpha) dY_s,$$
$$\tag{8}$$
$$v_t^n(Y) = \int_{[0,t[} F_n(t,s) d\xi_s = \sum_{\alpha=1}^{\bar{N}_n} \bar{h}_t^n(\alpha) \int_{[0,t[} \bar{g}_s^n(\alpha) d\xi_s$$

and let $(X^n, Y^n) = (X_t^n, Y_t^n)_{t\geq 0}$ be the process satisfying (1), (2) with the admissible control processes correspond to the kernels G_n and F_n.

Lemma 2. There exist equalities

a) $(X,Y) = \underset{n}{\text{l.i.m.}}\ (X^n, Y^n)$,

b) $(u(Y), v(\xi)) = \underset{n}{\text{l.i.m.}}\ (u^n(Y^n), v^n(\xi))$, (9)

c) $J_{uv}(u(Y), v(\xi)) = \lim_n J_{uv}(u^n(Y^n), v^n(\xi))$,

where l.i.m. stands for the limit in the mean square sense.

Proof of this lemma is analogously to the proof of Lemma 2 in (E.Ya.Rubinovich, 1997) and it is based on the representation (7).

3-rd step - transition to the Gaussian analogs of observable and unobservable processes.

Parallel with the processes X, Y from (1), (2) let us consider their Gaussian analogs, i.e., Gaussian processes $\widetilde{X} = (\widetilde{X}_t)_{t\geq 0}$, $\widetilde{Y} = (\widetilde{Y}_t)_{t\geq 0}$ with the same mathematical expectation and correlation. To obtain the representation for $(\widetilde{X}, \widetilde{Y})$ the martingale (ξ, η, ζ) is replaced by Gaussian martingale $(\widetilde{\xi}, \widetilde{\eta}, \widetilde{\zeta}) = (\widetilde{\xi}_t, \widetilde{\eta}_t, \widetilde{\zeta}_t)_{t\geq 0}$, with characteristics of it's components

$$\langle\widetilde{\xi}\rangle_t = \mathbf{E}\langle\xi\rangle_t, \quad \langle\widetilde{\eta}\rangle_t = \mathbf{E}\langle\eta\rangle_t, \quad \langle\widetilde{\zeta}\rangle_t = \mathbf{E}\langle\zeta\rangle_t,$$

and take Gaussian initial condition $\widetilde{X}_0$ which have mean value and covariance as X_0. Let as above $(\widetilde{X}^n, \widetilde{Y}^n)$ be the Gaussian analog corresponds to the control process with kernels $G_n(\cdot,\cdot)$, $F_n(\cdot,\cdot)$.

Lemma 3. The system $(\widetilde{X}, \widetilde{Y}, u(\widetilde{Y}), v(\widetilde{\xi}))$ are Gaussian.

Proof follows from the representations (6), (7) and Lemma 2, since

$$(\widetilde{X}, \widetilde{Y}) = \underset{n}{\text{l.i.m.}}\ (\widetilde{X}^n, \widetilde{Y}^n),$$
$$(u(\widetilde{Y}), v(\widetilde{\xi})) = \underset{n}{\text{l.i.m.}}\ (u^n(\widetilde{Y}^n), v^n(\widetilde{\xi})).$$

Lemma 4. On the approximating controls the values of performance index for primary processes X, Y and for their Gaussian analogs coincide, i.e.

$$J_{uv}(u^n(Y^n), v^n(\xi)) = J_{uv}(u^n(\widetilde{Y}^n), v^n(\widetilde{\xi})). \quad (10)$$

Proof of this lemma is analogously to the proof of Lemma 3 in (E.Ya.Rubinovich, 1997).

4-th step - solution of the minimax Gaussian problem.

Theorem 1. (about triple separation). Solution of the minimax generalized linear-quadratic stochastic control problem with incomplete information in Gaussian's case is reduced to the solution of three independent auxiliary problems:

1) the filtering problem for the process $\widetilde{X}$ by using the observation process $\widetilde{Y}$; this is a problem of conditional expectation $\pi_t = \pi_t(\widetilde{Y}) = \mathbf{E}(\widetilde{X}_t / F_t^{\widetilde{Y}})$ finding;

2) the deterministic control problem of Riccati type matrix equations with mixed restrictions;

3) the stochastic control problem with complete information for the variable π_t . In this case the optimal control has the form

$$u_t^o = u_t^o(\widetilde{Y}) = -L_t \pi_{t-}(\widetilde{Y}), \quad (11)$$

where L_t satisfies the generalized matrix Riccati equation with measure.

Proof of this theorem is in the Appendix.

5-th step - solution of the primary problem.

Theorem 2. Optimal admissible control in the primary problem has the form (11), where process $\widetilde{Y}$ is replaced by Y (i.e. by real observation process) in Kalman's filter equations for variable π_t.

Proof of this theorem is in the Appendix and bases on lemmas 1-3.

4. APPENDIX

Remark 1. For simplicity and without loss of generality let us assume that dimensionalities of the vectors $\widetilde{Y}_t$ and $v_t(\widetilde{\xi})$ coincide ($k = r_1$) that is formally the matrix B_t is missing in the equations.

Proof of Theorem 1.

1° Let us fix any admissible control $v_t = v_t(\widetilde{\xi})$ with kernel $F(t,s)$ and let $F_n(t,s) = h_t^n g_t^n$ be an approximating sequence for this kernel. Introduce a process $Z^n = (Z_t^n)_{t\geq 0} \in \mathbf{R}^{M_n}$, $(M_n = l_n N_n)$ with Ito's differential

$$dZ_t^n = g_t^n d\widetilde{\xi}_t, \quad Z_{0-}^n = 0$$

and consider an extended unobservable process

$$\theta^n = (\theta_t^n)_{t\geq 0}, \quad \theta_t^n = (\widetilde{X}_t^n \ \vdots\ Z_t^n)^* \in \mathbf{R}^{m+M_n}$$

in a block form with Ito's differential

$$d\theta^n = \overline{a}_t^n \theta_{t-}^n d\mu_t + b_t^n u_t(\widetilde{Y}^n) d\mu_t + D_t^n dW_t. \quad (A1)$$

In this case the observation process takes the form

$$d\widetilde{Y}_t^n = \overline{A}_t^n \theta_{t-}^n d\mu_t + d\widetilde{\zeta}_t, \quad (A2)$$

where $\overline{a}_t^n \in \mathbf{R}^{(m+M_n)\times(m+M_n)}$, $\overline{A}_t^n \in \mathbf{R}^{m\times(m+M_n)}$, $b_t^n \in \mathbf{R}^{(m+M_n)\times r}$, $D_t^n \in \mathbf{R}^{(m+M_n)\times(m+l)}$ are the next block-matrices:

$$\overline{a}_t^n = \begin{pmatrix} a_t & \vdots & 0_{m\times M_n} \\ \dots & \vdots & \dots \\ 0_{M_n\times m} & \vdots & 0_{M_n\times M_n} \end{pmatrix}, \quad b_t^n = \begin{pmatrix} b_t \\ \dots \\ 0_{M_n\times r} \end{pmatrix},$$

$$D_t^n = \begin{pmatrix} E_m & \vdots & 0_{m\times M_n} \\ \dots & \vdots & \dots \\ 0_{M_n\times m} & \vdots & g_t^n \end{pmatrix},$$

$$\overline{A}^n = (A_t \ \vdots\ h_t^n) \text{ and } W_t = (\widetilde{\eta}_t \ \vdots\ \widetilde{\xi}_t)^*.$$

Here $0_{m\times l} \in \mathbf{R}^{m\times l}$ – null matrix, $E_m \in \mathbf{R}^{m\times m}$ – unit matrix. In these variables the performance index takes the form

$$J_{uv}(u(\widetilde{Y}^n), v^n(\widetilde{\xi})) =$$
$$= \mathbf{E}\int_{[0,T]} [(\theta^n_{t-})^* \overline{K}^n_t \theta^n_{t-} + u^*_t(\widetilde{Y}^n) R_t u_t(\widetilde{Y}^n)] d\mu_t, \quad (A3)$$

The problem (A1)–(A3) was considered in (E.Ya.Rubinovich, 1997). For this problem an optimal minimizing control has the form

$$u^o_t(\widetilde{Y}^n) = -\overline{L}^n_t \overline{\pi}^n_{t-}(\widetilde{Y}^n), \quad (A4)$$

Here $\overline{L}^n_t \in \mathbf{R}^{r\times(m+M_n)}$ a deterministic matrix; conditional expectation $\overline{\pi}^n_t(\widetilde{Y}^n) = \mathbf{E}(\theta^n_t/\mathcal{F}^{\widetilde{Y}^n}_t)$ satisfy the equation

$$d\overline{\pi}^n_t(\widetilde{Y}^n) = (\overline{a}^n_t - \overline{b}^n_t \overline{L}^n_t)\overline{\pi}^n_{t-}(\widetilde{Y}^n)d\mu_t + \overline{\Gamma}^n_t d\overline{\nu}^n_t, \quad (A5)$$

where

$$\overline{\Gamma}^n_t = [E_{m+M_n} + \overline{a}^n_t \Delta\mu_t]\overline{\gamma}^n_{t-}(\overline{A}^n_t)^* \times$$
$$\times[\beta_t + \overline{A}^n_t \overline{\gamma}^n_{t-}(\overline{A}^n_t)^* \Delta\mu_t]^+, \quad (A6)$$

$\Delta\mu_t = \mu_t - \mu_{t-}$, and the process $\overline{\nu}^n = (\overline{\nu}^n_t)_{t\geq 0}$ with Ito's differential

$$d\overline{\nu}^n_t = d\widetilde{Y}^n_t - \overline{A}^n_t \overline{\pi}^n_{t-}(\widetilde{Y}^n)d\mu_t \quad (A7)$$

is a Gaussian martingale with respect to the family

$$\mathbf{F}^{\widetilde{Y}^n} = (\mathcal{F}^{\widetilde{Y}^n}_t)_{0\leq t\leq T} \text{ with } \mathcal{F}^{\widetilde{Y}^n}_t = \sigma\{\widetilde{Y}^n_s,\ s\leq t\}.$$

Here a covariance matrix $\overline{\gamma}^n_t$ satisfy the equation

$$d\overline{\gamma}^n_t = [\overline{a}^n_t \overline{\gamma}^n_{t-} + \overline{\gamma}^n_{t-}(\overline{a}^n_t)^*]d\mu_t +$$
$$+ D^n_t(d\langle W^n\rangle_t)(D^n_t)^* + \overline{a}^n_t \overline{\gamma}^n_{t-}(\overline{a}^n_t)^* \Delta\mu_t d\mu_t - \quad (A8)$$
$$-\overline{\Gamma}^n_t[\beta_t + \overline{A}^n_t \overline{\gamma}^n_{t-}(\overline{A}^n_t)^* \Delta\mu_t](\overline{\Gamma}^n_t)^* d\mu_t,$$

$$\text{with } d\langle W^n\rangle_t = \begin{pmatrix} \alpha_t & \vdots & 0_{m\times M_n} \\ \cdots & \vdots & \cdots \\ 0_{M_n\times m} & \vdots & g^n_t \kappa_t (g^n_t)^* \end{pmatrix} d\mu_t.$$

2° Next, introduce the notations

$$\widehat{Z}^n_t = \mathbf{E}(Z^n_t/\mathcal{F}^{\widetilde{Y}^n}_t),$$
$$v^n_t = h^n_t Z^n_{t-},$$
$$\widehat{v}^n_t = \mathbf{E}(v^n_t/\mathcal{F}^{\widetilde{Y}^n}_t) = h^n_t \widehat{Z}^n_{t-},$$
$$\overline{v}^n_t = v^n_t - \widehat{v}^n_t = h^n_t(Z^n_{t-} - \widehat{Z}^n_{t-}),$$
$$\pi^n_t = \pi^n_t(\widetilde{Y}^n_t) = \mathbf{E}(\widetilde{X}^n_t/\mathcal{F}^{\widetilde{Y}^n}_t),$$
$$\widehat{\gamma}^n_t = \mathbf{E}[\pi^n_t(\pi^n_t)^*],$$
$$m^n_t = m^n_t(\widetilde{Y}^n_t) = \widetilde{X}^n_t - \pi^n_t(\widetilde{Y}^n_t),$$
$$\gamma^n_t = \gamma^n_t(xx) = \mathbf{E}[m^n_t(m^n_t)^*],$$
$$\gamma^n_t(xz) = \mathbf{E}[m^n_t(Z^n_t - \widehat{Z}^n_t)^*],$$
$$\gamma^n_t(zz) = \mathbf{E}[(Z^n_t - \widehat{Z}^n_t)(Z^n_t - \widehat{Z}^n_t)^*].$$

$$\text{From this} \quad \overline{\gamma}^n_t = \begin{pmatrix} \gamma^n_t(xx) & \vdots & \gamma^n_t(xz) \\ \cdots & \vdots & \cdots \\ \gamma^n_t(zx) & \vdots & \gamma^n_t(zz) \end{pmatrix}.$$

3° It follows from the theorem of normal correlation (R.S.Liptser and A.N.Shiryaev, 1978)

$$\mathbf{E}[m^n_{t-}(\overline{v}^n_t)^*] = \mathbf{E}[m^n_{t-}\mathbf{E}((\overline{v}^n_t)^*/m^n_{t-})] =$$
$$= \mathbf{E}[m^n_{t-}(m^n_{t-})^*(S^n_t)^*] = \gamma^n_{t-}(S^n_t)^*, \quad (A9)$$

where $S^n_t \in \mathbf{R}^{k\times m}$ uniformly bounded on [0,T] and $d\mu$-intgrable deterministic matrix-function. It follows from Gaussian nature of $\overline{v}^n_t$ and m^n_t that the matrix S^n_t defines orthoprojector of $\overline{v}^n_t$ on m-dimensional subspace $\mathcal{L}^m_2(m^n_t)$ (defined analogously $\mathcal{L}_2(Y_t)$) with scalar product $x\cdot y = \mathbf{E}[x^*y]$.

4° From (A9) by virtue of introduced notations, it follows

$$h^n_t\gamma^n_{t-}(xz) = h^n_t\mathbf{E}[m^n_{t-}(Z^n_{t-} - \widehat{Z}^n_{t-})^*] =$$
$$= \mathbf{E}[m^n_{t-}(\overline{v}^n_t)^*] = \gamma^n_{t-}(S^n_t)^*, \quad (A10)$$

$$h^n_t\gamma^n_{t-}(zz)(h^n_t)^* =$$
$$= h^n_t\mathbf{E}[(Z^n_{t-} - \widehat{Z}^n_{t-})(Z^n_{t-} - \widehat{Z}^n_{t-})^*](h^n_t)^* =$$
$$= \mathbf{E}[\overline{v}^n_t(\overline{v}^n_t)^*]. \quad (A11)$$

5° Rewrite the equations (A3)–(A8) in the explicit block form and (on the basis of lemmas 2-3) pass to the limit (as $n\to\infty$) in the first block component of each equation (the limit transition technique see in (E.Ya.Rubinovich, 1997), (B.M.Miller and E.Ya.Rubinovich, 1995)). This yields in view of (A10), (A11)

$$d\pi_t = (a_t - b_t L_t)\pi_{t-}d\mu_t + \Gamma_t d\nu_t, \quad (A12)$$

where a limit gain matrix equals

$$\Gamma_t = (E_m + a_t\Delta\mu_t)\gamma_{t-}(A_t + S_t)^* \times$$
$$\times[\beta_t + (A_t\gamma_{t-}A^*_t + \mathbf{E}(\overline{v}_t\overline{v}^*_t) + \quad (A13)$$
$$+ A_t S_t\gamma_{t-} + \gamma_{t-}S^*_t A^*_t)\Delta\mu_t]^+,$$

and a limit covariance matrix satisfies the equation

$$d\gamma_t = (a_t\gamma_{t-} + \gamma_{t-}a^*_t)d\mu_t + a_t\gamma_{t-}a^*_t(\Delta\mu_t)^2 +$$
$$+\alpha_t d\mu_t - \Gamma_t[\beta_t + (A_t\gamma_{t-}A^*_t + \mathbf{E}(\overline{v}_t\overline{v}^*_t) + \quad (A14)$$
$$+ A_t S_t\gamma_{t-} + \gamma_{t-}S^*_t A^*_t)\Delta\mu_t]\Gamma^*_t d\mu_t.$$

The process $\nu = (\nu_t)_{t\geq 0}$ with Ito's differential

$$d\nu_t = d\widetilde{Y}_t - (A_t\pi_{t-} + \widehat{v}_t)d\mu_t$$

is a Gaussian martingale with respect to the family

$$\mathbf{F}^{\widetilde{Y}} = (\mathcal{F}^{\widetilde{Y}}_t)_{0\leq t\leq T} \text{ with } \mathcal{F}^{\widetilde{Y}}_t = \sigma\{\widetilde{Y}_s,\ s\leq t\}.$$

Here L_t is the limit first component of the block-matrix $\overline{L}^n_t$ (see (A48), (A49) in (E.Ya.Rubinovich, 1997)). Eqn. (A4) transforms to the next

$$u^o_t(\widetilde{Y}) = -L_t\pi_{t-}(\widetilde{Y}). \quad (A15)$$

Substitution (A15) in (4), on the basis of the conditional expectation property, gives

$$J_{uv}(u^o(\widetilde{Y}), v(\widetilde{\xi})) = J_{uF}(u^o(\widetilde{Y}), F) =$$
$$= \mathbf{Sp}\int_{[0,T]} [K^{1/2}_t\gamma_{t-}K^{1/2}_t + \widehat{K}^{1/2}_t\widehat{\gamma}_{t-}\widehat{K}^{1/2}_t]d\mu_t, \quad (A16)$$

where $\widehat{K}_t = K_t + L_t R_t L^*_t$ and the equality $J_{uv}(u^o(\widetilde{Y}), v(\widetilde{\xi})) = J_{uF}(u^o(\widetilde{Y}), F)$ emphasizes that the control v_t is defined by kernel $F(\cdot,\cdot)$. For the variable $\widehat{\gamma}_t$ It is easy to obtain from (A12) by Ito's formula the next equation

$$d\widehat{\gamma}_t = (\widehat{a}_t\widehat{\gamma}_{t-} + \widehat{\gamma}_{t-}\widehat{a}_t^*)d\mu_t + +\widehat{a}_t\widehat{\gamma}_{t-}\widehat{a}_t^*(\Delta\mu_t)^2 + \Gamma_t d\langle\nu\rangle_t\Gamma_t^*, \quad (A17)$$

where $\widehat{a}_t = a_t - b_t L_t$ and

$$d\langle\nu\rangle_t = [\beta_t + (A_t\gamma_{t-}A_t^* + \mathbf{E}(\bar{v}_t\bar{v}_t^*) + +A_tS_t\gamma_{t-} + \gamma_{t-}S_t^*A_t^*)\Delta\mu_t]d\mu_t. \quad (A18)$$

It is seen from (A14)–(A18), that the values of variables γ_t and $\widehat{\gamma}_t$ depends only on the component $\bar{v}_t$ of the control $v_t = \bar{v}_t + \widehat{v}_t$. It follows from this, in view of restriction (3), that it is sufficient to take a control v_t with component $\widehat{v}_t = 0$. In this case the restriction (3) takes the form

$$\mathbf{E}(\bar{v}_t^*\bar{v}_t) \le V_t^2.$$

It is obvious that the orthogonal projection of the element $\bar{v}_t$ (defined by the matrix S_t) satisfy this equation too, i.e.

$$\mathbf{E}[(S_tm_{t-})^*S_tm_{t-}] = \mathbf{Sp}\ S_t\gamma_{t-}S_t^* \le V_t^2. \quad (A19)$$

One checks directly that for maximization (A16) It is need to set

$$\mathbf{E}(\bar{v}_t^*\bar{v}_t) = V_t^2 \text{ or } \mathbf{E}(\bar{v}_t\bar{v}_t^*) = V_t^2E_m.$$

6° Let us consider a maximization control problem (A16)–(A19) with matrix phase variables γ_t, $\widehat{\gamma}_t$ and performance index (A16). The matrix variable S_t takes the role of control satisfying the mixed restriction (A19). Let a control S_t^o gives a solution of the problem (A16)–(A19) and $J_{uS}(u^o(\widetilde{Y}^o), S^o)$ is an optimal value of the functional (A16). Here a superscript "o" in $\widetilde{Y}^o$ means that observations are realized under the control $v_t^o(\widetilde{\xi})$ generating by the kernel $F^o(\cdot,\cdot)$, which corresponds to the matrix S_t^o. That is

$$J_{uS}(u^o(\widetilde{Y}^o), S^o) = J_{uF}(u^o(\widetilde{Y}^o), F^o) = = J_{uv}(u^o(\widetilde{Y}^o), v^o(\widetilde{\xi})).$$

Due to optimality of u_t^o and S_t^o the next inequalities take place

$$J_{uS}(u^o(\widetilde{Y}^o), S) \le J_{uS}(u^o(\widetilde{Y}^o), S^o) \le \le J_{uS}(u(\widetilde{Y}^o), S^o),$$

or

$$J_{uF}(u^o(\widetilde{Y}^o), F) \le J_{uF}(u^o(\widetilde{Y}^o), F^o) \le \le J_{uF}(u(\widetilde{Y}^o), F^o),$$

that is

$$J_{uv}(u^o(\widetilde{Y}^o), v(\widetilde{\xi})) \le J_{uv}(u^o(\widetilde{Y}^o), v^o(\widetilde{\xi})) \le \le J_{uv}(u(\widetilde{Y}^o), v^o(\widetilde{\xi})),$$

and hence

$$J_{uv}(u^o(\widetilde{Y}^o), v^o(\widetilde{\xi})) = J_u(u^o(\widetilde{Y}^o)). \quad (A20)$$

The proof of equality (A20) completes the proof of theorem.

Proof of Theorem 2.

Let the kernels $G^o(\cdot,\cdot)$ and $F^o(\cdot,\cdot)$ correspond to the controls $u^o \in \mathcal{U}$ and $v^o \in \mathcal{V}$ of the form (5) for the Gaussian analog of the original problem, i.e.

$$u_t^o(\widetilde{Y}^o) = -L_t\pi_{t-}^o(\widetilde{Y}^o) = \int_{[0,t[} G^o(t,s)d\widetilde{Y}_s^o,$$

$$v_t^o(\widetilde{\xi}) = \int_{[0,t[} F^o(t,s)d\widetilde{\xi}_s,$$

Show, that for arbitrary admissible control $v \in \mathcal{V}$ with some kernel $F(\cdot,\cdot)$

$$J_{uv}(u^o(Y), v(\xi)) \le J_{uv}(u(Y), v(\xi)) \quad (A21)$$

for any admissible control $u \in \mathcal{U}$ with kernel $G(\cdot,\cdot)$. Let as before G_n^o, F_n^o, G_n and F_n be arbitrary approximating sequences of the form (7), corresponding to the kernels G^o, F^o, G and F respectively. Now let us assume the contrary, namely, that there exist a control $\bar{u} \in \mathcal{U}$ that

$$J_{uv}(\bar{u}(Y), v(\xi)) < J_{uv}(u^o(Y), v(\xi)). \quad (A22)$$

By Theorem for Gaussian analogs

$$J_{uv}(u^o(\widetilde{Y}), v(\widetilde{\xi})) \le J_{uv}(\bar{u}(\widetilde{Y}), v(\widetilde{\xi})). \quad (A23)$$

By Lemma 4

$$J_{uv}(\bar{u}^n(Y^n), v(\xi)) = J_{uv}(\bar{u}^o(\widetilde{Y}^o), v(\widetilde{\xi})). \quad (A24)$$

Transition to the limit in (A24) by lemmas 4 and 2 gives

$$J_{uv}(\bar{u}(Y), v(\xi)) = J_{uv}(\bar{u}(\widetilde{Y}), v(\widetilde{\xi})). \quad (A25)$$

Comparison of (A22)–(A25) provides a contradictory chain of inequalities

$$J_{uv}(\bar{u}(\widetilde{Y}), v(\widetilde{\xi})) = J_{uv}(\bar{u}(Y), v(\xi)) < < J_{uv}(u^o(Y), v(\xi)) = = J_{uv}(u^o(\widetilde{Y}), v(\widetilde{\xi})) \le J_{uv}(\bar{u}(\widetilde{Y}), v(\widetilde{\xi})).$$

Therefore inequality (A21) is true. The next inequality is proved analogously.

$$J_{uv}(u^o(Y), v^o(\xi)) \le J_{uv}(u^o(Y^o), v^o(\xi)). \quad (26)$$

From (A21) and (A26) follows the assertion of Theorem.

5. REFERENCES

B.M.Miller and E.Ya.Rubinovich (1995). Regularization of a generalized kalman filter. *Math. and Computers in Simulation* **39**, 87–108.

E.Ya.Rubinovich (1997). Generalized linear-quadratic stochastic control problem with incomplete information. *Automatika i Telemechanika (in Russian)* **7**, 243–266, English translation is to appear in *Automation and Remote Control.*

M.H.A.Davis (1977). *Linear Estimation and Stochastic Control.* Chapman and Hall. London.

R.S.Liptser and A.N.Shiryaev (1978). *Statistics of Random Processes.* Springer. Berlin.

AUTHOR INDEX

www.ingramcontent.com/pod-product-compliance
Lightning Source LLC
Chambersburg PA
CBHW072058220326
41598CB00068BA/4454
* 9 7 8 0 0 8 0 4 2 9 3 2 8 *